T0270946

Paul K. Feyerabend
Physics and Philosophy
Philosophical Papers Volume 4

This collection of the writings of Paul Feyerabend is focused on his philosophy of quantum physics, the hotbed of the key issues of his most debated ideas. Written between 1948 and 1970, these writings come from his first and most productive period. These early works are important for two main reasons. First, they document Feyerabend's deep concern with the philosophical implications of quantum physics and its interpretations. These ideas were paid less attention in the following two decades. Second, the writings provide the crucial background for Feyerabend's critiques of Karl Popper and Thomas Kuhn. Although rarely considered by scholars, Feyerabend's early work culminated in the first version of *Against Method*. These writings guided him on all the key issues of his most well-known and debated theses, such as the incommensurability thesis, the principles of proliferation and tenacity, and his particular version of relativism, and more specifically on quantum mechanics.

Stefano Gattei is Assistant Professor at the IMT Institute for Advanced Studies, Italy, and Sidney Edelstein Fellow at the Chemical Heritage Foundation, Philadelphia, Pennsylvania. He has been working extensively on contemporary issues in the philosophy of science (Karl Popper's critical rationalism, Thomas Kuhn, the dynamics of theory-change and conceptual change, the incommensurability thesis, relativism) and on the history of early-modern astronomy and cosmology (especially Johannes Kepler and Galileo Galilei). He is the author of *Thomas Kuhn's Linguistic Turn and the Legacy of Logical Positivism* (2008) and *Karl Popper's Philosophy of Science: Rationality Without Foundations* (2009), as well as of several articles in the learned press.

Joseph Agassi is Professor Emeritus at Tel Aviv University and at York University, Toronto. He is the author of about twenty books and editor of about ten, as well as author of over 500 contributions to the learned press in the humanities, in diverse natural and social sciences, as well as in law and in education.

Paul K. Feyerabend
Physics and Philosophy

Philosophical Papers
Volume 4

Edited by

STEFANO GATTEI
Chemical Heritage Foundation, Philadelphia

JOSEPH AGASSI
Tel Aviv University

CAMBRIDGE
UNIVERSITY PRESS

CAMBRIDGE
UNIVERSITY PRESS

32 Avenue of the Americas, New York, NY 10013-2473, USA

Cambridge University Press is part of the University of Cambridge.

It furthers the University's mission by disseminating knowledge in the pursuit of education, learning, and research at the highest international levels of excellence.

www.cambridge.org
Information on this title: www.cambridge.org/9780521881302

First published 2016

A catalog record for this publication is available from the British Library.

Library of Congress Cataloging in Publication Data
Feyerabend, Paul, 1924–1994.
[Works. Selections. 2016]
Physics and philosophy / Paul Feyerabend; [edited by] Stefano Gattei, IMT Institute for Advanced Studies, Italy, Joseph Agassi, Tel Aviv University.
pages cm. – (Philosophical papers; volume 4)
"The present volume, the fourth collection of Paul Feyerabend's philosophical papers, is especially focused on physics, and more specifically on quantum mechanics. All papers fall within 1948 and 1970, thus covering the first period of Feyerabend's production"–Introduction.
Includes bibliographical references and index.
ISBN 978-0-521-88130-2 (Hardback: alk. paper)
1. Science–Philosophy. 2. Philosophy and science. 3. Quantum theory.
I. Gattei, Stefano. II. Agassi, Joseph. III. Title.
Q175.F425 2016
530.12–dc23 2015012678

ISBN 978-0-521-88130-2 Hardback

To Marcello Pera,
a close friend and appreciative colleague
of Paul Feyerabend

Contents

Contents

Acknowledgments

We are grateful to Grazia Borrini Feyerabend and to all the copyright holders (see pp. XXV-XXVII) for permissions to reprint the diverse works that reappear in this volume. We are grateful to John R. Wettersten for his draft of the translation of the papers originally published in German (Chapters 1–4, 8, 10, 12, and 24 here); we are also grateful to John L. Heilbron and to Concetta Luna for their careful revision of the translations of these papers and for their many useful suggestions.

Grazia Borrini Feyerabend has kindly provided constant and generous support during the whole lengthy process of the preparation of this volume. Brigitte Parakenings, administrator of the Philosophisches Archiv der Universität Konstanz, offered generous assistance to Stefano Gattei in his work at the *Paul Feyerabend Nachlaß*. Matteo Collodel kindly supplied some hard-to-access material included here. Ciro D'Amato patiently redrew all the illustrations that appear here. Roberta Corvi, Robert P. Farrell, John M. Preston, and Carlo Tonna discussed with Stefano Gattei many issues regarding the sizable output of Paul Feyerabend throughout these years. They all have our sincere gratitude.

Stefano Gattei
Chemical Heritage Foundation, Philadelphia, USA and
IMT Institute for Advanced Studies, Lucca, Italy

Joseph Agassi
Tel Aviv University, Israel and
York University, Toronto, Canada

Introduction

What does not satisfy me in [the statistical quantum theory], from the standpoint of principle, is its attitude towards that which appears to me to be the programmatic aim of all physics: the complete description of any (individual) real situation (as it supposedly exists irrespective of any act of observation or substantiation).

Albert Einstein

The question is not whether by a subtle and highly scholastic argument we may continue to uphold an untenable position. The question is whether we should think critically and rationally in physics, or defensively and apologetically.

Karl R. Popper

Only be aware what you will face in Copenhagen and how careful you will have to be. You will face a metaphysics. And metaphysicians usually are very dogmatic; but they are even more dogmatic when they believe their metaphysics to be truly factual.

Paul K. Feyerabend[1]

The present volume, the fourth collection of Paul Feyerabend's philosophical papers, is especially focused on physics, and more specifically on

---

[1] The sources of these three quotes are, respectively: Albert Einstein, "Reply to Criticisms", in Paul A. Schilpp (ed.), *Albert Einstein: Philosopher Scientist*, Evanston, IL: The Library of Living Philosophers, 1949, pp. 663–688: p. 667; Karl R. Popper, *Postscript to The Logic of Scientific Discovery*, edited by William W. Bartley, III, vol. III: *Quantum Theory and the Schism in Physics*, London: Hutchinson, 1982, p. 150; and Paul K. Feyerabend's second letter to Kuhn, in Paul Hoyningen-Huene, "Two Letters of Paul Feyerabend to Thomas S. Kuhn on a Draft of *The Structure of Scientific Revolutions*", *Studies in History and Philosophy of Science*, 26, 1995, pp. 353–387: p. 380.

quantum mechanics. All papers fall within 1948 and 1970, thus covering the first period of Feyerabend's production and his most elaborate products, beginning with his very first publication (prior to his having submitted his doctoral dissertation), an unsigned article in the *Veröffentlichungen des Österreichischen College*. They are significant for at least two reasons. First, they document his keen concern with the philosophical implications of quantum theory and its interpretations (a concern that never faded throughout his life, but that in the following two decades occupied a considerably lesser space in his published output). Second, they provide the crucial background to Feyerabend's critiques of Karl Popper (beginning with "Explanations, Reduction, and Empiricism," 1962) as well as of Thomas Kuhn ("Consolations for the Specialist," 1970). Indeed, some understanding of his less familiar works on physics here republished is necessary for understanding his more familiar concern with issues that soon became central to his later production. Although rarely considered by Feyerabend scholars, these early works of his represent the concerns that steered his work between the early 1950s and the late 1960s, work that culminated in the first version of his most famous book, *Against Method: Outline of an Anarchistic Theory of Knowledge* (1970), and that guided him to all the key issues of his most familiar and much-debated theses, such as his incommensurability thesis, his principles of proliferation and tenacity, and his peculiar version of relativism.

The present collection comprises all of Feyerabend's early works on the philosophy of physics with these exceptions: the works readily available in the previous collections of his papers, his doctoral dissertation (unpublished; it should appear as a separate volume, in the original or in translation), and two brief posthumous entries in *The Oxford Companion to Philosophy* (1995) that we could not receive permission to republish. Our making the bulk of his works on physics easily available – to scholars, philosophers, and historians of philosophy alike – should lead to a better general understanding of his heritage. To facilitate this, we offer here some historical, philosophical, and scientific background material that may help in putting his reflections into context, understanding his critical targets, and seeing their role in forging key elements of his later works.

The fundamental theories of nineteenth-century (classical) physics divided the agencies of the physical world into two distinct categories. There were material substances: the chemical elements, each comprising distinct, unchangeable atoms; the various compounds, molecules formed by the combination of atoms; and the macrobodies that comprise conglomerates of these molecules – as described by Isaac Newton, Leonhard

Euler, John Dalton, Augustin-Louis Cauchy, and their followers. To another category belonged electromagnetic fields that contain energy and radiate it as light waves, visible and invisible – as described by Michael Faraday, James Clerk Maxwell, John Henry Poynting, Heinrich Hertz, and their followers. The bridge between these two distinct categories was the ether that supposedly housed the energy of the fields. Faraday and his followers, including Einstein, dismissed the ether; also, Einstein declared that matter is concentrated energy. This inaugurated the era of relativist physics, namely of physics that abides by Einstein's principle of relativity rather than by Galileo's.

Another route that led to the collapse of classical physics came from a separate front that did not link up to the principle of relativity until 1925. In 1800 Joseph von Fraunhofer discovered spectral lines, namely, the fact that each element radiates and absorbs light differently. In 1859 Gustav Kirchhoff argued that, nevertheless, the ratio of the emission and the absorption that are characteristic to each element depends (not on the element but) only on the temperature of the radiating or absorbing atom. After a few failed attempts by various physicists to find what that dependence is, in October 1900 Max Planck advanced as a hypothesis a formula that fit experimental evidence. He then provided an explanation of that formula that included the assumption that the total energy in a field is made up of discrete quantities of energy, the size of each *quantum* being proportional to the frequency of the wave whose energy is concerned.[2] In 1905 Einstein offered the hypothesis that all radiation exists only in discrete energy packets, which he christened photons. (At the time Planck found this hypothesis much too wild.) According to classical electromagnetic theory, the fact that many metals emit electrons when light shines upon them (the photoelectric effect) can be attributed to the transfer of light energy to an electron in the metal; and the rate of this emission should depend on the amplitude or intensity of the light that falls on the piece of metal. Experimental results, by contrast, show that electrons are only dislodged if light reaches or exceeds a threshold frequency, below which no electron is emitted from metals. Einstein suggested that energy is exchanged only in discrete amounts, which perfectly fitted Planck's earlier discovery of the relation between energy and frequency.[3] The amplitude of the wave indicates the rate of the total amount

---

[2] Max Planck, "Über das Gesetz der Energieverteilung im Normalspektrum", *Annalen der Physik*, 309, 1901, pp. 553–563.

[3] Albert Einstein, "Über einen die Erzeugung und Verwandlung des Lichtes betreffenden heuristischen Gesichtspunkt", *Annalen der Physik*, 322, 1905, pp. 132–148.

of energy it conveys, but the frequency (or the wavelength) indicates the amount of energy that a single photon carries. Single photons, said Einstein, are responsible for single electron emission.

That was in 1905; in 1913 another bold step was taken. Building on the work of Joseph J. Thompson and his pupil Ernest Rutherford, the latter's pupil, Niels Bohr, published a revolutionary work on the structure of the hydrogen atom, according to which most of the atom is in its compact, central, positively charged nucleus with an individual negatively charged electron traveling in a circular orbit around it.[4] Arnold Sommerfeld improved Bohr's model, to render it compatible with the principle of relativity, thus also allowing for elliptical orbits as the effects of external magnetic fields[5] – the electrons are held in their orbits around the nucleus by electrical attraction, similar to the gravitational attraction that holds the planets in their orbits around the sun in our solar system (the so-called Rutherford-Bohr planetary atom, or old quantum theory). Unlike our solar system, however, Bohr's atom has discrete orbits: the energies of the electrons can occur only in certain fixed amounts, which correspond to certain fixed orbits, as the absorption of one photon should make the electron jump to a higher orbit, and the emission of one photon should cause a jump to a lower orbit. Contrary to classical electromagnetic theory, within the atom moving electrons do not radiate, except when they move from a higher to a lower orbit. This move was called quantum jump: an electron is not permitted to move between orbits but can disappear in one and simultaneously appear in another. Although the new quantum theory dispenses with these jumps,[6] they have survived to this day, having fired the imaginations of science fiction writers.

[4] Niels Bohr, "On the Constitution of Atoms and Molecules, Part I", *Philosophical Magazine*, 26, 1913, pp. 1–25. The paper was part of a trilogy, which also included "On the Constitution of Atoms and Molecules, Part II: Systems Containing Only a Single Nucleus", *Philosophical Magazine*, 26, 1913, pp. 476–502; and "On the Constitution of Atoms and Molecules, Part III: Systems Containing Several Nuclei", *Philosophical Magazine*, 26, 1913, pp. 857–875.

[5] Arnold Sommerfeld, "Zur Theorie der Balmerschen Serie", *Sitzungsberichte der mathematisch-physikalischen Klasse der K. B. Akademie der Wissenschaften zu München*, 1915, pp. 425–458; and "Die Feinstruktur der Wasserstoff-und der Wasserstoff-ähnlichen Linien", *Sitzungsberichte der mathematisch-physikalischen Klasse der K. B. Akademie der Wissenschaften zu München*, 1916, pp. 459–500. A revised version of these papers appeared as "Zur Quantentheorie der Spektrallinien", *Annalen der Physik*, 51, 1916, pp. 1–94 and 125–167, and later in his *Atombau und Spektrallinien*, Braunschweig: Vieweg, 1919.

[6] Erwin Schrödinger, "Are There Quantum Jumps?", *The British Journal for the Philosophy of Science*, 3, 1952, pp. 109–123.

The odd thing was not so much the jumps as the fact that two theories were here closely related but separate all the same: Planck and Einstein spoke of the quantized energy of fields, and Bohr spoke of matter absorbing and emitting these quanta. The unification of the two ideas began with Einstein's 1918 theory of absorption and emission (that was neglected until the advent of lasers). The most enigmatic aspect of the situation was the central formula of the theory, Planck's formula that correlates the energy of the photon with the wavelength associated with it: $E = hv$ (E is the energy of the photon, h is Planck's constant, and v is the frequency of the wave). What is this association? In 1924 Louis de Broglie used relativistic considerations to suggest that not only fields of energy but also material particles, such as electrons, have wave-like features, quite like photons.[7] He combined Einstein's formula ($E = mc^2$), relating matter to energy, with Planck's ($E = hv$), thereby obtaining $mc^2 = hv$. Since, for every photon, $v = c/\lambda$, he derived $\lambda = h/mc$, and assuming the same equation to hold for a particle moving with speed v, he obtained $\lambda = h/p$. Testing this idea, Clinton Davisson and Lester Germer found that electron beams behave like wave-fronts and can show interference patterns. This seemed so bizarre that Einstein had to insist on taking their work seriously. With de Broglie, the wave-particle duality became universal in fundamental physics. The question is, what does this amount to? What does it mean? The simplest answer is that the particle is a concentration of field energy. The trouble with this answer is that the wave theory allows for no stability of particle-like energy-concentration, whereas the received information is that every kind of particle has its characteristic degree of stability. The next simplest answer goes the opposite way: the particles are real; what makes for their quantum characteristics are waves that guide them this way. This is the pilot-waves theory of de Broglie. The difficulty with this answer may be smaller, as the particles in this case need not dissipate, but the waves should dissipate, and they do not.

A breakthrough came in 1925–1926, with the appearance of both versions of the classical quantum theory: the matrix mechanics of Werner Heisenberg, Max Born, and Pascual Jordan,[8] as well as Schrödinger's

---

[7] Louis de Broglie, *Recherches sur la Théorie des Quanta*, PhD dissertation, Université de Paris, 1924; published in *Annales de Physique*, 3, 1925, pp. 22–128.

[8] Max Born and Pascual Jordan, "Zur Quantenmechanik", *Zeitschrift für Physik*, 34, 1925, pp. 858–888; Max Born, Werner Heisenberg and Pascual Jordan, "Zur Quantenmechanik II", *Zeitschrift für Physik*, 35, 1926, pp. 557–615.

equation for the electron.[9] Both versions offered precise descriptions of the energy transfer involved in quantum transitions (of emission or absorption of light). Most physicists were slow to cope with matrix mechanics, due to its abstract nature and unfamiliar mathematics; instead, they welcomed Erwin Schrödinger's alternative wave mechanics, since its equation was more familiar, as it did away with all quantum jumps and discontinuities, although the waves it describes are of a new kind, material waves that have no analogue in classical mechanics and that were introduced by de Broglie only a short time before.

The success of Schrödinger was in that he viewed Bohr's electron as a standing wave – of vibrations with fixed end-points, like those of a musical instrument – and drew on the fact that such waves are harmonic, namely that the higher frequencies of the vibrating string are multiples of the base frequency. The novelty in his equation is that he showed that in quantum mechanics the resonance that depends on frequencies is the law of conservation of energy, since Planck's equation equates the frequency of a photon with its energy. Schrödinger thus viewed Bohr's electron as a "matter-wave" except that each wave has a fixed quantum of energy: whereas in the vibrating string the energy varies with the strength of the vibrations (their amplitudes), in quantum mechanics it depends on frequency. Schrödinger's equation led to much easier calculations and more familiar visualizations of atomic events than did Heisenberg's matrix mechanics, where the energy was found in an abstruse calculation. Schrödinger published then a proof that matrix and wave mechanics gave equivalent results: mathematically, they were the same theory. Wolfgang Pauli, who calculated the matrix for the energy levels of the hydrogen atom using matrix mechanics, advocated the use of Schrödinger's equation as a shortcut for

---

[9] Erwin Schrödinger, "Quantisierung als Eigenwertproblem", *Annalen der Physik*, 384, 1926, pp. 361–376. In this paper he presented what is now known as the Schrödinger equation, and gave a derivation of the wave function for time-independent systems, also showing that it gave correct energy eigenvalues for a hydrogen-like atom (such as the one described by Bohr in his own model). He later published, under the same title, three follow-ups to this paper (*Annalen der Physik*, 384, 1926, pp. 489–527; 385, 1926, pp. 437–490; and 386, 1926, pp. 109–139), in the second of which he showed the equivalence of his approach to Heisenberg's. The last of these papers, in which Schrödinger introduced a complex solution to the wave equation, marked the moment when quantum mechanics switched from real to complex numbers, never to return. Nevertheless, the equivalence is only partial: whatever matrix mechanics explains wave mechanics explains too, but not vice versa, since the wave equation also applies to continuous systems.

calculations within matrix mechanics (as means of transformation known as the diagonalization of matrices).

The advocates of the matrix formalism of quantum mechanics nevertheless faced the question, what is the wave function doing in the theory? To this Max Born came with a new idea: the waves carry not energy but probabilities – quantum mechanics is essentially statistical.[10] Heisenberg then insisted that the discontinuous quantum transitions give his version an edge over Schrödinger's. The intense debates that followed showed that both interpretations of atomic events are unsatisfactory. Both sides started searching for a more satisfactory theory, and began with the interpretation of the quantum mechanics equations in line with their own preferences.

One obvious defect of the theory in all versions was that it operated within Newton's framework, rather than within Einstein's. Paul Dirac tried to remedy this: he created a variant of Schrödinger's equation that abides by the demands of the principle of relativity; Jordan applied this to both matter and fields, in unified equations known as "transformation theory": these formed the basis of what is now regarded as the orthodox quantum mechanics.[11] The task then became a search for the physical meaning of these equations, for the ability to show the nature of physical objects in terms of waves or in terms of particles, or both. This was called quantum electrodynamics; it was an ambitious effort to include classical and quantum effects, and to allow for both waves and particles.

Next came the most philosophically seminal part of the theory, the Heisenberg inequalities or the Heisenberg uncertainty principle: at any given moment, Δp times Δq is bigger or equal to h – that is, the product of the inexactness of the measurement of the momentum of a particle multiplied by the inexactness of the measurement of its position is proportional to Planck's constant, so that the smaller the imprecision of one of these two variables, the bigger the other.[12] Heisenberg offered a sort of

---

[10] Max Born, "Zur Quantenmechanik der Stoßvorgänge", *Zeitschrift für Physik*, 37, 1926, pp. 863–867; and 38, 1926, pp. 803–827.

[11] Paul A. M. Dirac, "The Physical Interpretation of the Quantum Dynamics", *Proceedings of the Royal Society of London A*, 113, 1926, pp. 621–641; Pascual Jordan, "Über eine neue Begründung der Quantenmechanik", *Zeitschrift für Physik*, 40, 1926, pp. 809–838, and "Über eine neue Begründung der Quantenmechanik, II", *Zeitschrift für Physik*, 44, 1927, pp. 1–25.

[12] Werner Heisenberg, "Über den anschaulichen Inhalt der quantentheoretischen Kinematik und Mechanik", *Zeitschrift für Physik*, 43, 1927, pp. 172–198. The "observer effect", according to which measurements affect the measured items, is not under debate. Heisenberg went further and advanced the thesis that at the quantum level the effect cannot be

derivation of this principle and a popular explanation of it, both too inexact to count. The most direct way to derive it is, again, by taking the idea of a material particle as a wave of sorts, since the inexactness occurs in the general theory of waves.[13]

The uncertainty relations had far-reaching implications. First, Heisenberg advocated operationalism, the doctrine that in science every concept has a meaning only in terms of the experiments used to measure it, so that things that in principle cannot be measured have no scientific meaning. Thus, since the simultaneous values of a particle's position and momentum prescribe it a path, the uncertainty of the particle's path amounts to the concept of its path having no meaning. Operationalism is untenable, though: a basic assumption of modern physics, ever since Galileo and Newton, has been that the real world exists independently of us, regardless of whether or not we observe it. In Heisenberg's view, such concepts as orbits of electrons, or paths of particles, do not exist in nature, unless – and until – we observe them.

Heisenberg also drew profound implications for the concept of causality, or the determinacy of future events. Schrödinger had earlier attempted to offer an interpretation of his equation in which the electron waves represent the density of charge of the electron in the orbit around the nucleus. In his reading, every electron fills the whole universe, yet most of it is present in a reasonably small location. Born showed that the wave function of Schrödinger's equation represents not the density of charge or of matter but the probability of the location of the particle. In Born's reading, then, the results of quantum mechanics are not exact but statistical. Heisenberg took this one step further, challenging the notion of simple causality in nature: the future is not predetermined in the real world, not even the trajectory of an electron.

Schrödinger tried to refute this. Quantum theory asserts that a photon hitting a certain transparent filament has equal probability of passing and of not passing through it. Suppose that if and only if the photon passes through the transparent body, it triggers a gun that hits a cat. Suppose

---

reduced below a certain limit – in experiment or even in thought: the uncertainty principle is inherent in the properties of all quantum objects – of systems that are both wavelike and particlelike. In other words, Heisenberg postulated that the uncertainty principle states a fundamental property of quantum systems that can never be eliminated from science. Bohr viewed this as the destruction of the traditional barrier between us and the world, which he deemed a major philosophical consequence of quantum mechanics.

[13] Óscar Ciaurri and Juan L. Varona, "An Uncertainty Inequality for Fourier-Dunkl Series", *Journal of Computational and Applied Mathematics*, 233, 2010, pp. 1499–1504.

that all that happens in a closed opaque box. Then, according to Heisenberg, the cat is half alive and half dead until we open the box and look at the cat. This, said Schrödinger, is absurd.

In 1927, in the famous lecture he gave in Como, Niels Bohr stated his complementarity principle, which takes waves and particles as equally unavoidable in quantum-theoretical accounts: the wave and particle pictures complement each other. They are mutually exclusive, yet jointly both are essential for a complete description of quantum events; the uncertainty principle prevents them from coming together and conflicting with each other. By choosing either the wave or the particle picture, scientists influence the outcome of experiments, thereby causing a limitation in what we can know about nature "as it really is".[14] Complementarity, uncertainty, and the statistical interpretation of Schrödinger's wave function formed together the orthodox reading of quantum mechanics, known as the "Copenhagen interpretation".[15] In October 1927, at the Solvay conference in Brussels, its upholders went so far as to declare quantum mechanics complete, and the hypotheses upon which it rested as no longer in any need of modification. Dirac's austere and influential book, *The Principles of Quantum Mechanics*, first published in 1930 (fourth edition, 1958), provided the standard presentation of the theory for the decades to come. Philosophically its position was relatively clear; Dirac refused to ask the questions about the path of the electron that in principle theory and experiment do not answer. He also admitted that a central principle of the theory, the principle of superposition, does not yield to a simple, clear statement (p. 9; it is usually presented by examples).

Quantum theory discarded two central axioms of classical physics. First, it treated the basic material particles and energy as fields. Second, it rejected all "clockwork" pictures of nature: according to quantum mechanics, questions about future behavior of physical systems can be

---

[14] Niels Bohr, "The Quantum Postulate and the Recent Development of Atomic Theory", *Nature*, 121, 1928, pp. 580–590; reprinted in Niels Bohr, *Atomic Theory and the Description of Nature: Four Essays*, Cambridge: Cambridge University Press, 1934, pp. 52–91.

[15] Feyerabend said, "there is no such thing as the 'Copenhagen interpretation'" (see Chapter 6 of this volume). Indeed, Bohr and Heisenberg never agreed on all details of the reading of the mathematical formalism of quantum mechanics. The "Copenhagen interpretation" is a label that critics introduced, to denote Bohr's idea of complementarity plus Heisenberg's interpretation of the uncertainty relations, and Born's statistical interpretation of the wave function. At times, they added to this the correspondence principle that Bohr advanced in 1920: the behavior of systems described by (the old) quantum theory reproduces classical physics in the limit of large quantum numbers.

answered only statistically: Heisenberg's uncertainty principle replaces the rigid causality of classical physics with probabilities.

Physicists reacted to the new theories in different ways. To some, the abandonment of causality seemed a small price to pay for the great extension of understanding that quantum mechanics offers. Yet, as John von Neumann soon argued, there was an inescapable price to pay. In his 1932 *Mathematical Foundations of Quantum Mechanics* (originally published in German), he developed a mathematical framework for quantum mechanics, as that of special linear operators in Hilbert spaces. (Heisenberg's uncertainty principle, for instance, was translated in the noncommutability of two corresponding operators). Von Neuman's treatment allowed him to confront the foundational issue of determinism versus nondeterminism, and in the book he tried to dissuade researchers from seeking a causal system underlying the extant quantum-mechanical one without altering the system in some manner. (This search is known as the search for hidden variables; his argument was later refuted: see the following.) "It is therefore not, as is often assumed, a question of a re-interpretation of quantum mechanics", wrote von Neumann about his (alleged) proof, "the present system of quantum mechanics would have to be objectively false, in order that another description of the elementary processes than the statistical one be possible".[16] Physicists and philosophers of science readily and almost universally accepted von Neumann's claim.[17]

The question remained, and appeared in the following wordings: Is quantum mechanics complete? Does it apply to single particles or only to ensembles of single particles?[18] Einstein, dissatisfied with probabilities and, more fundamentally, taking for granted that nature exists independently of the experimenter, sought a theory that describes the behavior of particles as precisely determined. It is the task of research to uncover comprehensive yet nonstatistical laws of nature. He took quantum

---

[16] John von Neumann, *Mathematische Grundlagen der Quantenmechanik*, Berlin: Springer, 1932; English translation by Robert T. Beyer, *Mathematical Foundations of Quantum Mechanics*, Princeton, NJ: Princeton University Press, 1955, p. 325.

[17] See, for example, Max Born: "No concealed parameters can be introduced with the help of which the indeterministic description could be transformed into a deterministic one. Hence if a future theory should be deterministic, it cannot be a modification of the present one but must be essentially different" (*Natural Philosophy of Cause and Chance*, Oxford: Oxford University Press, 1949, p. 109).

[18] This refers to the fact that a single electron interferes with itself, that in an electron-beam that shows interference patterns the interaction between the different electrons with each other is negligible.

mechanics in Born's reading as satisfactory as far as it goes, but he denied that it goes all the way. In other words, he deemed quantum mechanics as incomplete. The debate was fierce. It was argued that the precise measurement of one particle in two different places and times allows conclusions about its past path. Heisenberg dismissed this argument, declaring only predictions, not retrodictions, to be of concern for science. One might object that the interest in dinosaurs proves this wrong, but Heisenberg's apologists might answer that this is in the macroworld where classical science applies, and thus is irrelevant to the quantum world.

In 1935 Popper's *magnum opus*, his *Logik der Forschung*, appeared and advocated the view of science as the set of empirically testable theories, to wit, refutable ones. In effort to present quantum mechanics as testable, Popper offered an attempt to refute it by planning an experiment that might go beyond the limit of precision allowed by Heisenberg's principle. Heisenberg, Einstein, and others found a mistake in Popper's plan, since in it an electron passes a barrier that thus reduces the precision of its position or momentum.[19] Very soon afterward, Einstein, Boris

---

[19] Karl R. Popper, *Logik der Forschung: Zur Erkenntnistheorie der modernen Naturwissenschaft*, Vienna: Springer, 1935, pp. 172–181; English translation, *The Logic of Scientific Discovery*, London: Hutchinson, 1959, pp. 236–246; see also Appendix *XI, especially pp. 444–445. Popper's thought experiment may be derived from the Einstein, Podolsky, and Rosen thought experiment by imagining a film with a hole in it that the electron goes through, assuming that the film's position and momentum remain unchanged. Popper's error – first noted by Carl Friedrich von Weizsäcker (see Popper and von Weizsäcker's exchange in "Zur Kritik der Ungenauigkeitsrelationen", *Die Naturwissenschaften* 22, 1934, pp. 807–808), by Heisenberg (in private letters), and by Einstein in a letter reprinted in Appendix *XII ("The Experiment of Einstein, Podolsky, and Rosen") of Popper's *The Logic of Scientific Discovery*, cit., pp. 457–464 – was in ignoring the fact that in transition the electron gets "smeared" unpredictably. As a consequence, Popper withdrew his thought experiment, later to propose improved versions of it throughout the 1980s, at times in collaboration with physicists. Beginning in 2000, Popper's thought experiment appeared prominently in several papers published in journals of theoretical physics, giving rise to a heated discussion that leaves it as one of the open questions in the contemporary philosophical debate on quantum physics. In fact, more than realism is at stake: alongside with Einstein, Podolsky, and Rosen (as well as Vigier, Bohm, and Bell), Popper strongly opposed to the claim to finality and completeness of the standard interpretation. In his opinion – just as in that of Feyerabend after him – such claim was anathema, as it clashes with the realism of the critical attitude as expounded in *Logik der Forschung*. His aim was not to provide "a crucial experiment of quantum mechanics but only of its (subjectivist) Copenhagen interpretation (which they call 'the standard interpretation')" ("Popper versus Copenhagen", *Nature*, 328, 1987, p. 675). Michael Redhead said ("Popper and the Quantum Theory", in Anthony O'Hear (ed.), *Karl Popper: Philosophy and Problems*, Cambridge: Cambridge University Press, 1995, pp. 163–176: p. 176), "Popper's carefully argued criticisms won the support of a number of admiring and influential physicists. He has done a great service to the

Podolsky, and Nathan Rosen proposed a similar experiment, in which the barrier is replaced with a particle and the effect of the collision between the initial particle and its obstacle is measurable. (Their argument is known as the "EPR paradox".)[20] They argued that the uncertainty principle forbids certain precise measurements of the two particles, because the theory allows conclusions from results of one measurement of characteristics on one particle on the other and vice versa, even though the measurements can be performed when the two are at a great distance from each other. In the imaginary experiment with two particles speeding away from each other, but with correlated properties, to be precise, an observer could choose to find the position of the first particle by merely observing the second, and the momentum of the second particle by merely observing the position of the first. This way, the observer will find the precise position and momentum of both without violating the theory, yet while violating the principle of uncertainty. Hence, the EPR paradox does not refute the theory but only shows, or is claimed to show, that the theory is incomplete.

Bohr answered Einstein, Podolsky, and Rosen. He reaffirmed – at least twice – the assertion that the uncertainty of the measurement of characteristics of quantum objects is due to "the impossibility of controlling the reaction of the object on the measuring instruments, if these are to serve their purpose";[21] in other words, Bohr declared the EPR thought experiment not performable. He argued not so much for this claim as against Einstein's realist views.

By 1939 most of the younger theoretical physicists were convinced that Einstein's objections sprang only from nostalgia for the apparent certainties of nineteenth-century physics. Since 1945, however, a few physicists

philosophy of quantum mechanics by emphasizing the distinction between state preparation and measurement and trying to get a clearer understanding of the true significance of the uncertainty principle, but above all by spearheading the resistance to the dogmatic tranquilizing philosophy of the Copenhagenists. Because some detailed arguments are flawed, this does not mean that his overall influence has not been abundantly beneficial".

[20] Albert Einstein, Boris Podolsky, and Nathan Rosen, "Can Quantum-Mechanical Description of Physical Reality be Considered Complete?", *Physical Review*, 47, 1935, pp. 777–780. Bohr's reply – Niels Bohr, "Can Quantum-Mechanical Description of Physical Reality be Considered Complete?", *Physical Review*, 48, 1935, pp. 696–702 – left Einstein, Podolsky, and Rosen unconvinced. See his "Reply to Criticisms", cit., especially pp. 666–674.

[21] Bohr, "Can Quantum-Mechanical Description of Physical Reality be Considered Complete?", cit., p. 697; Bohr's reply is repeated in his "Discussion with Einstein on Epistemological Problems in Atomic Physics", in Schilpp (ed.), *Albert Einstein: Philosopher-Scientist*, cit., pp. 199–241.

once again have begun to criticize orthodox Copenhagen view, arguing that the statistical nature of quantum mechanics implied that it is only really applicable to ensembles of particles. The question, then, is, can research go further and find causal theories that accord with quantum mechanics, exactly or approximately? This is the question of hidden variables. The best known hidden-variables theory was that of the American physicist and philosopher David Bohm,[22] who in 1952 offered a detailed version of de Broglie's pilot-waves theory of a quarter of a century earlier. This showed von Neumann's celebrated proof wanting. As Bohm presented it, he managed to avoid the presuppositions of the proof. Heisenberg was swift to respond: the theory is metaphysical and so irrelevant. This response is obviously a rescue operation: irrelevant though Bohm's response may be, it rendered Einstein's dream of a causal completion of quantum theory possible again. Bohm distinguished between the quantum particle and a hidden pilot-wave that governs its motion; whether these waves exist or not, they represent the possibility of hidden variables thus rendering the principles of quantum mechanics somewhat less durable.

Bohm claimed[23] that theoretical speculation about subquantum physics is called for, von Neumann's arguments notwithstanding; he marshaled physical, historical, and philosophical arguments for this claim. In his view, alternative theories about the subquantum world will give observably different results, particularly concerning very high energies or the internal structure of the atom's nucleus. In presenting his own outline of a possible subquantum theory, Bohm employed explanatory models of the sort that Heisenberg had rejected. He compared the wave/particles of the quantum world to clouds or tidal waves, thereby representing transient configurations with blurred edges, continually forming, dissolving, and traveling across an underlying substratum (or "field") of energy. Accordingly, the statistics of orthodox quantum theory can once again be treated as statistics of a familiar kind that do not preclude causality. Finally, Bohm suggested to reconsider certain assumptions on which all physical theories have rested ever since the seventeenth century: in his view, just as Einstein had rejected some of Newton's assumptions in the

---

[22] In his first book, *Quantum Theory* (New York: Prentice-Hall, 1951), Bohm defended the Copenhagen interpretation, but soon thereafter he rejected his own former view and became one of the leading defenders of the hidden-variables theory.

[23] David Bohm, *Causality and Chance in Modern Physics*, London: Routledge & Kegan Paul, 1957.

theory of relativity, so we are compelled to reject Cartesian assumptions about space and geometry, drawing new concepts from topology.

The EPR thought experiment appeared impossible to perform. In 1957 Bohm and Yakir Aharonov presented a variant of it that is performable.[24] Nevertheless, the situation was not very clear, and it was in 1964 that John S. Bell helped dispel much of the fog that Bohr, Heisenberg, and von Neumann had created.[25] He used a simple case of a theorem that Maurice Fréchet had published in 1935 (Fréchet's inequalities),[26] known as Bell's inequalities. He applied these to quantum cases on the supposition that Einstein, Podolsky, and Rosen had used for their thought experiment: the assumptions of locality (of the proximity of cause and effect) and realism (the physical system is independent of its observer). These assumptions, together, are called "local realism", or "local hidden variables". He showed that no physical theory of local hidden variables can explain all of the predictive success of quantum mechanics.[27] The assumptions of local realism then prove quantum mechanics incomplete. In a previous paper (published only later, in 1966), Bell addressed the impossibility proofs that hidden variables are impossible, as von Neumann's proof contains a conceptual error (it relied on an assumption that is inapplicable to quantum theory: the probability-weighted average of the sum of observable quantities equals the sum of the average values of each of the separate observable quantities).[28] Alongside Einstein, Schrödinger, de Broglie,

[24] David Bohm and Yakir Aharonov, "Discussion of Experimental Proof for the Paradox of Einstein, Rosen, and Podolsky", *Physical Review*, 108, 1957, pp. 1070–1076.

[25] "Bohm showed explicitly how parameters could indeed be introduced, into nonrelativistic wave mechanics, with the help of which the indeterministic description could be transformed into a deterministic one. More importantly, in my opinion, the subjectivity of the orthodox version, the necessary reference to the 'observer', could be eliminated": John S. Bell, "On the Impossible Pilot Wave", *Foundations of Physics*, 12, 1982, pp. 989–999; reprinted in John S. Bell, *Speakable and Unspeakable in Quantum Mechanics: Collected Papers on Quantum Philosophy*, Cambridge: Cambridge University Press, 1987, pp. 159–168: p. 160.

[26] Maurice Fréchet, "Généralisations du théorème des probabilités totales", *Fundamenta Mathematicae*, 25, 1935, pp. 379–387.

[27] John S. Bell, "On the Einstein-Podolsky-Rosen Paradox", *Physics*, 1, 1964, pp. 195–200; reprinted in John S. Bell, *Speakable and Unspeakable in Quantum Mechanics*, cit., pp. 14–21.

[28] John S. Bell, "On the Problem of Hidden Variables in Quantum Theory", *Reviews of Modern Physics*, 38, 1966, pp. 447–452; reprinted in John S. Bell, *Speakable and Unspeakable in Quantum Mechanics*, cit., pp. 1–13. The supposed flaw had already been discovered by Grete Hermann in 1935, but her refutation remained nearly unknown for decades, until Bell rediscovered it. The alleged theorem had a strong influence.

and Bohm, Bell rejected the received interpretation of quantum theory, and called attention to the fact that empirical evidence does not at all force us to renounce realism. The long and the short of it is that due to Bell's clarifications, the EPR thought experiment was performed and its result corroborates the incredible quantum prediction that both Einstein and Bohr had deemed impossible. This is known as quantum entanglement: no matter how distant the two entangled particles are, the choice of variable of the one to measure limits the possible choice of the other one to measure.

This is the background of the physical and philosophical debate to which the papers collected in this volume belong. Paul Feyerabend's interests in the physical sciences – particularly astronomy, mechanics, and quantum theory – were deep. As a teenager, he attended Vienna's high school (*Realgymnasium*), at which he learned Latin, English, and science. His physics teacher was Oswald Thomas, a famous astronomer known for his works on popular astronomy. He was widely read in Austria and in Germany, and triggered interest in physics, especially in astronomy. Helped by his father, Paul "built a telescope from a bicycle and an old clothing stand", and "became a regular observer for the Swiss Institute of Solar research".[29]

I was interested in both the technical and the more general aspects of physics and astronomy, but I drew no distinction between them. For me, Eddington, Mach (his *Mechanics* and *Theory of Heat*), and Hugo Dingler (*Foundations of Geometry*) were scientists who moved freely from one end of their subject to the other.[30]

After the war he went back to Vienna with the intent to study physics, mathematics, and astronomy. Instead, he chose to read history and sociology, but he soon became dissatisfied with them and returned to theoretical physics. His teachers were Hans Thirring, Karl Przibram, and Felix Ehrenhaft. The last was a critic of all orthodoxy in physics; many physicists considered him a charlatan. Feyerabend much appreciated his fearless iconoclasm. He must have been successful as a student, since in 1948 and in 1949 he was offered grants to attend the international summer seminar of the Austrian College Society in Alpbach. In 1948 he met Karl Popper there, and impressed him sufficiently to receive his help

---

[29] Paul K. Feyerabend, *Killing Time*, Chicago–London: The University of Chicago Press, 1995, p. 29.
[30] Ibid., p. 30.

to attain a scholarship to go to England. Before reaching England, in 1949, he witnessed a much-expected clash between Ehrenhaft and representatives of "the orthodoxy". Years later, in *Science in a Free Society*, he reported:

Ehrenhaft gave a brief account of his discoveries adding general observations on the state of physics. "Now gentlemen" he concluded triumphantly, turning to Rosenfeld and Pryce who sat in the front row – "what can you say?". And he answered immediately. "There is nothing at all you can say with all your fine theories. *Sitzen müssen sie bleiben! Still müssen sie sein!*".

The discussion, as was to be expected, was quite turbulent and it was continued for days with Thirring and Popper taking Ehrenhaft's side against Rosenfeld and Pryce. Confronted with the experiments the latter occasionally acted almost as some of Galileo's opponents must have acted when confronted with the telescope. They pointed out that no conclusions could be drawn from complex phenomena and that a detailed analysis was needed.[31]

At the time, Feyerabend continues, such heated discussions had little effect on him:

None of us was prepared to give up theory or to deny its excellence. We founded a Club for the Salvation of Theoretical Physics and started discussing simple experiments. It turned out that the relation between theory and experiment was much more complex than is shown in textbooks and even in research papers. ... We continued to prefer abstractions if the difficulties we had found had not been an expression of the nature but could be removed by some ingenious device, yet to be discovered. Only much later did Ehrenhaft's lesson sink in and our attitude at the time as well as the attitude of the entire profession provided me with an excellent illustration of the nature of scientific rationality.[32]

Ehrenhaft's lesson sunk in much later – after attending Popper's lectures and seminars at the London School of Economics, the best school for sharpening one's critical acumen – and then Feyerabend started publishing on the philosophy of quantum mechanics. He found the dominance achieved by the Copenhagen interpretation quite undeserved; he found it incredible that this interpretation was considered the last word on the matter – by scientists and philosophers of science alike. His early works are the products of his study with Popper, whose unorthodox views on the philosophical interpretation of quantum theory are the concern of chapter 7 of *Logik der Forschung* (1935) and more so in the *Postscript* to its English edition that Popper was working on then, as well as in a few

---

[31] Paul K. Feyerabend, *Science in a Free Society*, London: New Left Books, 1978, p. 111; see also *Killing Time*, cit., pp. 65–67.

[32] Paul K. Feyerabend, *Science in a Free Society*, cit., p. 111.

other works Popper published in the 1950s.[33] In his very first publications, Feyerabend focuses on quantum theory as one of the most interesting examples of the way in which philosophical speculation, empirical research, and mathematical ingenuity jointly contribute to the development of physical theory. He sides with Popper – as well as with Einstein, de Broglie, Bohm and Vigier – in challenging the orthodoxy of the Copenhagen interpretation and advocating a realistic interpretation of quantum mechanics. As its orthodox interpreters, especially the "logical" positivists among them, tried to strip it of its metaphysical features, they rendered it a mere prediction device, no longer requiring researchers to provide an account of the atomic world as it exists independently of observation and experiment. (Reichenbach went so far as to exclude from scientific theory events that take place between observations, which he called "inter-phenomena".) Feyerabend openly distances himself by most logical positivists.[34]

---

[33] Feyerabend made extensive annotations throughout a copy of Popper's *Logik der Forschung*, particularly in the chapters devoted to probability and quantum mechanics. Later, he had access to the manuscript of *Postscript to The Logic of Scientific Discovery* that was published in a completely reworked form some thirty years later, in three volumes. The first two volumes of the *Postscript* were in galley proofs in the mid-1950s and the third volume in the late 1950s. In the published version, the first two volumes would be devoted to realism and indeterminism – two of the key issues repeatedly discussed in the papers collected in the present volume, and the third volume to the quantum paradoxes and Popper's propensity interpretation of probability as applied to the interpretation of quantum physics. See his *Postscript to The Logic of Scientific Discovery*, edited by William W. Bartley, III, vol. 1: *Realism and the Aim of Science*, vol. 2: *The Open Universe: An Argument for Indeterminism*, and vol. 3: *Quantum Theory and the Schism in Physics*, London: Hutchinson, 1982–1983. The central theme of the third volume was presented by Popper at the Ninth Symposium of the Colston Research Society, in Bristol, which Feyerabend attended, too (see *PP1*, pp. 207–218, as well as Chapter 16 of the present collection: here Feyerabend introduces the traditional thesis that would be central to his later work, namely, that observations are theoretically biased, and inevitably so). Feyerabend's own work developed in close parallel with Popper's: in addition to the latter's *The Logic of Scientific Discovery*, London: Hutchinson, 1959, chs. 8–9, see his "The Propensity Interpretation of the Calculus of Probability, and the Quantum Theory", in Stephan Körner and Maurice H. L. Pryce (eds.), *Observation and Interpretation: A Symposium of Philosophers and Physicists*, New York: Academic Press Inc., Publishers, and London: Butterworths Scientific Publications, 1957, pp. 65–70 and 88–89, his "The Propensity Interpretation of Probability", *The British Journal for the Philosophy of Science*, 10, 1959, pp. 25–42; and his "Philosophy and Physics: The Influence on Physics of Some Metaphysical Speculations on the Structure of Matter", in *Atti del XII Congresso Internazionale di Filosofia (Venezia, 12–18 Settembre, 1958)*, Venice: G. F. Sansoni, 1960, vol. 2, pp. 367–374.

[34] Feyerabend ridicules Reichenbach's interphenomena thesis: see his "Reichenbach's Interpretation of Quantum Mechanics", *Philosophical Studies*, 9, 4, 1958, pp. 47–59; reprinted in *PP1*, pp. 236–246.

Most importantly, Feyerabend was struck by the attitude of orthodox physicists toward the gaps of quantum theory. Although they admitted that it would have to undergo some decisive changes in order to cope with some new discoveries and that the future new theories will introduce new concepts for the description of the new facts, they insisted that the basic elements of current theory would remain unchanged. The basic structure of the theory did not require a revision, and any modifications would not affect its indeterminist framework. By contrast, in Feyerabend's view the Copenhagen interpretation was but one possible interpretation of the quantum formalism. He upheld a pluralistic approach, as opposed to Thomas Kuhn's advocacy of conformism within a scientific research community. So Feyerabend came to defend the right of "hidden-variables" theorists, such as David Bohm, whom he admired. (They were both at the University of Bristol at the time.) Only a realistic interpretation, he said, can reveal the revolutionary potential of scientific theories.

Bohm called attention to some aspects of microphysics that he deemed problematic and most physicists deemed settled. This was a clash of ideas that intrigued Feyerabend. Whereas it is often assumed – both in philosophy and in the sciences, not to mention the community of scholars at large – that within the sciences theories are (almost) uniquely determined by facts, so that speculation and ingenuity have a limited role to play, Bohm's (and, later on, Bell's) questions indicated that the notorious divide between the sciences and the humanities is due to this very erroneous picture of science. Bohm opposed the received view

that complementarity, and complementarity alone, solves all the ontological and conceptual problems of microphysics; that this solution possesses absolute validity; that the only thing left to the physicist of the future is to find, and to solve equations for the prediction of events which are otherwise well understood.[35]

The claim that the Copenhagen interpretation of quantum mechanics is the only possible interpretation allowed by experimental results, then, is downright dishonest. Feyerabend said this in 1960 in a few letters to Kuhn, upon reading the first draft of what would be published in 1962 as *The Structure of Scientific Revolutions*. The issue, in this case, is historical reconstructions, not interpretations of a physical theory, but the argument is exactly the same, and in his letters quantum physics is often referred to:

What you are writing is not just history. It is *ideology covered up as history* ... points of view *can* be made explicit, and it *is* possible to write history in such a

---

[35] Feyerabend, "Professor Bohm's Philosophy of Nature", *The British Journal for the Philosophy of Science*, 10, 1960, pp. 321–338; reprinted in *PP1*, pp. 219–235: p. 219.

manner that the reader is always aware of one's ideology or point of view *as well as of the possibility of an alternative interpretation of the historical facts.* That is, history can be written in such a manner *that what is factual and what is reasonable appear as two clearly distinct affairs.* ... What I do object to most emphatically is the way you present this belief of yours; you present it not as a *demand*, but as something that is an obvious consequence of historical facts.[36]

According to Popper, experiment does not impose the strange consequences drawn from quantum theory; an erroneous philosophical approach to physics does that. It is positivism: Bohr and Heisenberg, Popper claimed, were seduced by traditional positivists such as Ernst Mach as well as by the new ones, the "logical" positivists, including the members of the Vienna Circle. Their theory was not logically true but hypothetical, and erroneous. Feyerabend disagreed. He claimed that the Copenhagen theorists had some perfectly good physical arguments for thinking that their view alone was compatible with the observed results of experiments, and he offered a defence of their instrumentalist interpretation. Ultimately, however, he argued for the necessity that the observed results of experiments themselves be challenged, thereby using the case of quantum theory (as, in other contexts, he appealed to Galileo's case, or to other cases from the rich history of science) to push for a reconsideration of the methodological rules by which researchers are abiding or declare to be abiding. Here we may find, *in nuce*, Feyerabend's pluralistic test model,[37] in which theories are contrasted with one another as well as with experience: "the methodological unit to which we must refer when discussing questions of test and empirical content is constituted by a

---

[36] Paul Hoyningen-Huene, "Two Letters of Paul Feyerabend to Thomas S. Kuhn on a Draft of *The Structure of Scientific Revolutions*", cit., p. 355; see also pp. 356, 360, 367–368 and 379–380. In another letter, Feyerabend describes the Copenhagen interpretation not as a paradigm, as Kuhn did, but as "what remains of a former paradigm (the classical theories) when this has been freed from anything that goes beyond experience. ... They have not simply added *another* theory to the theories of the past which at some future time may be replaced by again another theory. ... From now on we have entered a new age of scientific activity. There will be no more revolutions, there will be only accumulation" (ibid., p. 379). See also Paul Hoyningen-Huene, "More letters by Paul Feyerabend to Thomas S. Kuhn on *Proto-Structure*", *Studies in History and Philosophy of Science*, 37, 2006, pp. 610–632.

[37] Feyerabend's theoretical pluralism (scientific progress is enhanced by the simultaneous presence of a sufficiently large number of competing theories), advocated in his early works, is not to be confused with his later methodological pluralism (science has no distinctive method, therefore anything goes). "Theoretical pluralism (that is, Feyerabend's pluralistic methodology)", writes John Preston (*Feyerabend: Philosophy, Science and Society*, Cambridge: Polity Press, 1997, p. 139), "is intended to be a single methodology for all scientific inquiry. It sponsors the proliferation of theories, but not of methods for evaluating theories".

whole set of partly overlapping, factually adequate, but mutually inconsistent theories".[38] Otherwise, he suggested, there would be no more arguing or judging among disciplines: criticism, evaluation, and explanation would no longer be the aims of proper philosophical discourse. All philosophers would be left with, then, would be descriptions of the logics, grammars, or first principles of the various kinds of discourse, and the many sorts of language games and forms of life in which they are embedded. Philosophical critique would no longer be of content, but of criteria application; as Feyerabend put it, all that would be left are "consolations for the specialist".[39]

Stefano Gattei

Joseph Agassi

---

[38] Feyerabend, "How to Be a Good Empiricist: A Plea for Tolerance in Matters Epistemological", in Bernard Baumrin (ed.), *Philosophy of Science: The Delaware Seminar*, vol. 2, New York: Interscience, 1963, pp. 3–39; reprinted in *PP3*, pp. 78–103: p. 92. This was but an extension of ideas Popper had already formulated in *The Logic of Scientific Discovery* and elsewhere. In the 1962 original version of "Explanation, Reduction, and Empiricism" (in Herbert Feigl and Grover Maxwell (eds.), *Scientific Explanation, Space and Time*, Minneapolis: University of Minnesota Press, 1962, pp. 28–97: pp. 31–32), Feyerabend readily acknowledged this; later he withdrew the acknowledgement – after Popper had called attention to it. See Popper, *Objective Knowledge: An Evolutionary Approach*, Oxford: Clarendon Press, 1972, p. 205; and Feyerabend, "Explanation, Reduction, and Empiricism", in *PP1*, pp. 44–96: p. 47, footnote 6. This implies that Popper never advocated a monistic model, according to which a single theory is tested against "experience". Although there is hardly any passage in which Feyerabend explicitly associated Popper with this thesis, a number of Feyerabend scholars assume that he did.

[39] Paul K. Feyerabend, "Consolations for the Specialist", in Imre Lakatos and Alan Musgrave (eds.), *Criticism and the Growth of Knowledge*, Cambridge: Cambridge University Press, 1970, pp. 197–230; see also his "Kuhns Struktur wissenschaftlicher Revolutionen: Ein Trostbüchlein für Spezialisten?", in Paul K. Feyerabend, *Der wissenschaftstheoretische Realismus und die Autorität der Wissenschaften*, Braunschweig: Vieweg, 1978, pp. 153–204.

Editorial Note

Two volumes of collected philosophical papers were edited by Feyerabend himself and appeared in 1981; he published another collection in
1987. In 1999, five years after his demise, John Preston edited a third
volume of his collected papers, and Bert Terpstra saw through the press
Feyerabend's last (unfinished) manuscript, to which he attached a number
of previously published essays dealing with its main themes. These books
will be referred to as follows:

PP1   Realism, Rationalism and Scientific Method: Philosophical Papers,
      vol. 1, Cambridge: Cambridge University Press, 1981.

PP2   Problems of Empiricism: Philosophical Papers, vol. 2, Cambridge:
      Cambridge University Press: Cambridge 1981.

FR   Farewell to Reason, London: Verso/New Left Books, 1987.

PP3   Knowledge, Science and Relativism: Philosophical Papers, vol. 3,
      edited by John M. Preston, Cambridge: Cambridge University
      Press, 1999.

CA   The Conquest of Abundance: A Tale of Abstraction versus the
      Richness of Being, edited by Bert Terpstra, Chicago–London: The
      University of Chicago Press, 1999.

Discussions of specific issues and detailed analyses of problems related
to contemporary physics are scattered throughout Feyerabend's
works. They appeared in various forms: as journal articles, book
chapters, reviews, and comments, as well as in book form. Unlike the
previous volumes, which cover a variety of issues in the philosophy of
science, the present collection focuses on Feyerabend's papers on the

philosophy of physics, especially on quantum mechanics. It was compiled with a dual purpose in mind. First, the editors aim at providing the community of scholars with the ultimate collection of Feyerabend's papers on physics, including a few papers originally published in German (among them Feyerabend's first published work, unsigned) and that are now available in English for the first time. Second, we exclude items that are available in his earlier collections (except for Chapter 7; see following sources). The most important of these are listed here:

"On the Quantum-Theory of Measurement", in Stephan Körner and Maurice H. L. Pryce (eds.), *Observation and Interpretation: A Symposium of Philosophers and Physicists*, New York: Academic Press Inc., Publishers, and London: Butterworths Scientific Publications, 1957, pp. 121–130; reprinted in *PP1*, Part 2, Ch. 13 (pp. 207–218). A preliminary version of this paper appeared as "Zur Quantentheorie der Messung", *Zeitschrift für Physik*, vol. 148, no. 5 (October 1957), pp. 551–559.

"An Attempt at a Realistic Interpretation of Experience", *Proceedings of the Aristotelian Society*, New Series, vol. 58 (1957–1958), pp. 143–170; reprinted in *PP1*, Part 1, Ch. 2 (pp. 17–36).

"Reichenbach's Interpretation of Quantum Mechanics", *Philosophical Studies*, vol. 9, no. 4 (June 1958), pp. 47–59; reprinted in *PP1*, Part 2, Ch. 15 (pp. 236–246).

"On the Interpretation of Scientific Theories", in *Atti del XII Congresso internazionale di filosofia: Venezia, 12–18 settembre 1958*, vol. V: *Logica, gnoseologia, filosofia della scienza, filosofia del linguaggio*, Florence: G. C. Sansoni, 1960, pp. 151–169; reprinted in *PP1*, Part 1, Ch. 3 (pp. 37–43).

"Das Problem der Existenz theoretischer Entitäten", in Ernst Topitsch (ed.), *Probleme der Wissenschaftstheorie: Festschrift für Victor Kraft*, Vienna: Springer, 1960, pp. 35–72; English translation by Daniel Sirtes and Eric Oberheim, "The Problem of the Existence of Theoretical Entities", in *PP3*, Ch. 1 (pp. 16–49).

"Professor Bohm's Philosophy of Nature" (review of David Bohm, *Causality and Chance in Modern Physics*, Foreword by Louis de Broglie, London: Routledge & Kegan Paul, and Princeton, NJ: Van Nostrand, 1957), *The British Journal for the Philosophy of Science*, vol. 10, no. 40 (February 1960), pp. 321–338; reprinted in *PP1*, Part 2, Ch. 14 (pp. 219–235).

"On a Recent Critique of Complementarity: Part I", *Philosophy of Science*, vol. 35, no. 4 (December 1968), pp. 309–331, and "On a Recent Critique of Complementarity: Part II", *Philosophy of Science*, vol. 36, no. 1 (March 1969), pp. 82–105; reprinted together in *PP1*, Part 2, Ch. 16, under the title "Niels Bohr's world view" (pp. 247–297).

"Zahar on Mach, Einstein and Modern Science", *The British Journal for the Philosophy of Science*, vol. 31, no. 3 (September 1980), pp. 273–282; reprinted in *PP2*, Ch. 6, under the title "Mach, Einstein and the Popperians" (pp. 247–297).

"Realism and the Bohr-Rosenfeld Condition", Appendix (1981) to "Consolations for the Specialist", in *PP2*, Ch. 8 (pp. 162–167).

"Mach's Theory of Research and Its Relation to Einstein", *Studies in History and Philosophy of Science*, vol. 15, no. 1 (March 1984), pp. 1–22; reprinted in *FR*, Ch. 7 (pp. 192–218).

"Quantum Theory and Our View of the World", *Stroom. Mededelingenblad Faculteit Natuur-en Sterrenkunde*, vol. 6, no. 28 (1992), pp. 19–24; reprinted in *CA*, Part 2, Ch. 3 (pp. 161–177).

Related works, still absent from this series, are Feyerabend's Ph. D. dissertation: *Zur Theorie der Basissätze*, University of Vienna, 1951 (unpublished), as well as two very short entries on Niels Bohr and Ernst Mach that appeared posthumously in *The Oxford Companion to Philosophy*, edited by Ted Honderich and published by Oxford University Press in 1995 (on pp. 98 and 516, respectively) and republished in a new edition in 2005 (see pp. 102 and 549, respectively).

For a complete list of Feyerabend's works, see the constantly updated online chronological and annotated bibliographies by Matteo Collodel at www.collodel.org/feyerabend/.

The papers collected in the present volume vary considerably in size, purpose and style: their features reflect the time in which they were written and the context in which they were originally inserted. For these reasons, they have been organized into three sections: (1) original papers and book chapters, (2) reviews and comments in conference proceedings, and (3) encyclopedia entries. Within each section, works follow a chronological order. They are here reprinted as they originally appeared. Editorial interventions were limited to tacit correction of typographical errors, as well as to slight modifications of punctuation, for reasons of uniformity. A few bibliographical references, mainly to the republication of Feyerabend's works, as well as a few other editorial interventions, in the body of the texts or in the footnotes, are clearly marked by curly brackets (the only exception being Chapter 15, where Feyerabend himself used curly brackets in the body of the text).

In order to facilitate references to the original publications, page numbers there are given here within square brackets in bold superscript fonts. All notes appear as footnotes. Endnotes turned into footnotes include added references to their original pagination – within square brackets, in bold superscript fonts.

SOURCES OF THE WORKS COLLECTED
IN THE PRESENT VOLUME

1. "Der Begriff der Verständlichkeit in der modernen Physik", in *Veröffentlichungen des Österreichischen College*, Vienna: Österreichischen College, 1948, pp. 6–10.

2. "Physik und Ontologie", *Wissenschaft und Weltbild: Monatsschrift für alle Gebiete der Forschung*, vol. 7, nos. 11–12 (November–December 1954), pp. 464–476.

3. "Determinismus und Quantenmechanik", *Wiener Zeitschrift für Philosophie, Psychologie, Pädagogik*, vol. 5, no. 2 (1954), pp. 89–111.

4. "Eine Bemerkung zum Neumannschen Beweis", *Zeitschrift für Physik*, vol. 145, no. 4 (August 1956), 421–423.

5. "Complementarity", *Proceedings of the Aristotelian Society, Supplementary Volumes*, vol. 32 (1958), 75–104.

6. "Niels Bohr's Interpretation of the Quantum Theory", in Herbert Feigl and Grover Maxwell (eds.), *Current Issues in the Philosophy of Science: Symposia of Scientists and Philosophers, Proceedings of Section L of the American Association for the Advancement of Science, 1959*, New York: Holt, Rinehart & Winston, 1961, pp. 371–390.

   The paper is followed by Feyerabend's "Rejoinder to Hanson", ibid., pp. 398–400.

7. "Problems of Microphysics", in Robert G. Colodny (ed.), *Frontiers of Science and Philosophy* (University of Pittsburgh Series in the Philosophy of Science, vol. 1), Pittsburgh, PA: University of Pittsburgh Press, 1962, pp. 189–283.

   A preliminary (almost final) Polish translation appeared as "O Interpretacij Relacyj Nieokreslonosci [On the Interpretation of the Uncertainty Relations]", *Studia Filozoficzne*, vol. 19, no. 4 (1960), pp. 21–78.

This paper was partially (§§4–11) reprinted, in a slightly altered version, in *PP1*, Ch. 17, under the title "Hidden Variables and the Argument of Einstein, Podolsky and Rosen" (pp. 298–342).

8. "Über konservative Züge in den Wissenschaften, insbesondere in der Quantentheorie, und ihre Beseitigung", in *Club Voltaire: Jahrbuch für kritische Aufklärung*, vol. 1, edited by Gerhard Szczesny, Munich: Szczesny Verlag, 1963, pp. 280–293.

9. *Problems of Microphysics*, The Voice of America. Forum Philosophy of Science Series, no. 17, Washington: U. S. Information Agency, 1964.

10. "Eigenart und Wandlungen physikalischer Erkenntnis", *Physikalische Blätter*, vol. 21, no. 5 (May 1965), pp. 197–203.

11. "Dialectical Materialism and the Quantum Theory", *Slavic Review*, vol. 25, no. 3 (September 1966), pp. 414–417.

12. "Bemerkungen zur Verwendung nicht-klassischer Logiken in der Quantentheorie", in Paul Weingartner (ed.), *Deskription, Analytizität und Existenz: Forschungsgespräche des internationalen Forschungszentrums für Grundfragen der Wissenschaften Salzburg, drittes und viertes Forschungsgespräch*, Salzburg-Munich: Anton Pustet, 1966, pp. 351–359.

13. "On the Possibility of a Perpetuum Mobile of the Second Kind", in Paul K. Feyerabend and Grover Maxwell (eds.), *Mind, Matter, and Method: Essays in Philosophy and Science in Honor of Herbert Feigl*, Minneapolis: University of Minnesota Press, 1966, pp. 409–412.

14. "In Defence of Classical Physics", *Studies in History and Philosophy of Science*, vol. 1, no. 1 (May 1970), pp. 59–85.

15. Review of Alfred Landé, *Foundations of Quantum-Mechanics: A Study in Continuity and Symmetry* (New Haven, CT: Yale University Press, 1955), *The British Journal for the Philosophy of Science*, vol. 7, no. 28 (February 1957), pp. 354–357.

16. "Discussions with Léon Rosenfeld and David Bohm (and others)", excerpts from Stephen Körner and Maurice H. L. Pryce (eds.), *Observation and Interpretation: A Symposium of Philosophers and Physicists*, New York: Academic Press Inc., Publishers, and London: Butterworths Scientific Publications, 1957, pp. 48–57, 112–113, 138–147, and 182–186.

17. Review of John von Neumann, *Mathematical Foundations of Quantum Mechanics*, translated by Robert T. Beyer (Princeton, NJ: Princeton University Press, 1955), *The British Journal for the*

Philosophy of Science, vol. 8, no. 32 (February 1958), pp. 343–347.

18. Review of Hans Reichenbach, *The Direction of Time* (Berkeley–Los Angeles: University of California Press, 1956), *The British Journal for the Philosophy of Science*, vol. 9, no. 36 (February 1959), pp. 336–337.

19. "Professor Landé on the Reduction of the Wave Packet", *American Journal of Physics*, vol. 28, no. 5 (May 1960), pp. 507–508.

20. "Comments on Grünbaum's 'Law and Convention in Physical Theory'", in Herbert Feigl and Grover Maxwell (eds.), *Current Issues in the Philosophy of Science: Symposia of Scientists and Philosophers, Proceedings of Section L of the American Association for the Advancement of Science, 1959*, New York: Holt, Rinehart & Winston, 1961, pp. 155–161.

21. "Comment on Hill's 'Quantum Physics and Relativity Theory'", ibid., pp. 441–443.

22. Review of Norwood R. Hanson, *The Concept of the Positron: A Philosophical Analysis* (New York: Cambridge University Press, 1963), *The Philosophical Review*, vol. 73, no. 2 (April 1964), pp. 264–266.

23. Review of Hans Reichenbach, *Philosophic Foundations of Quantum Mechanics* (Berkeley–Los Angeles: University of California Press, 1965), *The British Journal for the Philosophy of Science*, vol. 17, no. 4 (February 1967), pp. 326–328.

24. "Naturphilosophie", in Alwin Diemer and Ivo Frenzel (eds.), *Das Fischer Lexikon: Enzyklopädie des Wissens*, vol. 11: *Philosophie*, Frankfurt am Main: Fischer Bücherei, 1958, pp. 203–227.

25. "Philosophical Problems of Quantum Theory" (1964), Philosophisches Archiv der Universität Konstanz, *Paul Feyerabend Nachlaß* (PF 11-12-3), 38 pages.

    As with the next few chapters, this was meant to be an entry to *The Encyclopedia of Philosophy*, edited by Paul Edwards and published in 1967: see Chapter 12, footnote 1.

26. "Boltzmann, Ludwig", in Paul Edwards (ed.), *The Encyclopedia of Philosophy*, New York-London: Macmillan, 1967, vol. 1, pp. 334–337.

27. "Heisenberg, Werner", ibid., vol. 3, pp. 466–467.

28. "Planck, Max", ibid., vol. 6, pp. 312–314.

29. "Schrödinger, Erwin", ibid., vol. 7, pp. 332–333.

PART ONE

PAPERS AND BOOK CHAPTERS
(1948–1970)

The Concept of Intelligibility in Modern Physics
(1948)

[6](Written as an antithesis, after a discussion at the philosophical and physical scientific study-groups of the *Collegegemeinschaft* of Vienna, to Prof. Schrödinger's essay, "Die Besonderheit des Weltbildes der Naturwissenschaften".)[1]

In modern physics, it is often said that it is utterly impossible to grasp what a philosopher understands by the concept of outside world. By contrast, it is maintained that some sort of regularity forces the physicist to hold fast to the phenomena, and remove from his mental picture all those elements that display no references to the phenomena. This purely descriptive attitude has become known as positivism, and has rapidly fallen into discredit among a large number of philosophers. Therefore, we have to try to provide a description of the foundations of this peculiar method, as well as of its epistemological assumptions.

We may proceed in two different ways. We may either let physics and philosophy have their say together, and listen to the discussion that emerges. Yet, I do not believe that much edifying may come from this, not the least because in the past century there has been a very important conceptual change in both disciplines.

The second approach, which will be followed here, is more indirect and leads through concepts that are generally common in everyday life (which, indeed, is the common point of departure of the exact sciences as well as of

[1] Erwin Schrödinger, "Die Besonderheit des Weltbildes der Naturwissenschaften", *Acta Physica Austriaca*, 1, 1948, pp. 201–245; translated into English as "On the Peculiarity of the Scientific World-View", in Erwin Schrödinger, *What is Life? And Other Scientific Essays*, Garden City: Doubleday, 1956, pp. 178–228.

philosophy). We therefore replace the [2]pair of concepts, real external world/phenomenon, which represents our difficulty in philosophical terms, with the pair of concepts intelligible/abstract, and see what comes from it. First, the concept of intelligibility itself; in the natural sciences, we have always sought to resolve all phenomena available to the senses into simple visual models, and in so doing to make the mechanism intelligible. Such models explain macroscopic regularity, but do not themselves require any further explanation. They are immediately clear, evident, vivid, if we wish to put it this way. At this point, we can already see that the concept of intelligibility often almost coincides with that of vividness. In most cases, however, it is not just about whether this or that model can be pictured (we will indicate this with I), but rather, it is essentially about what laws it obeys (indicated with II). So, for example, a chair, which changes its size when we bring it to different places in a room, is vivid in the sense of no. I. It is, to be sure, not usual for chairs to behave in this way, but the processes at work can be observed, measured, and, in short, pictured. A vivid rendition in the second sense additionally means, however, that we expect the pictured object to behave like the familiar things we are accustomed to. In the case of Greek atomists, such a vivid rendition presupposes, quite primitively, that everything that happens can be traced back to collisions; whereas in the case of classical mechanics, to the motion of attracting masses. In the one case, it is a model that became plausible through the behavior of things in the immediate environment; in the other, it is a conception that comes from the regularity of planetary orbits, which was already understood. From the first point of view, the motion of the planets and the law that underlies it seems to be incomprehensible, absurd, and from the very beginning they tried to replace it with the strain properties of the intervening medium. The abundance of theories that appeared at the time, to which Newton was by no means the least contributor, is a psychologically interesting indication of how is it possible to grasp the concept of picturing. The dictum about the absurdity of action at a distance tells us nothing about the forces working in the universe. Today we know that very well. Rather, it tells us something about the way of thinking of those who cannot imagine something other than push, pull, or pressure, since these were the only kinds of forces in the immediate surroundings known at the time. Independently of that, Newton analyzes the relationships of the motion of the planets, and ascertains the law from which all the planetary orbits can be simply derived, by superposing a constant velocity factor. It was the first practical application of a way of thinking that today has become known as positivist.

It is very instructive, by contrast, to look at physics in the 18th and 19th centuries, which specifically analyzes and resolves, by way of models, the phenomena under consideration from the opposite side. It is the era of celestial mechanics, in which action at a distance became plausible to the extent that they even sought to trace immediately evident phenomena, such as the elastic reflection of a billiard ball from a wall, back to complex structured forces acting at a distance. Laplace's theory of capillary pressure is the best example of the extent to which the concept of intelligibility is subject to change, and how little the failure to picture a theory can be used as an argument against its content – an argument that certain physicists still [8]put forward against the modern development of the natural sciences.

To summarize:

Intelligible is any such regularity to which we have become accustomed through long use, whose structure is understood from itself. Thus, first, the regularities of the local environment; and next, those of distant environments which are directly accessible to us (celestial mechanics).

Toward the end of the previous century, they even tried to obtain models for the regularity of atoms on the basis of celestial mechanics. But from the beginning we can say that this approach cannot be in any way epistemologically justified, other than by the (heuristically important) principle of continuity, and that in the end success is but a matter of chance. For, in this case, we expect that atoms behave no more and no less than the objects of the world to which we have been so far accustomed; that the laws that our tables, chairs and bathtubs as a whole obey may also account for the emission of spectral lines or for the structural relations of atomic nuclei. If we look at this hypothesis in its full significance, we will not be surprised if, in the long run, there are certain divergences, which, with all good will, cannot be made sense of so primitively.

But now as to the approach itself:

First, we know the atomic weight through measurements of the density of the elements. The velocity, and maybe also the rough structure (spherical, elliptical, dumbbell-shaped) of molecules, follow from thermodynamics and the molecular theory of gases. Thomson's experiments established the fact that every atom contains the same amount of positive and negative electricity. We knows the necessity of repulsive forces from the law of the attraction of electricity. In analogy to the regularity of the planetary orbits, Bohr constructed a model in which the centrifugal force of the revolving electron takes over the role of the repulsive force required. So far, everything is quite clear and satisfactory. Yet, the matter gets disputable as soon as we further consider the conditions of motion in atoms according to the classical laws: first, each nonuniformly moving charge is a source of electromagnetic radiation; secondly, according to the law of the conservation of energy, the radiation process produces a loss of the motion, so that the atom would eventually collapse under continuous emission.

Against this we have:

   1. The stability of atoms.

   2. The sharpness of spectral lines.

As a consequence, the model appears to be useless. But now (principle of continuity) begin the attempts to rescue it (auxiliary hypotheses):

  (a) The electron, to be sure, goes around the core; but it is at the same time unable to emit radiation.

  (b) It circles only on certain orbits (consequence of the quantization principle); hence, it does not have the capability of planets to describe arbitrary orbits.

  (c) The emission and absorption process produces a sudden change in the direction of the electrons, which – though not bound to any specific orbit – leads the electron discontinuously to the next energy level.

[9]Each of the points (a), (b), and (c) eliminates a classical law. The model, to be sure, is still clear, but by now only in the sense of no. I. It resembles more a haunted house than a physical edifice.

We see what is going on here: the approach of the classical model will necessarily be transformed until nothing remains of it. (Last remnant: Sommerfeld's smeared electron [*Wimmelelektron*].)[2]

This raises the question whether this anticipation of classical laws is in any way an appropriate route to a satisfactory conception of the structure of atoms, whether we do not here stand in front of what is in principle a new field, which cannot be grasped with pictures from the world of tables and chairs, and whether it is not methodologically more effective first to simply register those regularities without immediately referring them to a structured carrier, "*atom*". This is the position of modern physics. We might find it unsatisfactory, just as the Cartesians found the idea of direct action at a distance unsatisfactory, although shortly thereafter they thought they understood these regularities. Now, just as then, we are dealing with a transitional phase, at the end of which we will think that

---

[2] *Wimmelelektron* seems to be a coinage of Feyerabend's. Johannes Stark used the term *Wimmelbewegung* in his article "Die Kausalität im Verhalten des Elektrons", *Annalen der Physik*, 6, 1930, pp. 681–699: pp. 681, 684–686 (and elsewhere), to refer to Brownian motion; "smeared" gives the idea of swarming, and fits well with Feyerabend's use of the term.

a different way of thinking is clear and intelligible. But, in that case as in the present one, all elements of the earlier way of thinking must be resolutely removed, in order to allow for the new regularities to emerge. *This* is the position of today's positivism. It makes it possible to formulate the connections that will appear intelligible tomorrow.

Its line of action is radical: "atom" is not this or that thing, but the sum of phenomena known in a certain domain. This is to be understood in this way: the phenomenal differentiation of elements leads to a preliminary, primitive classification (periodic system of the first kind). Finer investigations, which admittedly presuppose the identity of the substrate, allow us to find the nuclear shell as the chief feature of comparison, according to which the sequence of elementary building blocks can be ordered. Classified in this way, these elements are examined with respect to their spectrum, of their behavior in a magnetic field, etc. In this way, a series of regularities comes to light, which are assigned to the elements of the corresponding atomic number. The sum of all these regularities is then "the atom X".

Now, from the laws themselves follow certain quantities, which turn out be largely independent of changing external conditions (mathematically speaking: of arbitrary transformations), and which themselves still retain such invariance, even if the building blocks of earlier physics have already been through several changes. A well-known example of such a quantity is the interval in the general theory of relativity. It may look as if the transformation of space and time coordinates (which so far had a meaning independently of velocity) opens the door to all sorts of unpredictability. Nevertheless, even here there is a quantity, admittedly not directly observable, which turns out to be completely independent of velocity and gravitational deformation. This shows that the objects of our perception cannot be the ultimate invariants, and therefore are also unsuited for an invariant representation of all laws of nature.

Once we have clarified these relations, which are mostly only mathematically formulable, for ourselves, we recognize [10]a simplicity of a completely different kind than was the case in the classical picture. We fare like a wanderer who, after many travels, sees right in front of him a region that was hitherto completely unknown and amazing to him. We understand the new area from its immanent regularity, and have thereby made more progress than we would have, had we built a model from sticks and hooks, which, after a few operations would have been doomed to stagnation. Admittedly, it is always and everywhere possible to transfer already known relations to newly discovered areas, and in practice – for the sake of continuity (= convenience) – one will initially proceed like that. But there is no principle that could permanently guarantee the success of this method. For the so-called "unity of natural forces" only persists in a given worldview, and may be reduced ad absurdum by any new discovery.

Two problems remain to be addressed:

(1) The question of the causal determination of atomic processes; and
(2) The question of the possibility of metaphysical constructions.

As to (1), it can also be rephrased in this way: does causality hold at the atomic level? Given the previous discussion, the answer is clear: if by causality we understand the relation that allows the motion of ponderable particles to depend on one another or on certain forces, then something of this kind cannot be found among atoms, not because there are no laws in that case, but merely because we can no longer get by with the picture of ponderable particles and the representations we know from the macroscopic domain. Or, more explicitly: there is no position where a mass point must be located with absolute precision, because mass points and positions no longer are the essential descriptive notions. The strict connection, which on the large scale we call causality, exists between certain mathematical quantities, and no longer between objects of our perception (particle A and particle B). If we regard those particles as essential, then admittedly all those problems quantum theory offers to a primitive and vivid explanation arise.

As to (2): after what has been said so far, there is no difficulty, from now on, to transferring the newly discovered invariances to real things, and to set up a metaphysics on this basis. For, from (1), the argument of the unknowability of the so-called external world ceases to apply. If we focus on ponderable particles, then, admittedly, it is problematic to see how the external world is to be constructed. If, however, we use the new concepts, there is no reason why we should not speak of the external world in this context as well. To determine this is, indeed, already a task for philosophy itself.

2

Physics and Ontology
(1954)

1.

[464]One goal that philosophers and scientists of all times set themselves was to get to know the world, so that it would become possible, on the basis of a *picture* of the world, to explain in a satisfactory way at least its most important phenomena. The history of the efforts to obtain such a picture of the world shows three clearly separated stages: the mythological stage, the metaphysical stage, and the stage of natural science. We have to move the beginning of the metaphysical stage to the time of the Ionic natural philosophers, when we began for the first time to seek tradition-independent reasons for the construction of a particular system to explain the world. Later, above all with Plato's *Timaeus* and Aristotle, a return to a more abstract form of mythical [465]stage occurred. We will talk about this in what follows. However, the beginning of the scientific stage took place only recently, and was initiated by the so-called revolution in physics. The meaning of this revolution for philosophy – not only for the content of its teachings, but also for the methods it employed – cannot be emphasized enough. In this essay, therefore, I will focus above all on this third stage.

2.

For what follows it is important to see that, ever since Plato and Aristotle, a prejudice has been spread in philosophy, which from now on will be referred to as the ontological prejudice. According to the ontological prejudice, knowledge is necessarily linked to certainty. On the other hand, it is assumed that only claims that are certain say something about the world: as a consequence, we must be able to formulate a picture of the world by way of propositions whose truth stands once and for all, and whence doubt is forever removed. Statements about being are absolutely true. Statements whose truth we may doubt do not concern being, even when they are capable of providing good service as tools for the prediction of interesting phenomena.

The ontological prejudice leads to a division of the originally unitary attempt to create a unitary and satisfying picture of the world. The Presocratics still discuss the validity and usefulness of a cosmological description by appealing to the same arguments about the validity of a statement about the fundamental parts of the world. Empirical arguments and speculation go hand in hand. As soon as absolute truth is demanded only for statements providing a picture of the world, we have a sharp distinction between "mere" descriptions – which, being uncertain, do not convey knowledge, and, as a consequence, do not touch being, either – or "opinion", on the one side, and knowledge on the other. We may attain opinion with experiment in conjunction with speculation. But in order to achieve knowledge, we need a different method.

These methods were developed by Plato and Aristotle and were further cultivated by their followers. One organ of knowledge is intellectual intuition – a direct capacity, a direct insight into the principles that lay the basis of the course of the world. Knowledge, which originates in intellectual intuition, is absolutely certain. Intellectual intuition is especially suited for philosophers, and the return to it distinguishes philosophy more than other disciplines (with the single exception of mathematics, perhaps). Philosophers look at the world as it really is. Thereby, the separation between philosophy and science is sharp, and drawn once and for all: the birth of philosophy as an autonomous discipline, which is not only independent from scientific considerations, but insists that it is the only one capable of providing a picture of the world, goes together with the ontological prejudice, with the idea of absolutely certain claims, and with the assumption that only such claims are "ontologically relevant", that is, only they show us a picture of the world, as it "really" is. The conflict between knowledge directed at reality and opinion, which is appropriate to prediction, clearly emerges in St. Thomas. In *Summa* I, 32 he calls attention to two differing ways of giving an account of something.

The first is that one proves a certain principle in an adequate way. So in cosmology one offers an adequate reason for believing that the motion of the heavens is uniform. According to the second type, one introduces no reason that grounds the principle in an adequate way, [466]but rather shows that, when the principle is assumed, its consequences are in agreement with the facts. So, in astronomy one used the hypothesis of epicycles and eccentric circles because, on the basis of this hypothesis, the observable phenomena of the heavenly motions can be surely portrayed. But this is no sufficient argument, because perhaps, on the basis of another hypothesis, they could also be just as surely portrayed.

All philosophers of the Platonic-Aristotelian tradition also have *a positivistic philosophy of science*. That means, according to their conception, the laws of science are *appropriate instruments for making predictions*. But the existence of the laws does not allow us to infer a conclusion about the real construction of the world. Reason: the laws are not absolutely certain.

3.

A second version of a positivistic philosophy of science can be traced back to Berkeley. The considerations that lead Berkeley to deny ontological relevance to scientific statements, in his case, above all the laws of Newtonian mechanics, are closely related to his theory of meaning.[1] According to this theory, a descriptive expression is meaningful if and only if it denotes (in an unambiguous ways) an idea or a complex of ideas. A sentence is meaningful if and only if it is put together out of meaningful expressions of the type just defined and logical signs according to certain grammatical rules. But now, the sentences of Newtonian mechanics contain expressions such as "force", "absolute motion", etc., to which no idea corresponds. From this, we can draw two different conclusions. (a) General laws of the given type are meaningless, because they contain "absurd and unintelligible expressions" – as Berkeley puts it.[2] (b) The meaning of these laws is different from what we tend to assume *prima facie*. A philosopher who defends position (a) will not deny the *usefulness* of physical laws. So Berkeley repeatedly emphasizes that Newton's laws are "mathematical hypotheses", which allow predictions in an excellent way.[3] But he denies that they teach us something about the world: they are useful instruments for prediction – but, being meaningless sentences, they have *no descriptive content*.

[476][1] For what follows, see K. R. Popper's excellent article, "Berkeley as a Precursor of Mach", *The British Journal for the Philosophy of Science*, IV/13. Lenin emphasized the close relationship between Berkeley's philosophy and the modern positivists in *Materialism and Empirio-Criticism*.
[2] *Principles*, 32.   [3] *De Motu*, 66.

Today this view of the nature of physical laws has become the philosophical creed, above all through the authority of philosophers such as Wittgenstein[4] and physicists such as H. Hertz,[5] Ph. Frank,[6] and above all Niels Bohr,[7] of nearly all those physicists who in some way contributed to the progress of quantum theory. It counts as proven that, for example, the quantum mechanical formalism "must be viewed as a tool, which allows us to deduce predictions of a certain or statistical nature".[8]

So has the postivistic, or "formalistic", philosophy of science today spread everywhere. Among traditional philosophers, who devalue science because they think they possess the key to knowledge of reality, and among positivists, who have given up the hope of knowing "reality", who hold the problem of knowledge of reality to be a pseudo-problem and content themselves with possessing tools, which, more or less, allow them to make predictions. However hostile positivists and traditional philosophers may otherwise be to each other, on one point they concur: that from science, especially from physics, we cannot expect any picture of the world.[9] What follows is dedicated to criticizing this position.

4.

I begin with the criticism of the positivist version, that is, Berkeley's version, which found an excellent advocate, above all, in E. Mach. In order to clearly present the criticism, I [467]employ a diagram that depicts the presentation of the formalistic theory provided by H. Hertz, which is unsurpassed in clarity.[10] This diagram shows that the positivistic version, too, proceeds from a definite theory of "that which is", that is, from a definite ontology, from an ontology of things (sense data, objects of the everyday world, classical situations), about whose existence predictions should be made. In order to keep the description as general as possible from the beginning, I name these things *elements*. Those who are more

---

[4] See Schlick, "Die Kausalität in der gegenwärtigen Physik", *Die Naturwissenschaften*, XIX, 1931.

[5] *Mechanik*, "Einleitung", p. 1.   [6] *Das Kausalgesetz und seine Grenzen*, I.

[7] "Causality and Complementarity", *Dialectica*, 7/8.   [8] *Loc. cit.*, p. 314.

[9] The unity goes still further. Lacking philosophical principles, many physicists, when they began philosophizing clung to an idealistic metaphysics (Y-function, as knowledge function) and thereby made enthusiastic followers among philosophers, who held to the bad philosophy of the physicists, but not to their good physics (which they by no means understood). So it then came to philosophical judgments, which so meaningfully begin with "Today's physics says…".

[10] *Loc. cit.*

comfortable with illustrations may thereby easily think of sense data, simple feelings – in short, of all phenomena which are subjectively present in observations.

Associated with the ontology of elements is a specific *theory of meaning*. According to it, an expression is meaningful if, and only if, it refers to an element, or else it belongs to the class of logical expressions. According to the kind of philosophy formalists advocate, the assignment of descriptive signs to elements is viewed as justified either by convention or by being in the essence of the signs. Sentences have a meaning when the meaningful descriptive signs in them appear to be combined in specific ways with logical expressions. We name sentences of this kind *elementary sentences*. Therefore, the function of a theory is the following. We have a series of elementary sentences $E_1, E_2, \ldots E_n$; by mapping the theory to these elementary sentences we obtain a series of other elementary sentences, $F_1, F_2, \ldots F_n$. We test the usefulness of the theory by checking whether the situations described by the generated sentences are actually the case (see the sketch below).

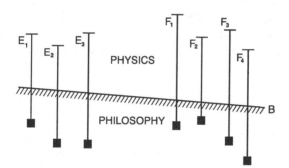

It is the task of *physics* to discover useful connections between elements. Connections between elements are useful, when they (a) are simple and (b) lead to correct predictions. It is the task of *philosophy* to determine the meaning of expressions of the elementary sentences as well as the nature of elements. This task must be carried out before the physicist begins to work. For, before he can create instruments, which, given some elementary sentences, produce other elementary sentences, he needs to know which sentences are elementary sentences – which means, what is an element and what is not. The whole ontology must already be determined. Therefore, no consequences can be deduced from physical theories that were not already known before the establishment of the theories. Border B is impenetrable. No communication occurs between the field of physics and the field of philosophy. The positivistic theory shares this feature with traditional philosophy.

This last determination is decisive for the theory presented. When it becomes clear, namely, that the invention of a new theory leads us to the assumption of the existence of new elements, which, under some conditions, belong to a completely different category from the elements presumed at the beginning, then three consequences follow: (1) the initial assumptions notwithstanding, scientific theories are ontologically relevant; (2) as a consequence, the construction of an ontology is not something a philosopher can perform independently of scientists; (3) the conditions of meaning, too (which, indeed, are relative to a given ontology), *change* as science progresses: what is meaningful follows from the theories we apply – in short, there can be no definite philosophical criterion of meaning, which would be valid *once and for all*: a criterion of meaning is descriptive, not normative.

Let us now consider the way in which the breakthrough of border B can take place.

(1) Let us assume the elements are observed planetary positions. Elementary sentences are sentences of the form $(\alpha\delta)(P, T) = (a, d)$, where $(\alpha\delta)$ is a functor (right ascension, declination), [468]whose value is pairs of numbers; P, a definite, individual planetary position (a definite element); T, a moment in time. The criterion for the presence of an element at a definite position (for the correctness of a definite elementary sentence) is (K_1): agreeing statements of trustworthy observers under good conditions (good vision, etc.). When all conditions are fulfilled, an element is in the position where the observers report it.

Let us now assume that laws were found, which allow us to predict the movements of the elements. An example is Kepler's laws. With their help we can predict the position of a planet after a twofold transformation (from Earth-centered system to Sun-centered, back to Earth-centered), provided we know its position at a certain time. If, now, observers have enough trust in the law, the following may occur: let us assume that according to K_1 the positions of P at T_1, T_2, T_3, T_4, are those that are shown in the drawing below. Arch b indicates the orbit computed on the basis of the law. In this case observers would attribute the divergence in T_3 to unknown conditions, which means, they would move P to T_3 on the basis of a systematic mistake, whereas the planet is, in fact, in S. But this means that, from now on, we are no longer talking about elements with the meaning initially attributed to them. For, since in the position S there is *no* determined position of the planet according to K_1, then there is no element either. Nevertheless, we say that something passes through S. From now on, then, we do not merely speak of elements, but also of other objects (planets, or however we wish to name them in the present case). In other words: we are induced by a series of simple laws to expand our ontology. We are no longer directly interested in the prediction of the orbit of the elements: rather, we are concerned with the prediction of the orbit of newly constructed objects.

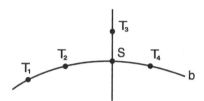

(2) This is still clearer if we consider the transition to the theory of gravitation. According to this theory, the presence of a planet has disturbing effects on neighboring material. It may happen that relevance is denied to a series of otherwise reliably appearing observations of the movement of a point of light, because no observable effects of the presence of material can be proved. In this case, then, the points of light are directly interpreted as light effects of distant masses, and what matters is the movement of these masses and the [470]question whether our direct interpretation may not rest on a mistake. Therefore, with the theories we adopt, the circle of our criteria, by which we determine what happens at a particular position, and the complexity of the objects thus determined, expand. But this has the following consequence: first, whenever the conditions C_1, C_2, ... C_n are fulfilled, we regard an object as present. It is not a matter of further conditions. Later we speak of a systematic mistake, an illusion, although C_1, C_2, ... C_n are not fulfilled – which shows that from now on we demand the fulfillment of further conditions, if we have to speak of an object in the proper sense of the word, and not of an illusion. The relationship of such new objects to those previously considered may be best elucidated by appealing to a psychological example: let us assume that an experimental subject (ES) makes optical observations. For this person there exists no difference between the optical hallucination of table and a real table, since the only objects he is able to grasp are optical impressions. These optical impressions are the elements of our ES. The ES finds himself, then, in the same position as an

observer who strives to determine relationships between planet positions. Now, the ES is also allowed to have tactual impressions of the elements. He will then soon experience a new kind of objects, sight-touch objects. The earlier elements are only a way in which these objects appear.[11] In this moment the difference between real object and hallucination (or appearance) is relevant. A sight-touch object is for the ES a new object in a completely different meaning, as is a crocodile for someone, who, until then, has not yet seen any. Indeed, the crocodile can be set in a reasonable relationship to already-known objects (river, countryside). But in a hallucinated glass no sight-touch water can be poured. But then it follows that the change in ontology, which is tied to the introduction of laws, is far more lasting than it first appeared. It is not the case that new elements are discovered in addition to the elements at hand (touch impressions to sight impressions). Putting forward theories sometimes leads to the fact that *all* the elements that served as a starting point are moved to a lower level of reality. What really is no longer the sighted positions of the planets, but the attracting planets themselves. It comes down not only to an enrichment, but to a complete *rearrangement of the ontology.*

(3) This implies a criticism of a theory of interpretation of scientific calculations, which, among others, is defended by Carnap.[12] He distinguishes between elementary and nonelementary expressions in scientific calculations. Elementary expressions are expressions of the following kind: the truth value of syntactically correctly formed sentences, which contain elementary expressions as the only descriptive expressions, can be determined by observation. Let us define the elementary expressions (expressions such as, say, "red", "hard", etc.) on the basis of nonelementary expressions of the calculation. The interpretation of the system occurs, then, whenever we indicate the semantical meaning of the elementary expressions. In a special case, we interpret the Newtonian theory by assigning to the expression "planet position", which was defined as "Earth-centered coordinates of the planet", "observed planet location", that is, an element in the sense explained at the beginning of Section 4.

---

[11] The terminology "way in which these objects appear" and, in general, the way of observing here introduced go back to E. Tranekjaer-Rasmussen. See *Virkelighed og beskrivelse*, Copenhagen, 1954. "Psychologie und Gesetzerkenntnis"; *Gesetzt und Wirklichkeit*, ed. Moser, 1948, 91ff.

[12] *Foundations of Logic and Mathematics*, in *International Encyclopedia of United Science*, Vol. I/3, 64.

Whenever all that is given, we only need, once again, to follow the movements of the elements, in order to check the theory. But:

(a) The movement of the elements is very well described by Kepler's laws. However, these laws *contradict* Newton's theory (for they [471]are valid only for an infinitely heavy Sun, and for the planets with negligible masses).

(b) Due to systematic disturbances, the movement of the elements often varies from the movement of the Earth-centered coordinates of the planets, with which it should be in accord, because of the interpretation given (the expression "Earth-centered coordinates of the planet P" acquires a meaning as long as it is assigned a certain element, a certain point of light). That shows that the assignment is more complicated than we might initially think.

(c) Systematic mistakes, disturbances, which depend on the imperfection of the observation instruments, which are produced by atmospheric influences, etc., are constantly newly discovered. That shows that the assignment cannot be given once and for all, but it must change according to the knowledge of systematic mistakes. *But the assignment first should give a meaning to the expression of the calculation, and thereby to the whole theory!* Without it we have, depending on the point of view we adopt, only an uninterpreted calculation with meaningless signs. But then, how is it possible to switch from one assignment to another, and to justify this switch? Does that not show that a theory must already have a meaning independently of an assignment of the given kind? And thus, that it is mistaken to think that the semantic interpretation of the elementary expressions first provides it with a meaning?

Certainly, we cannot deny the assignment of any function. Let us consider how we can teach a lay person what the theory is about. We show him the calculations, the rules according to which it is calculated, and we also explain to which observations the statements in the calculations are connected. We tell him, for example: "You can determine the Earth-centered coordinates of Mercury by following the path of this point of light in heaven". By this statement we affirm that the observation of the point of light represents a generally trustworthy method for the determination of the Earth-centered coordinates of Mercury, but not that the expression "Earth-centered coordinates of Mercury" really receives a meaning by the connection with the given point of light. As a consequence, the point of light behaves, with respect to the Earth-centered coordinates, as the perception of an object with respect to the object itself – the statement of the perceptions from which we can move to the description of an object, provides us, in general, with a trustworthy means to find an object and to determine its properties, which are at hand; but, on account of perceptual illusions, which are always possible, it can never be seen as constituting the object, not even partially. We express this relationship by saying that the statement of perceptions, of elements, in brief the whole so-called interpretation of a calculation by reference to observations, has a pragmatic, not a semantic, function. In general, it provides a trustworthy help for the determination of truth values of certain statements in the calculation, but it cannot be recognized as constituting the *meaning* of the expressions of the calculation.[13]

We cannot end this section without dealing with a possible criticism of our exposition. The criticism is the following. Let us assume, it is said in the first case, that something moves through S, although there was no element to see (to determine) at the position S. Let us assume that in the Newtonian theory scientists are interested in the positions of planets, not in points of light. In so doing, however, a new ontology is only apparently introduced. For these sentences, without being referred to elements, have no independent meaning. They are abbreviations[472]. Spelled out, they have the same form as the hypotheses we discussed in the first place. Just like them, they merely say which elements occur when other elements are present at any given time. Their meaning is exhausted by this relationship to the elements. However, only the elements exist.

There are two ways of meeting this objection. First: as in Section 4, point 3, we assume here that the meaning of all scientific statements is given by their relationship to elements. When it turns out that they can have a meaning independently of a *given* assignment, the argument breaks down. But this is exactly what happens, as was set out in Section 4, point 3c. From this follows that each "translated version" (see point 3b) is doomed to failure, and that we must take scientific statements, especially existence statements, as they *prima facie* offer themselves: as statements, which communicate to us movement, change, etc., of things

---

[13] Philosophers who have recognized the condition mentioned under (c), but who do not want to abandon attributing a semantic function to the assignment of elementary expressions to the elements, change this assignment to a probability assignment (on the detour through the so-called reduction sentences). But, thereby, they achieve only an apparent improvement. For there are conceivable conditions under which it is completely improbable that a certain point of light collapses with the Earth-centered coordinates of a given planet. The probability version is particularly defended by A. Pap.

which were not in question in the ontology upheld before the formulation of the laws. Second: if hypotheses are used to produce elementary statements, why should we not try the same with the elementary statements themselves? It would then follow that the elements do not play the role that was attributed to them, and that there are really no elementary components. The only way we can avoid this consequence is by using elementary statements, for which such an analysis seems to be pointless from the very beginning – absolutely certain elementary statements, so-called basic sentences[14] or determinations.[15] The positivist theory of science fell back on such statements early on. While in the traditional philosophy the general principles are attributed with absolute and unbreakable security, the absolute security is put by many positivists – in accordance with their inductivistic theory of science – in the basic sentences. Both are obsessed by the ideal of security.[16]

5.

In the last section I showed, by way of examples, that we must ascribe ontological relevance to the statements of physical theories. That these theories, according to their wording, as well as to their meaning, imply the existence of objects that were not in question before the theories were put forward, and that are different from the objects external to theory not only according to their appearance – as a dog from a cat – but also categorically – as the hallucination of a table from a real table. We now want to prove that the most modern development of physical sciences allows us to remove *complete* physical ontology from physical theories. *Consequently, a philosophical ontology from now on has not less, but nothing more, to say.*

In order to make this claim plausible, we must distinguish between various sentence forms within physical sciences. The two most important we name *regularity claims* and *universal theories*. An example of a regularity claim is the sentence stating that matter expands when heated. An example of a universal theory is Newton's theory. Regularity claims indicate what happens in a system given certain circumstances. Regularity claims are not connected with one another but should not be in

---

[14] See A. J. Ayer, *Basic Propositions*; *Philosophical Analysis*, ed. Black.

[15] See B. v. Juhos, "Theorie empirischer Sätze", *Archiv für Rechts-und Sozialphilosophie*, XXXVII/1. M. Schlick, "Das Fundament der Erkenntnis", *Erkenntnis* Bd. III.

[16] A detailed criticism of the idea of absolutely secure basic statements is found in my thesis *Zur Theorie der Basissätze*, Wien 1951.

contradiction, either. When we discover in a system a new type of regularity (say, electrical conductivity), then we add it to the already known regularities. We say, then, that we have discovered a new *property* of the system. Regularity claims teach us the properties of physical [473]systems. How these systems are themselves made does not follow from them. The reason why they have just these properties cannot be explained in a physical way. Therefore, as long as we, in physics or in sciences in general, only possess regularity claims, it is philosophical ontology that must rack its brain about the nature of the systems that behave in this way. Physics was still in this state of affairs at the time of Galileo. One of its most important principles, the causality principle, obviously counts for systems and not for properties of these systems. Therefore, at this time its justification could not follow be grounded physically, but was accepted on the basis of general philosophical considerations. Philosophy and science shared the task of obtaining a picture of the world. The sciences provided the descriptions, philosophy speculated about the essence of things. But these speculations were far too vague to be able to offer an explanation of the processes described by physics.

A *universal theory*, such as Newtonian mechanics, has, by contrast, a completely different structure. First of all, it is misleading to speak of a *theory*, since the equations of Newtonian mechanics cannot be immediately and directly applied to the systems being investigated. In order to be able to predict the behavior of a system on the basis of Newtonian mechanics, we need a *model* of this system, that is, a division of this system into material parts, and the details of the forces at work on these parts. When a certain precision is given, then it is possible that various models are equally appropriate for the predictions: the movement of the planets in the solar system can be equally well predicted, whether we start with mass points of spheres or spherical parts tied together by cohesion forces. Let us now assume that we discover a new phenomenon in a system. As long as we only have regularity claims at hand, we will attempt to determine this phenomenon more exactly and to describe it with a new regularity claim. But the possession of a universal theory gives us other duties. It forces us to analyze the system and to show that the phenomenon is predictable, when we (a) take a certain model as a basis, (b) predict the behavior of this model on the basis of a universal theory, and (c) translate the final stage of this model back into the normal language that we normally use to describe the system. By explaining the differing, observed regularities on a system in a unified way, the universal theory establishes a connection between these phenomena. It explains why precisely these regularities count, and how, in its formulation, the emergence of each event can be explained. A universal theory is, then, *a possible picture of the world*, and we only have to make sure that it is correct, in order to know how the systems, whose behavior we have so far described with the help of regulatory claims, are *really* made. From this it follows that from now on all philosophical problems, whose object is the nature of physical systems, can be divided into two partial problems, which are no longer specifically philosophical: (a) in the problem of the truth of a given universal theory T: this problem is decided by physics; (b) in the question, what is the situation of a world correctly described by T. That is a logical-cum-mathematical problem. The causal problem, for example, falls into the two following partial problems: (a) the truth value of quantum theory; (b) the structure of quantum theory. The last problem was superbly solved by *von Neumann*'s researches.[17]

[474]If a universal theory turns out to be false, then a peculiar situation arises, which is often completely misunderstood by modern philosophers of physics. Let us look at the situation that was at hand after the refutation of the classical theory of the movement of particles. It was clear that the classical picture of a particle moving along a continuous trajectory could not explain appearances in the microscopic field. But the movements of the planets could be predicted, just as before, with the best possible precision on the basis of the classical theory. As a consequence, it followed that we, indeed, did not make a simple mistake when assuming that matter is made of discrete particles that move on their way along continuous paths. We draw the same conclusion from the assumption of moving parts in the Wilson chamber. Here it seems, the particle nature of matter stands directly in front of our eyes! But then we forget, in this case, that we are supporting this conclusion with a very familiar, but nevertheless very complicated, interpretation, in which the laws of classical mechanics play an essential role. But it is these laws which are no longer at our disposal! It is true that we can use them to predict phenomena in a wide area, just as we did before. But the fact that they are no longer universally valid prevents us from drawing ontological consequences from them. All we can say is that material particles *appear* under certain conditions as compact particles; but we cannot say that they *are* compact particles. (Certainly, a wave can also appear as a particle, under certain conditions.)

---

[17] See, in this regard, my essay "Determinismus und Quantenmechanik", *Wiener Zeitschrift für Philosophie, Psychologie und Pädagogik*, 1954 {reprinted as Chapter 3 of the present collection}.

Think, for example, of looking at a wave crest on the surface of a sea from a great distance. A refuted universal theory allows us then to say how physical systems *appear* to us under certain conditions; it allows no further conclusion about the *nature* of such systems. This state of transition from one theory to another was settled once and for all by Niels Bohr's doctrine of complementarity – although the new theory, which was certain to replace the classical theory, already existed (or at least existed in outline), that is, quantum theory. Instead, then, of taking seriously the new picture of the world, the new theory, we hold fast, in a typically positivistic way, to the classical aspects (which in this case represent Berkeley's sense data) and deny that the nature of quantum mechanical systems can ever be known. Here I lack the space to show how other features and difficulties of the new theory, too, arise only because the advocates of this theory do not hold on to the theory, but more to their positivistic philosophy.

6.

The arguments we have so far advanced, criticizing positivists and showing that their ontology has long been superseded in the course of scientific progress, must suggest the following objection from the exponents of traditional philosophy, who swear allegiance to the ontological prejudice: by now, what physics is to provide are universal theories. Universal theories are possible pictures of the world, that is, they are, according to their form, capable of providing an explanation for each occurring appearance in the universe. It must be already acknowledged that the analysis of a picture of the world uncovers the nature of the world. But as long as we are not *certain* that this picture of the world, which is has just been offered for analysis, is the correct one, then we cannot draw from it consequences about the world. The analysis of an illusion explains nothing to us about reality. Most theories, past pictures of the world, which physics had to offer, turned out to be illusions. Who guarantees us that we are not at present once again victims of a mistake?

Answer: no one can give us such a guarantee. But from this follows that [475]we should now formulate our theories more carefully; that we, as some physicists advise us,[18] should insert such clauses as "apparently", or "it seems, that...", etc. Such a procedure would make it impossible to

---

[18] See von Neumann, *Mathematische Grundlagen der Quantenmechanik*, p. 224 {English translation by Robert T. Beyer, *Mathematical Foundations of Quantum Mechanics*,

carry out the most important tasks for natural scientists: to correct theories. Those who claim nothing cannot be contradicted. Those who make no attempt cannot be corrected. As a consequence, we have to apply our theories as if we felt completely certain of their correctness; we have to speak as if the world is really constructed as our theory says it is, and we must not allow ourselves the smallest spark of cleverness in saying: "Of course, the theory will not work, but let's have a look and see how far we can go!" For that would belittle the possibility of a correction. In short: we have to understand physical theories as they *prima facie* offer themselves – as statements about the world.

What physicists are not allowed – surrounding their theories with cautionary clauses – seems to be permitted philosophers; indeed, it seems to be imperative. Philosophers know that the statements of physicists do not count as necessary, that they have a hypothetical character. Does it follow that they cannot be used as components for a description of the world, as components for a picture of the world? This question divides into two partial problems: (a) Can hypotheses have a descriptive function? (b) Can hypotheses be true? The first question is certainly to be answered positively. For, the descriptive function of a statement does not depend on the truth value of that statement, and a statement lacking a descriptive function cannot be refuted. The second question is answered differently, according to the way we understand it. It is certainly possible do define a hypothesis as "true" in a reasonable way. This possibility is the prerequisite of the application of scientific methods. These methods consist precisely in assuming that a theory, a statement is true, and to see how far we can get with it. But from this very method it follows that we can never be certain of the truth of a hypothesis. But does not the opposite idea, that only certain statements are ontologically relevant, proceed from an over-simple idea? That is, from an idea of the world against which Heraclitus exercised the strongest criticism? This idea is the following: the world can be locked with respect to knowledge. It is a well-ordered building, a cosmos, which those who know can, examine fully from beginning to end. Certain knowledge is, then, at hand, when the cosmos has been examined and when we have assured ourselves that nothing has escaped our attention. Now, does the progress of the natural sciences not show that this idea is naïve and childish, and that it carries a whole ontology based on the ontological prejudice? How should we now decide

Princeton: Princeton University Press, 1955, p. 420; Feyerabend's review of the English edition is reprinted as Chapter 17 of the present collection}.

in the conflict between traditional ontology and ontology that follows from physical theories? Here we must simply follow our interests. We have to compare the achievements of the two. On the one side, the universal theories of the natural sciences, which show us fantastic connections and lead us in constant progress to discovery of new traits of the world, to penetrate into ever new levels of the world, against which the now abandoned fields still can act as first approximations, which are exact enough to make it possible for us to discover mistakes in our conceptions; universal enough to answer all relevant questions about the physical universe, which were earlier left to philosophy and its speculations. On the other side, the vague and empty speculations about the world that traditional philosophies, [476]with their mark of certainty, provide us with. By which it should be clear that this security, should it really be attained some time, would mean the end of knowledge – and to what extent this would mean the end of human beings on this Earth, is yet to be seen. For knowledge, and advancement of knowledge, is one of the essential functions, one may say, of human beings.

7.

Summary of the theses: (1) The aim of ancient philosophers was to attain a unified picture of the world, which could serve to explain the phenomena in our world. (2) They used empirical (scientific) and philosophical considerations in the same way. (3) The ontological prejudice led to a division between science, which advances hypotheses, and philosophy, which is capable of achieving absolutely certain knowledge. (4) Positivist theory of science leads to the same division, even when its philosophy is not formulated in such detail. Both see in hypotheses only tools for making predictions. (5) This works to some degree, as long as physics only has at its disposal claims of regularity. But, with the emergence of universal theories, it acquired the means of description, which show the universality and explanatory power of world pictures, but with much greater detail. (6) Such universal theories are the modern pictures of the world, of which we create more and more, in order to understand the course of the events around us.

3

Determinism and Quantum Mechanics
(1954)

[88]Let us imagine doing the following experiment with a mechanical system S (a pocket watch, for example): we bring S into a certain state, Z (winding up the pocket watch, setting the hands at a certain position), and then leave it alone. S will either remain in that state, or will change in a certain way. In our example: the springs will set the wheels of the pocket watch in motion, and will make the hands rotate into a different position. After a certain time, T, a new state, Z_1, has emerged. We can now attempt to restore the state Z: we once again wind up the pocket watch and bring the hands into the same position in which they were in state Z. If we leave the system alone, two things may happen: after the time T, Z_1 has once again come to be, or after the time T, a different position, Z_2, has come to be. In the first case, we would conjecture that we have discovered a regularity. We can describe this regularity by saying that the state Z *caused* the state Z_1. However, it is not certain how we would behave in the second case. For a determinist, it seems reasonable to argue in the following way: the position Z_2 *has* a certain cause, but we do not know it. We presume, indeed, that we have set the system S back in the same state, Z. Actually, this is not the case. If we look carefully, we will discover that the state Z', with which we began, differs slightly from the state Z'', in which we reset the state, so little, however, that we did not notice the difference and thought that at both times the same state, Z, was present. With our inexact means of observation we believed two states to be the same when they differed in a small respect. But everything depends on this difference! Indeed, we later learned that Z_1 comes from Z', and Z_2 comes

from Z''. It is indeed the case that Z_1 [89] is caused by Z – for Z is, according to the view just given, a disjunction of states under which the cause of Z_1 also occurs. But Z_1 is not the effect of Z, because, besides Z_1, Z can produce other effects, say Z_2.

Let us clarify this argument by appealing to the example of the pocket watch: when we have once again completely wound up the pocket watch, and put the hand back in the same position, it is still possible that, after the time T, another state will appear, as happened the first time. That would not be surprising. We could, for instance, make the assumption that the escapement of the watch sometimes subsides, so that every beat does not move the neighbouring cogwheel. How often this happens within T depends on tiny differences in initial positions, on the wear and tear that the watch experiences in a certain time interval, as well as on the fatigue of the spring. We can also mention here that the parameters affecting the escapement change with temperature, and that therefore a temperature difference in both cases must lead to a difference in the frequency of the escapement, etc. So, when we want to bring the watch back to its initial position, it is not sufficient that we simply once again wind it up and bring the hands back to the original position. We also have to make sure that equality of the given positions is preserved. When that is the case, then the same position will recur after T, as in the first attempt.

2.

If we try to grasp more abstractly what we have just explained with a graphic example, then we come to the following procedure. We generally describe the state of a system by stating that it has such and such features: the watch, say, is fully wound and shows 1:30. In physics, indication of results that have certain measurements replaces indication of the properties of the systems. An indication of a state in physics has, then, the following form: at time t the measurement of the quantity A yields the value a, the measurement of the quantity B the value b, etc. (In the case of the pocket watch: the tension of the spring is so and so high, the hands form this or that angle with the 6–12 line, etc.) Let us shorten these statements by [90] writing $((Aa, Bb, \ldots; t))$. For the sake of simplicity let us assume that the system S is described by indicating the results of only two measurements. We bring it into the position $((Aa, Bb; t))$ and determine that $((Aa', Bb'; t + T))$. We restore the first state and after the time span T we get $((Aa'', Bb''; t' + T))$, where $a'' \neq a'$, or $b'' \neq b'$, or both.

We name a system that behaves in this way an *undetermined system*. In the case of undetermined systems we generally assume that there is a series of measurable quantities, Q_1, Q_2, ... Q_n (whose values can be random functions of the time – we distinguish the different functions with upper indices), so that $((Aa, Bb; Q_1^{i_1}q_1, Q_2^{i_2}q_2, ... Q_n^{i_n}q_n; t))$ *always* leads to $((Aa', Bb'; Q_1^{i_1}q_1', ... Q_n^{i_n}q_n'; t + T))$. We call the Q_i the forces working on S, and a system, for which measurable quantities and forces are related as just described, a *definite system*. Should it turn out that the temporal course of the values of some of the Q_i (say, of Q_i, ... Q_t) can be computed using certain laws, and in case the remaining Q_i only depend on the times, their initial values, and the initial values of A and B as well as those of Q_1 to Q_t, then we label S, through A and B, as incompletely described and the class $((Q_1, Q_2, ... Q_t))$ an *extension* of the description of S. The class of the remaining Q_i, whose measurable quantities can also be a random function of time, we call the *completion* of the description of S. A definite system, whose measured values are extensions, is called a *complete system*.

We can now give the *assumption of strict determinism* the following physical sense: *for every indeterminate system, there exists a completion or an extension, or both, which renders it a determined system* (assumption K).

3.

It is important to be clear about the nature of assumption K. Let us imagine that someone proceeds as follows: he examines a system S, and determines that equal initial states do not lead to equal final states. He says, "From this follows, therefore, that the initial states, which I hold to be equal, were not, in fact, equal. I have, indeed, taken care that the measurable quantities A, B, ... L have the same value. But the fact that the same final state [91]has not been reached shows that there must be other measurable quantities, M, N, ..., which are so hidden that I have not yet discovered them, but which have different values in the states I regarded as equal". We name these measurable quantities thus defined "hidden parameters". We can summarize his argument in the following way: although the system was returned to the same initial state, from this initial state various final states followed. From this he concluded that, after all, the initial state was not, in fact, the same. It is true that no empirical difference could be determined – but this is due to the roughness of our observation instruments. In reality there are hidden parameters, which are not yet known, and whose difference in the initial states led to a difference in the final states. Under these conditions, what becomes of assumption K? That is easy to see: when we conclude, whenever identical initial states lead to different final states, that the initial states were, after all, not identical, then assumption K can by no means be refuted. For, in that case, we only *refer* to those states as initially identical that lead to identical final states. Statement K no longer *says* anything about *the way in which* things in nature behave; it only *shows* under what conditions we refer to the initial states as "identical".

We must therefore proceed differently. Instead of taking a difference of the final states as a *criterion* for the difference of the initial states, we must view them as an *opportunity* to seek a difference in the initial states. This means that the so-called hidden parameters cannot be defined by appealing to the class of deviations of final states from a given initial state, which we deem to be unequivocal; we should rather try to present them in a factual way, and bring them about by way of experiment. In this case we have *explained* the deviation (and we have not merely described it in a different way, by changing its name) and we have at the same time guaranteed that assumption K is not an uninteresting linguistic convention, but an empirical statement.[1] We should, then, reformulate assumption K as [92]*Assumption K'*: for any undetermined system there is an *independent* provable completion or extension or both, which renders it a determined system. When we formulate the principle like this, however, we encounter another difficulty: the principle now says, it is true, something about reality, but we do not see how it could be refuted. It has, namely, the form "for every *x* there is a *y*, such that...". Therefore, when we encounter a system that behaves so irregularly that from an initial determined state different final states arise repeatedly, and when in addition a search for independent and determinable completions or expansions is unsuccessful, we cannot conclude that there are no such coordinates – as they could, in fact, lead a very hidden existence.

---

[91][1] Many philosophers use the method we criticize in order to establish absolute certainty for the causal principle. They overlook that they can do this only at the cost of the interest that this principle has for us: if [92]"Each event is well-determined" is true only because under all conditions I can assume any (perhaps unobservable) event as a determiner of a given event, then this statement only mirrors how I proceed in the attribution of names. It says nothing about nature.

4.

When, then, quantum theorists explain to us that there is no causality, and when they also claim to have a *proof* of this contention, they must understand something other than assumption K' by the statement, "There is causality". For, it follows from the explanation outlined, there cannot be a *proof* of the falsehood of K'. In order to understand what statement in atomic theory is refuted when we say that there is no causality, we must first look more closely at the various types of regularities that occur in physics. Let us consider, for example, the statement that all heated bodies expand. This statement can be used to predict and explain the occurrence of an expansion in a particular case. It is refuted by a single counter-example – the question that may arise is only whether we have experimented accurately. The expansion of a given body is also *directly* explained by application of the statement that all bodies expand in certain conditions, together with the determination that these conditions do indeed hold.

Let us compare this, for example, to Newton's theory of motion of mechanical systems. The theory states the following: if I, II, III, IV, etc. are parts of a system, K_1, K_2, etc., the forces [93]affecting them, and p_1, p_2, ... the momenta of their centers, then $p_1 = K_1$, $p_2 = K_2$, etc. From knowledge of the initial situation, of the initial speeds as well as of the forces and their change in the course of time, we can calculate the subsequent movement of any such system. An example often mentioned of such a mechanical system is the planetary system: from knowledge of the position of the planets at a particular time, as well as of their velocity at that time, we can predict the rest of their movements. In the case of the planetary system, we also see that a division into parts is no unequivocal procedure. We may take the planets and the Sun as the elements of the system, but also just as well the planets and two halves of the Sun's sphere tied together by attractive forces, or else we may choose, instead of a planet, a smaller sphere and a surrounding spherical shell. In some cases such a portrayal is even adequate, for example, if we want to describe the phenomenon of the tides. Therefore, we can treat the Newtonian theory as a *theoretical scheme*, with whose help we can describe a mechanical system, now in one way and now in another, depending on the problems we want to deal with, and on the degree of precision necessary for our prediction. When we then find that a mechanical system does not behave as the theory had predicted, we cannot, on the basis of this experience, declare the theory refuted. The discrepancy between theory and observation may also be caused by the fact that we have not divided the mechanical system in its parts in a proper way, or that we have introduced false assumptions concerning the forces affecting the elements. Call in the pocket watch again: we can try to describe its behavior by symbolically referring to the spring as a constant working force, to its escapement as a brake device (it is, indeed, known that constant forces, when affecting bodies moving in a viscous medium, produce a terminal, nonaccelerated motion of these bodies) with certain friction coefficients, etc. When the watch does not behave as we assume on the basis of this description together with mechanical laws, the reason may lie in the fact that we [94]made too simple a picture of the inner structure of the watch, or that mechanics is false. On the other hand: in order to explain the behavior of a mechanical system on the basis of Newton's theory, we have to make assumptions about the structure of the system, or, to put it differently, we have to make a *model* of the way in which the system works.

5.

In order to obtain an adequate definition of the concept of "model", let us start with the following consideration: let A_1, A_2, A_3, ... be a series of measurable quantities, whose value is determined by observation or by a conjunction of observation and calculation. These quantities do not need to stand in any perceivable connection. Some of these quantities are determined by the same method (such as the velocity of the Moon and the velocity of Jupiter). We call the classes of such quantities *measurement classes*. Let M be the class of all quantities (a numerically infinite class), N_j some subclass of M. With the statement "$W(A_i, t) = a_i$" we express that the quantity A_i at time t has the value a_i. Let $N_j!(t) \equiv \pi_i [W(A_i, t) = a_i]$, where i runs through all indices of A_i that are part of N_j, and let also $N_j!$ $(\Delta; t) \equiv \pi_i [a_i + \Delta \geq W(A_i, t) \geq a_i - \Delta]$. We call P a Δ-*state with reference to the theory* H if, and only if, when

(a) there is a j so that $P = N_j$;

(b) $P!(t) \ \& \ H \prec P! \ (t + T)$ for every T within given borders with a_i clearly variable;

(c) there is no P' so that if $P' \subset P \ \& \ P' \neq P$, and (b) holds for P';

(d) $P! \ (\Delta; T + T)$, whereby a_i is the predicted value for $t + T$;

(e) the value of no element of P is calculated by applying an equation of the (b) kind;

and we name the spatial region, in which the measurements were performed, together with the elements contained in it, the system, whose state
has been measured with preciseness Δ.

We can therefore define "Δ-state with reference to H" independently of
any reference to a definite system.

When we then have a unit S of material and moving parts, for example
a pocket watch, we then, for the most part, call this unit a system, and ask
how its parts would move in the future. Thereby, there are certain
quantities that are evident from the very beginning – for example, in the
case of the pocket watch the angle of the hands with the 6–12 line.
Possibly, these quantities already constitute a Δ-state – in this case, we
call the system Δ-closed, outward and inward. Possibly, finer details are to
be taken into account, and only such details constitute a Δ-state. Then we
call the system an inward-open system, but lockable. It may also be the
case that no Δ-state can be found, but that there is a class of quantities
that satisfy the conditions from (a) to (e), whereby instead of (b) is P! (t) &
H & π_i [W(G_i, t) = g_i (t) $\prec$ P! (t + T) and the class of G_i represents a
completion, but not an [95]extension of the original description. Then S is
open outward. Let G/ be the class of G. We call P & G/ a Δ-*model of
S relative to H*. In classical mechanics G_i are for the most part forces that
influence the system from the outside.

Δ-states do not contain all the quantities that, in a system, are exactly
measurable. Some of the quantities not contained in them are dependent
on the elements of a Δ-state with reference to a certain theory H, and may
be defined on its basis. For example, in classical mechanics the rotation-
impulse of a particle with reference to a specific center as $n[r \; \dot{r}]$. Other
quantities belong to other states (so, for example, the temperature does
not belong to mechanical states), other quantities fall again below the
precision limits (for example, the space-coordinates of the planets of any
fixed star in the stellar astronomy). We call therefore a theory H *exhaustive* in reference to a class L of measurement classes, when, for every Δ and
for every system, there is a Δ-model PG/ of S relative to H, which satisfies
the following conditions. Let K be a class of quantities, which is constructed in the following way: from each measurement class of L we
choose at least one quantity, but in any case all those quantities, whose
measurement is necessary in order to determine a Δ-state with reference to
H. Then every element of K is either an element of P, or else it may be
defined on the basis of the elements of P.

The *claim of determinism* may therefore be formulated in the following
way: *the class of measurement quantities can be divided in measurement*

classes and the class of measurement classes can be again likewise divided into subclasses L_1, L_2, ... (whereby L_1 & L_2 & ... coincide with the class of all measurement classes – assuming a definite distribution of the quantities in measurement classes), so that there is an exhaustive theory for every L_1.

This claim shares irrefutability with K′. We can, however, also raise the question whether a certain theory U is exhaustive with regard to a previously given class L_1 of quantities, or not. *The answer to this question can be decided with a purely mathematical proof.*

First, we must notice that there is a trivial and a nontrivial answer to the question just posed: with reference to a class of measurement classes L, a theory may be nonexhaustive, because there is at least one measurement class in L, about whose elements, H makes no statement. So, for example, classical mechanics is nonexhaustive with regard to a class of quantities, which also contain, in a certain position, the forces of the electromagnetic field strength. But let us assume that all the measurement classes are understood as the basis of measurement classes, about whose elements the theory itself makes statements – which means, the class of all observable classes of the theory. We can *prove* that quantum theory is not exhaustive with regard to the class of its observable classes, that is, on the basis of quantum theory it is not possible to predict the value of every observable of this theory at a certain time, with any precision.

6.

[96]Let us confront this new formulation with the question, whether assumption K′ is true or not. Assumption K′ contains references to completely arbitrary quantities. Assumption K′ is fulfilled when I find a thermodynamic magnitude that would allow me to explain irregularities in the behavior of a mechanical system such as a pocket watch. On the other hand, in answering the question, whether a theory is exhaustive with regard to the class of its observable classes, such a liberty is not allowed. True, here we can also (once again using the example of classical mechanics) bring in all possible *forces* in the explanation of the movement of the elements of a system. But the movement of the elements themselves is *always* described through the indication of position and momentum coordinates, and predicted with the help of the fundamental equations of mechanics. However comprehensive the model is with whose help we want to make the behavior of a certain system explainable, the movement of elements of this system is always viewed as completely described by the

indication of position and momentum coordinates. And, therefore, well defined borders are set for the discovery of new parameters. (Is it not possible to attribute to the elements new, magical characteristics in order to explain their behavior? Such a procedure would draw with it a transition to a new theory, and thus neglect the question whether *classical mechanics* is exhaustive with regard to its observables.) We see that the new formulation of the problem is different from the old formulation in an important respect:[2] the old formulation asks whether the world is made so that K′ is valid. We have shown that for this question we will never be able to find enough empirical evidence. The new formulation asks whether K′ is fulfilled, if the world satisfies the postulates of quantum theory, and if the quantities brought in for extension are only the observables of quantum theory. It can be shown that this question must be answered with "no".

7.

Before I go on to introduce this proof in a condensed form, I still have to make some remarks about considerations [97]often used to prove the statement. The considerations are the following: in order to experience something about the behavior of our external world, we must take observations into account. These observations disturb the observed system, or, to be more precise, the observation of a quantity B in a system makes untenable the statement that the system remains in a state in which the magnitude A has the value a. Therefore, our knowledge of systems can never be complete. And, accordingly, we also cannot fulfil K′, although it is perhaps possible that the quantities introduced in K′ actually exist. What stands out in this argument is its carefreeness. First, we have to notice that observations *always* disturb. When we say, then, that events in quantum theory are not causally determined, while in the classical theory they are, this must lie in the difference between the theories, and not in the fact that was observed.

The argument may be refined in the following way:[3] with the classical theory it is possible to compute the influence of observation on the observed system, because we, at least in thought, can make the interaction between observer and observed system endlessly small. But in quantum

---

[2] G. Hermann overlooks this difference in her otherwise excellent portrayal *Die naturphilosophische Grundlagen der Quantenmechanik*, Berlin 1935.

[3] See, for example, Heisenberg, *Physikalische Grundlage der Quantentheorie*.

theory there is a lower limit to the effect. We cannot, therefore, still try to explain the transition of the quantum effect for its part, but we must take this transition as an indivisible single event.[4] But all we can say about indivisible individual events is that they happen or not, that is, for such events only statistical laws can be advanced. With this argument, it is already clear that a whole theory is involved. But it is not clear how. Then how does something behave when nobody is observing it? In this case no one could have put forth the quantum theory, but it could nevertheless be right. But now, how should we explain the indeterminacy involved in the quantum theory? The explanation of the indeterminate character of quantum theory, with reference to [98]the processes which occur during observation, can offer the impression that quantum theory only shows us the way in which the world appears when measured, and that it is not a portrayal of features of the world, just as it is independent of any observation. It leads to the assumption that the incompleteness of our means of observation prevents us from penetrating into a fully determined world, and this assumption – when we add that the concept of the "real outside world" is a concept with no meaning – is only a new version of Berkeley's philosophy: being is to be perceived – that is: being is to be empirically determined.[5] In fact, many philosophers, who have held on to the bad philosophy of the physicists, but not to their good science, have received this apparent consequence with enthusiasm. But we can now determine an important feature of quantum theory, the feature of being deterministic, *without* analysis of observation records and by appealing to pure mathematical considerations. Before we move on to these considerations, a few more comments about the incorrectness of the proof previously are in order:[6] this proof proceeded from the assumption that (a) each measurement involves an interaction, and therefore that (b) each measurement must lead to an essential vagueness. Now, neither (a) nor (b) is true. When a shooter fires his rifle and does *not* hit the target, then we can conclude that he shot past it, and we reached this conclusion by appealing to measurement, which involves no interaction. On the other hand, in case a measurement is made in a specific situation, we can predict the result with full precision *in spite of* the alleged interaction, which shows that the

[4] See Bohm D., *Quantum Mechanics*, Princeton 1950.
[5] See Einstein, "Reply to My Critics", in the Einstein volume of "The Library of Living Philosophers".
[6] For the following see Schrödinger, "Über die Unanwendbarkeit der Geometrie im Kleinen", *Naturwissenschaften*, XXII/341.

finite value of the quantum of action is not sufficient to infer indefiniteness of the measurement result.

8.

The proof we now present is due to J. von Neumann,[7] and proceeds in the following way: let us assume a (sufficiently large) class K of atomic systems S_i, with the same [99]structure S, all of which may be found in the same surroundings U (for example, a class of electrons in a certain electrical field, a class of hydrogen atoms in a well-defined field of light, etc.). The systems should not have any influence on each other; we only set them together in theory. ((S_I in U)) is an element of the class, which we have built as we have indicated. Let N be the number of these elements and A_1, A_2, A_3, ... the physical determination of parts of S, whose indication completely determines each S_I, according to the assumptions of quantum theory. We now measure one after the other the series A_1, A_2, ... with the help of the methods described above. We proceed as follows: A_I is measured in P $<<$ N elements from K (let the class of these elements be K_1) and the result written down. A_2 is measured in Q $<<$ N elements (class K_2), etc. We assume that K_1, K_2, ...K_i, ... have no common elements. This assumption is introduced in order to carry out all measurements in complete independence from one another (the assumption that single measurements disturb is, therefore, not considered). We obtain certain distributions of the values of the measured quantities. Since we have presupposed that all the systems under examination satisfy the same conditions, we can assume that this distribution is the same for all subclasses of K, and that it also holds for K itself (this assumption can be further scrutinized by undertaking sample checks). The result of the measurement is a statement of the following kind: the elements of K give in the measurement of A_I the distribution A_I (x), in the measurement of A_2 the distribution A_2 (x), etc., where A_I (x) represents the probability that the measurement of A_I on an element of K, on which no measurement is planned, gives the value x.

We are tempted to explain how the distribution comes about in the following way: each of the examined systems finds itself in a certain state, which is characterized by *sharp* values of the quantities of A_1, A_2, ... The

---

[7] In his book *Mathematische Grundlagen der Quantenmechanik*, Berlin 1930 {Feyerabend's review of the English translation of this book is reprinted as Chapter 17 of the present collection}.

systems, which are elements of K, build a *mixture*, that is, they are statistically divided over many states. It is therefore no wonder that we do not always obtain from successive measurements of the elements of the mixture the same measurement value, but a statistical distribution of measurements. The following is an example: [100]we consider all the bodies that have passed a certain sieve with openings of constant breadth as the systems under examination. The size of the bodies will then have upper limits, but will vary under these limits. The variation thus arises that the class on which we carry out our investigation is a mixture of elements, each of which has a definite magnitude.

As soon as this explanation is accepted, we can try to divide the class K into subclasses, whose elements have clear measurement values and which, mixed together in a definite relationship, give rise to a range of measurement values, which is identical to the range found in the original class. The failure of such an attempt can be explained in two ways: (1) there are subclasses with clear measurement values, but we have no possibility to separate them experimentally out of a class K, which is given in advance. This means that we have no possibility of producing classes of atomic systems, whose measurement values scatter across a region, which is smaller than a certain value given in advance. (2) There are no such subclasses. In the first case, it would be at least theoretically possible to find an explanation for the fact that the scattering comes about, in which the assumption of clear measurement values plays an important role. But in the second case we cannot attribute the scattering to the mixture of clearly determined systems; we must transfer them to *within the system itself*.

Cases (1) and (2) are jumbled together by some philosophers of science, particularly operationalists. Operationalists say approximately this: if we have no method to discover the difference between two cases, then this difference does not exist. If we have no method to isolate a certain case, then this case likewise does not exist. These principles contain a worth-while core – but, like nearly every general foundation principle, it hides behind the vagueness of the expressions used. By "method" we can under-stand a certain experimental procedure; but we can also understand such a procedure accompanied by the construction of hypotheses and, finally, the mere procedure of hypothesis building. The principle is false when "method" is understood in the first sense, true [101]in the other two cases. For example, it is no doubt possible that a certain chemical element has not been isolated yet although we have good reasons to assume its existence. These reasons are provided by a theory, which has already led to many fruitful discoveries and which we are, therefore, not inclined to abandon easily. In the case of quantum theory it can then be shown that there are no subclasses of the given features. The scattering of the measurement values is not a property of mixtures, nor does it belong to the elements of the mixture themselves, but it is just a property of these elements.

9.

The proof of this claim is relatively simple: we call a class K, in which the division associated with any measurable quantity is displayed in each of its subclasses by the *same* function, a pure instance. We further call a class K of systems, which, at least with regard to quantity, scatter, an unclear instance. Let us assume that there are unclear pure instances. Then a class of systems is available, which scatter with reference to certain values. But this class only contains subclasses, which show the same scattering – as a consequence, it is not possible to build the scattering of the total system by composition of nonscattering subclasses. Von Neumann has thus proven that there is (a) no clear instance in the quantum theory, and that (b) there are pure unclear instances. Thus, (A) we have said everything we need in order to prove the thesis that quantum theory is an indeterministic discipline. When, however, we go over to finer quantum theoretical models of a certain system, some of its measurement values will always turn out to be different from one time to another. But there is more that follows from the proof: (B) let us assume that someone poses the task of building a theory that has the same results as quantum theory, and that therefore is as well confirmed as quantum theory, but that, on the basis of an intricate mechanism, is capable of giving a causal explanation for each instance. The proof just given shows that such a theory, regardless of how simple or complicated the assumptions it uses may be, must contradict quantum theory and, therefore, since quantum theory is well confirmed by experience, also experience. [102]What we have shown is that: (1) if the world is made as quantum theory says it is, there are no predictable events in it; (2) quantum theory is well confirmed, so the world is as it says; therefore, (3) there are no predictable events in it. The first statement is provable; the other two are only probable.

10.

The argumentation at the end of the previous paragraph shows the advantage of applying comprehensive theories to decisions concerning relevant philosophical questions, such as, for instance, deciding whether all events in the world are predictable, or whether there are nonpredictable events in it. We have seen that statements such as K′ cannot be directly verified by experience. The possibility of their being directly verified is precluded by their logical structure. The considerations just introduced lead to an indirect verification of K′. This indirect verification seems to face a task more difficult still than the verification of the K′ statement: quantum theory is certainly logically more complex than K′, and there is no prospect of finding ways to bring its logical form into simpler terms.[8] We can nevertheless verify it, and agree whether it is to be accepted or rejected. Physicists undertake this duty. Philosophers are left with the task merely of deciding what follows from the truth of the theory – and this decision can be made with the help of pure mathematical considerations. The philosophical task of inquiring into the general features of the world (such as, for example, the question whether the world contains nonpredictable events) is thereby divided into two parts: the task of verifying a theory, for which specific physical methods are provided in advance, and the task of the logical-cum-mathematical analysis of this theory.

Various philosophers have denied that these two partial problems exhaust the corresponding philosophical problem (the causality problem). They say that the concept of predict[103]ability has no "ontological relevance".[9] That is to say that if an event cannot be predicted by a specific theory, this does not mean that such an event does not have a cause. That is to be conceded. But, it does not follow that we have to weaken the way in which theories express themselves, for instance that we have to say "perhaps this event does not have a cause" instead of "this event does not have a cause". For, if we provide all our natural scientific theories with precautionary clauses, such as "perhaps", "possibly", etc., they cannot in any way be refuted, because they do not state anything. Against this, it may be objected once again that natural scientific theories, even when they are still so definitively formulated, nevertheless do not refer to the world – at least we have no guarantee that they do. This argument confuses the truth of a theory with the objects it refers to: optics has to do with the phenomena of light – therein

---

[8] This is what is to be said about all attempts to describe the laws of physics in the system of *Principia Mathematica*.

[9] See H. Wein in the *Clausthaler Gesprächen*. Report in *Philosophia Naturalis*, I/1, p. 142. The circle of people working for this journal is especially known for its philosophical obstinacy.

lays the object it refers to. If a certain optical theory should be proven false, it does not therefore follow that this theory does not speak about phenomena of light, but about something else. It makes a false statement – about light. We shall call a theory ontologically relevant if it refers to an object, that is, if it speaks about the world. The confusion between truth and reference to an object leads philosophers, about whom we are talking here, to call a theory ontologically relevant only if it cannot be corrected. Let us see how this incorrigibility can be obtained: either we retain the same expressions, but give them a new meaning – and so what cannot be corrected, here, is writing down certain thoughts, not the thoughts themselves. We have already treated this case (Section 3). Or else, we keep repeating the statements of the theory that is being questioned, whatever the case may be. But can we call a doctrine ontologically relevant, in some reasonable sense of the word, if it can never be affected by anything, and always sticks to its claims? Hardly. That said, we must attribute ontological relevance to physical theories, and deny such relevance to philosophical theories. Philosophical analysis divides, as described, into an [104]empirical problem (the investigation of the truth of a theory) and a logical-cum-mathematical problem (the investigation of the structure of this theory).

11.

Essentially more interesting and fruitful are the objections that Bohm has recently offered against von Neumann's proof.[10] He thinks it is possible to provide a causal interpretation of the formulas of quantum theory that does not contradict them. One more remark, before I offer a sketch of Bohm's interpretation: physicists and philosophers who defend the idea that a causal interpretation of the formulas of quantum theory is possible are always very concerned that this interpretation does not contradict quantum theory. That is why von Neumann's proof seemed, for them, to represent an obstacle that could not be overcome. As a consequence, they overlook the fact that comprehensive theories, which unify a series of less comprehensive theories, almost invariably contradict them: Kepler's laws contradict Newton's theory, as they can be derived from it only approximately.[11] As a consequence, as long as the contradiction between

---

[10] "Quantum Theory in Terms of 'Hidden Variables'", *Phys. Review* 85 (1952), p. 166 ff.
[11] See K. R. Popper, "Naturgesetze und theoretische Systeme", *Jahrbuch der Hochschulwochen Alpbach*, 1948, ed. Moser.

quantum theory and its allegedly causal interpretation falls under the threshold of measurement, its existence cannot be used as an argument against the interpretation.

But Bohm takes up the task of constructing an interpretation that does *not* contradict quantum theory; accordingly, he proceeds from the following considerations: the time-dependent Schrödinger equation for the motion of a particle in the field V(x) has the form $i\hbar\frac{\partial\psi}{\partial t} = -\frac{\hbar^2}{2m}\nabla^2\psi + V(x)$. When we write ψ in the form $\sqrt{P}e^{\frac{iS}{\hbar}}$ we obtain

$$\frac{\partial P}{\partial t} + \nabla\left(P \cdot \frac{\nabla S}{m}\right) = 0 \tag{1}$$

$$\frac{\partial P}{\partial t} + \frac{(\nabla S)^2}{2m} + V(x) - \frac{\hbar^2}{4m}\left[\frac{\nabla^2 P}{P} - \frac{1}{2}\frac{(\nabla P)^2}{P^2}\right] = 0 \tag{2}$$

For $\hbar \to 0$, (2) is Hamilton-Jacobi equation of motion; and for $v = \frac{1}{m}\nabla S$, (1) is the continuity equation. [105]We can then view electrons as particles that have exact values of position and speed, but whose movement, in addition to the classical potential V(x), is controlled by the "quantum-theoretical potential" $U = -\frac{\hbar^2}{4m}\left[\frac{\nabla^2 P}{P} - \frac{1}{2}\frac{(\nabla P)^2}{P^2}\right]$, which is to be interpreted as the potential of a real wave field (comparable with Maxwell's electromagnetic field). In order to compute the exact motion of a particle, it is necessary that (a) the field for the problem at hand be calculated, (b) the initial situation and the initial speed of the particle be known. The calculation (a) is identical to the normal calculation of the ψ-function of the particle. It yields an expression for U that in most cases fluctuates temporarily in a very strong way – so strongly that only statistical predictions as to the paths of the particles are possible. Just like statistics in thermodynamics, however, this statistics is in principle avoidable, and necessary only due to practical reasons. In the case of a measurement we can, therefore, predict that one of the intrinsic values will turn up, but we cannot say which intrinsic value will turn up, nor, in detail, which path the particle will follow. The results for the system obtained from the observed intrinsic values are, as a consequence of the sensitivity of U to small variations in the surroundings, essentially determined by the initial positions of the measurement apparatus. What we measure is, in fact, a combination of the features of the system and of the measurement apparatus. The assumption that *they* could be elements of a deterministic interpretation would contradict quantum theory in its current form. This is, according to Bohm, the content of von Neumann's proof. But the variables that Bohm brings in are of another type. They turn up in quantum theory itself, and in principle – given the initial values and P – can themselves be calculated exactly (x-place; x-coordinate; x-momentum: ∇S). Since the measurement quantities that refer to them describe features of a combination, the value of these variables at a given time cannot be ascertained by measurement. This means that the questionable variables are not observables, nor are their values clearly computable on the basis of the results of the measurements. That also explains that, according to (a), the form of the wave function after the measurement is decisive for the position of the particle after the measurement. This allows us to draw conclusions with respect to only one [106]of two complementary observables. Therefore, we have to *guess* the initial values required for the prediction of the path of the particle. If we have these values, the path can be completely computed and in principle we can predict where, after a series of interactions, the particle will turn up. On account of the limits of the measurements, however, this prediction, too, can be controlled only through partial guesses. What we achieve, therefore, is the following: we obtain a description that interprets every quantum theoretical process as a continuous motion of a particle or a swarm of particles in a classical + quantum theoretical field of force. This alternative description does not make predictions possible, or statements about the past, since the required path values cannot be determined by measurement. But (by guessing) the initial and end values can be found, which (aa), insofar as measurement results are at hand, agree with them; and (bb) do not lead to contradictions when we use equations (1) and (2) to combine these partly guessed, partly measured values. By the way, the impossibility of completely determining by measurement any value of a particle's path exists only when we assume that the formalism of quantum theory in its present form provides the correct description of natural appearances (which Bohm assumes). But when we dispense with these assumptions, it becomes questionable whether (bb) – that is, the consistency of the particle motion – is secured.

From this description we see (1) that no contradiction with quantum theory is present, and that we have to restrict *von Neumann*'s proof to variables that are in principle (assuming the truth of quantum theory) accessible to observation. Furthermore, we have to replace (B), in Section 9, with the following: a causal interpretation of the results of quantum theory, whose conditions are in principle observable, contradicts quantum theory in its present form. In Section 6 we said that in these terms: quantum theory is not exhaustive with respect to its classes of observables. But (2): does this interpretation entail significant progress in the problem of determinism? Let us consider a given case. We compute in the normal way and we observe which of its values [107]obtains. *Next*, we try out variables for constructing a path that does not contradict measurement values, as far as they are available, and is also in agreement with equations (1) and (2). Now, the defenders of quantum theory have, to be sure, never disputed that it is possible to reconstruct the path of an electron that contradicts the uncertainty relation.[12] It is Bohm's contribution to have shown quite generally the consistency of such a reconstruction with the formalism of quantum theory. But, obviously, this does not answer the problem of determinism – for, this consists in the question, whether there are *nonpredictable* events, or to put it as the objective content of a given theory – *noncaused* events. But in Bohm's interpretation it is impossible to provide an exact *prediction*, since (a) it works with partially nonobservable variables and because (b) it works with statistical mixtures. The first limit stands in principle as long as we (like Bohm) deem quantum theory to be correct. Therefore, Bohm has not shown that it is possible to interpret quantum theory as a *deterministic* discipline.[13] The second limit is regarded by Bohm, as well as by a number of other physicists, as a merely practical one. I shall discuss the correctness of this view in the next section.

12.

The attitude of many physicists and philosophers toward classical statistics is roughly the following: in classical statistics (statistical thermodynamics, for example) we are concerned with particles whose paths can, in principle, be measured with any arbitrary precision. *In practice*, however, it is impossible to measure these paths. Therefore, we introduce statistical assumptions, which do not, as in quantum theory, express statistical features of nature itself, but are merely an expression of human limitations. We now consider, as we did in Section 8, a sufficiently large class K of statistical [108]collections with the same construction and in the same surroundings. The systems again should not interact with one

---

[12] See, for example, Heisenberg, *Physikalische Grundlagen der Quantentheorie*, p. 15.
[13] Briefly, Bohm shows that, with guesses, it is possible to find values on the basis of which the outcome of a given measurement can be predicted. But then we can leave the calculation aside and immediately try to guess the final values. Does that prove determinism? No indeterminist disputes that we can guess future events.

another; we only unite them as a thought experiment. A molecule of an ideal gas, enclosed in a container at room temperature, is an element of K. With tiny instruments we measure momenta at certain positions in the gas. The measurement variables in this case are the momenta in these positions. According to Maxwell's division, measurement values are generally scattered over an interval. According to the argument in Section 8, we prove that such scattering cannot be explained by mixture: thus, it represents a characteristic feature of the very systems under examination. It is a feature of nature, not an expression of our lack of knowledge. To this it is generally replied that such scattering, which is admittedly objective, does not affect the limit of the possibility of predicting the path of an individual molecule. Let us consider the following example: a molecule of ideal gas is enclosed in a container with two small openings. Through one opening we send a certain molecule, M, of the type of the enclosed gas, whose speed and place at a given time we assume as known. We ask at what time and direction will the molecule appear at the other opening. According to the idea we are now discussing, such a prediction should be possible in principle. Let us make, then, the following thought experiment: we measure the position and the speed of all molecules in the gas at the time of the entrance of M, and from this calculate the subsequent motion of this molecular planetary system. According to this prediction, M will appear at the second opening later, at a determined moment in time. In order to check the prediction, it must be possible to recognize M. According to our hypothesis, however, M is a molecule of the same type of the molecules enclosed in the container – therefore, we can perform the check only if we follow M on its path through the gas, and if at the same time we define a path for M through the gas. But this assumption contradicts the assumption that M is a part of the gas, that is, it interacts with the other molecules of the gas, which obey *only* the gas laws. As a consequence, we are back to [109]statistics, and the thought experiment we have performed turns out – as often is the case with thought experiments – to be empirically empty. From this and similar considerations[14] it follows that even in classical physics we take a point of view from which it is meaningless to ask what cause prompts a molecule to move in a given way. But the deterministic program proves more

---

[14] See the outstanding essay by A. Landé, "Continuity, a Key to Quantum Mechanics", *Philosophy of Science*, April 1953. Through reasoning very similar to von Neumann's, Landé shows that it would be contradictory to assume that causes can be given for the detailed behavior of each molecule.

seriously unworkable. Let us first take it in its strongest form: a condition of the universe completely determines each further condition, that is, every event in the universe is exactly predictable when we know the events occurring at a given moment in time. For the prediction we use a theory T. From T, together with a certain state S, follows, among other consequences, that in the future all scientists will hold a theory T_1, which contradicts T (see footnote 12) as true, and that they will also have reasons to hold it as true. But then *we* could also immediately use T_1 in place of T. Once again, the prediction will be that all scientists, at some given future time, will hold T_2 as true, and there will be reasons to hold T_2 as true. Therefore, we can immediately use T_2. And so, on the basis of the prediction of a *single* theory T, we immediately have all its improved and comprehensive successors – without any need, as it seems, to gather further experience. But the experience required for the correction of theories is now hidden in the current state of the universe. If we assume that we can exhaust this experience in a finite number of steps, then it seems, without any doubt, that all improvements of a theory are merely a matter of calculation, but no longer of observation. We can avoid this very implausible consequence only if we give up either strict determinism, or the assumption that experience is exhaustible (that the universe is finite). Let us keep hold of the first and limit [110]ourselves to complete systems, whose state relative to a certain theory and a certain model can, in general, be ascertained in a finite number of steps. If we (1) consider that the measurement of a system by way of another always has the consequence of a certain openness, that (2) this openness cannot be reduced to zero, because the parts of each measurement apparatus must have a finite extension, if they are to be strong enough to further transfer the effects transferred to it to a reading mechanism; if we, on the other hand, observe that for purely mathematical reasons a prediction system (made of experimental tools and calculation apparatus) can never predict its own state (by the way, this theorem can be immediately applied to cosmology – if we do not assume that the calculating physicist is outside of the universe, as a sort of demon: cosmology, the theological physics of the present!), then it follows that, even in a world that strictly obeys the laws of classical mechanics, there are unpredictable, and therefore undetermined, events.[15] When we take all this into account, we will increasingly tend to see the assumption of universal determinism as a

---

[15] See K. R. Popper, "Indeterminacy in Quantum-Mechanics and in Classical Physics", *British Journal for the Philosophy of Science*, 1951.

prejudice, and to view the world as it *prima facie* offers itself to us: on the one hand, there is a series of predictable events, but, on the other, there are also many events that are not predictable and that, therefore, if we take seriously our theories from the ontological point of view, we must refer to as uncaused. Then it also becomes clear that the discoveries of quantum theory look so surprising only because we were caught in the philosophical thesis of determinism, which, in this case, clashes, for the first time, against an obstacle that made all attempts at reinterpretations – in which philosophers always seek salvation – impossible. What we often refer to as a "crisis in physics" is also a greater, and long overdue, crisis of certain philosophical prejudices.

4

A Remark on von Neumann's Proof
(1956)

[421]The attacks that Bohm [1], de Broglie [2], Weizel [5], and others directed against von Neumann's proof of the impossibility of a deterministic interpretation of elementary quantum mechanics (QM) show that it has some weak points, but do not show where the weaknesses lie. The best discussion in this regard is that of Fenyes [3]. He shows that some consequences of QM considered to be among its characteristic features are consequences of interpreted parts P of its probability computations. These consequences hold for every theory containing P. But Fenyes' arguments are intricate, and may be replaced by a relatively simple consideration. This is what I would like to do in this note.

The aim of von Neumann's proof (NP) is to show that every deterministic theory of atomic phenomena necessarily contradicts QM, whether it works with observable or with "hidden" variables ([4], p. 171). We call this claim D. In order to prove D, two sentences are necessary and sufficient, A and B, which, as von Neumann shows, follow from QM ([4], pp. 169 ff.). We skip this step of the proof (whose validity has been incorrectly questioned by others) and focus our attention to the derivation of D from A and B. A and B refer to the infinite aggregate of atomic systems (von Neumann accepts Mises' interpretation of probability – see [4], pp. 158 ff., and note on p. 156) and state propositions about the relative frequency of measurement values within any whole. According to A, there are no scatter-free aggregates, that is, no aggregates whose elements agree as to the values of all variables; according to B, each aggregate of quantum mechanical systems can be produced by mixture of unitary aggregates (a unitary aggregate being one whose (infinite) subaggregates are characterized by the same frequency function as itself). Now, "the question" may be raised

whether the scattering of the unitary aggregates does not derive from the fact that these are not the real states, [422]but only mixtures of various states – while, besides the variables of the wave functions, still other variables would be required (these are the "hidden parameters") to identify the real state ... The statistics of the unitary aggregate would then arise through averaging over all real states on which it is built; that is, by averaging over the value field of the "hidden parameters" realized in these states. But this is impossible, for two reasons: first, because the relevant unified aggregate could then be described as a mixture of two different aggregates, in conflict with its definition. Second, because aggregates without scattering ... do not exist,

that is to say, because such an assumption would contradict A and B. This is basically von Neumann's proof of *D* ([4], pp. 170 ff.).

In order to come to a proper appraisal of the proof, let us now recall that *A* and *B*, on which the proof is based, follow from every interesting statistical theory, for example, from the theory of the game of dice.[1] According to *A*, in this case, none of the numbers of points will constantly turn out constantly; according to *B*, no game system, however cleverly thought out (when it leaves untouched only the conditions of chance of the game), can lead to subaggregates that show a different distribution of the results (number of points) as the initial aggregate (which, in the case of the game of dice, has in general the distribution $^1/_6, {}^1/_6, {}^1/_6, {}^1/_6, {}^1/_6, {}^1/_6$).[2] When this is the case, however, it should not be possible, on the basis of NP (by adding "hidden variables" to those used exclusively in the game), to come to more detailed, and, above all, to a deterministic description of the single throw of the dice. And the same would have to hold for the gas theory – although we all think (and von Neumann, too: [4], p. 109) that in the latter case hidden variables exist. But if von Neumann's argument on this point is not correct, it can also be incorrect in quantum mechanics, since the proof is based on exactly the same assumptions, namely, *A* and *B*. Which additional assumptions were smuggled in?

It is the assumption (S) (see [4] p. 171), that a deterministic description must always lead to scatter-free aggregates. But this assumption is false.

---

[1] I owe to Prof. K. R. Popper (London) the remark that NP did not reach its goal because it is valid not only for QM, but also for the game of dice.

[2] That is to say, *B* *follows* from the so-called principle of excluded game system, which, according to von Mises, is a necessary condition of randomness in a series of events. In a discussion with the author, Mr. J. Agassi has maintained the *identity* of *B* and the principle mentioned.

For determinism maintains that [423]events that chronologically derive from one another are clearly linked with each other (presuming that they were completely described), but it says nothing about the distribution of those events at a certain point in time. So, no one will doubt that the many small influences that a die experiences during the shaking process determine the outcome (together with other circumstances), although from that it by no means follows that by shaking we can repeat some of these influences as we might wish, and therefore produce sharp aggregates. Von Neumann believes that S in the classical case is fulfilled. Here he is in error – and his own proof should have taught him better. He correctly assumes that in quantum theory S is not fulfilled – but an examination of the classical case shows that this is no exception that distinguishes QM from the classical case. The correct conclusion is that S is never fulfilled wherever statistical assumptions are present, and that S has nothing to do with the existence or nonexistence of hidden parameters. All that follows from NP is this: hidden parameters, which were introduced for a more exact description of statistical processes, will likewise scatter.

Literature

[1] Bohm, D.: *Phys. Rev.*, 85, 180 (1952).
[2] Broglie, L. de: *La Physique quantique, restera-t-elle indéterministe?*, Paris 1953.
[3] Fenyes, I.: *Z. Physik*, 132, 81 ff. (1952).
[4] Neumann, J. v.: *Mathematische Grundlagen der Quantenmechanik*, Berlin 1930 {Feyerabend's review of this book is reprinted as Chapter 17 of the present collection}.
[5] Weizel, W.: *Z. Physik*, 134, 264 ff. (1953).

5

Complementarity
(1958)

1. INTRODUCTION

[75]Bohr's idea of complementarity originated as an attempt to devise a consistent and exhaustive picture of the behavior of microscopic systems. His arguments for this picture and against alternatives are based, partly upon empirical investigation, partly upon philosophical analysis. In the present paper, I have set myself the task of making explicit the assumptions upon which this analysis rests.

I shall proceed in the following way. In Section 2 I shall discuss a simple example of the "paradoxes" the idea of complementarity sets out to explain and I shall indicate the kind of solution proposed. In Sections 3 and 4 this solution will be discussed in more general terms. Bohr's views about the way in which the quantum of action is to be incorporated into (any) quantum theory will be developed, and some of its presuppositions investigated. The principle of complementarity will be stated and explained. Section 5 applies complementarity to the only coherent quantum theory of today, viz. elementary quantum mechanics. The Copenhagen interpretation is presented and criticized. Although this interpretation can be attacked on general grounds, it nevertheless seems to be supported by a closer analysis, not of its philosophical elements, but of the theory it is supposed to elucidate. In Section 6 I shall inquire how far this support goes and to what extent it may be regarded as a support of Bohr's more general ideas. Section 7 deals with the wider application of complementarity and with the question whether such a wider application is (a) justified, (b) enlightening. This section also contains a summary of the results of the paper.

[76]In writing this paper I have learned much from discussions with Profs. D. Bohm, S. Körner, and K. R. Popper, as well as from Prof. Popper's papers and lectures both on the subjects of quantum mechanics and of scientific method in general.

2. DUALITY

About light the following experimental facts are known:

(a_1) The phenomena of interference, diffraction, and of the propagation of light in various media can be completely accounted for by the wave theory (or by the electromagnetic theory).

(a_2) The interference patterns are independent of the intensity.

These facts have been firmly established, partly by the numerous successful applications of the wave theory in the construction of optical instruments, partly by more recent experiments of great accuracy which were prompted by doubts in the Copenhagen interpretation.[1]
Experiments have also shown that:

(b_1) the interaction of light with matter consists in localizable events, and that the energy transferred from the electromagnetic field to the absorbing material is concentrated in a small spatial volume.[2]

(b_2) Absorption and emission of light are individual processes which possess a definite direction.[3]

(b_3) The simultaneous fluctuations of intensity in different fringes of the interference pattern are statistically independent.[4]

[77]On the basis of (a) it was assumed that *light consists of waves* (A). "What is light?" writes H. Hertz in 1889.[5] "Since the days of Young and Fresnel we had known that it is a wave phenomenon. We know the velocity of the waves, their length, we know that they are transversal

---

[1] I am referring to experiments carried out by Jánossy and Náray. They are described in a booklet published in September, 1957, by the *Hungarian Academy of Sciences.*

[2] This is a consequence of the photo-effect. Cf. Richtmeyer-Kennard-Lauritsen, *Introduction to Modern Physics*, New York, 1955.

[3] This was first derived by Einstein on the basis of theoretical considerations. The experimental analysis which followed the discovery of the Compton effect showed the correctness of Einstein's results. Cf. the ref. in fn. 2 as well as Compton's own report in *Naturwissenschaften*, Vol. 17 (1929), pp. 507 ff.

[4] This was demonstrated by S. I. Vavilov; cf. his Микроструктура Свэта, Moscow, 1950. Cf. also the booklet referred to in fn. 1.

[5] *Gesammelte Werke* I., 340.

waves; in short, we know completely the geometrical properties of the propagation of light. With respect to these matters all doubt has been removed and for the physicist a refutation seems unthinkable. For all a human being can possibly know the wave theory of light is a certainty".

The experiments referred to under (b) have shown that (A) cannot be correct. First, because according to (A) the energy transported by light would have to be equally distributed over the whole wave field, which contradicts (b_1). Secondly, because according to the wave theory the field corresponding to a single act of emission would have no resultant momentum, which contradicts (b_2). And finally, because in a wave theory all events which could possibly happen along a wave front are co-ordinated by their phase, which contradicts (b_3). In order to explain the experiences described in (b), the particle theory of light has been revived: *light consists of particles* (B).

It is easy to see that this assumption also runs into difficulties. It cannot account for the coherence which is necessary for interference. Mechanisms bordering on telepathy are needed in order to explain the behavior of a single photon in an interferometer which is being rapidly rearranged.[6] Nor does it seem to be possible to introduce the field as a purely statistical entity as this would allow the particle to dwell sometimes in a state which is incompatible with the principle of the conservation of energy. And even the attempt to save the particle interpretation by admitting action at a distance does not eliminate the need for the [78]waves; for the behavior of the particle does not depend upon the whole universe, but only upon the situation along its wave front. This throws us back upon the wave theory. We may sum up these results as follows: the facts known about light (and, as we may add, about matter in general), divide into two classes, (a) and (b). The facts of the first class, while contradicting the particle theory, can be completely and exhaustively explained in terms of the wave theory. The facts of the second class, while contradicting the wave theory, can be completely and exhaustively explained in terms of the particle picture. No *known* system of physical concepts can provide us with an explanation which covers, and is compatible with, all the facts about light and matter. This has been called the *duality* of light and matter.

For a physicist who conceives of light (and of matter) as something which is fundamentally a single and objective entity, this situation

---

[6] Cf., e.g., Max Born, *Optik*, Berlin, 1933, p. 465, as well as W. Pauli, *Handbuch der Physik* (Geiger-Scheel), Vol. XXIII (1925), p. 81.

constitutes a challenge. For him this situation proves that the theories available at the moment are insufficient. He will consider as appropriate the search for a new conceptual scheme which is general enough to account for the undulatory and the discrete aspect of light and matter, without giving any one of these two aspects universal validity. He will demand that such a new theory, as indeed any theory, must satisfy two requirements. It must be empirically adequate, i.e., it must contain the facts, mentioned in (*a*) and (*b*) as approximately valid under mutually exclusive conditions (to be formulated in terms of a set of its parameters). And it must be universal, i.e., it must be of a form which allows us to say what light is rather than what light appears to be under various conditions. He could, of course, not guarantee that a theory satisfying these two demands would ever be invented. But in any case the existence of two nonexhaustive and complementary descriptions would seem to him to be a historical accident, an unsatisfactory intermediate stage of scientific development which should, and might, be overcome by new ideas. The [79]ideal of scientific procedure behind this attitude will be called, for reasons to be explained later, the *classical ideal*. It is clear that the classical ideal is closely connected with the position of *realism*.[7]

Bohr's point of view may be introduced by saying that it is the exact opposite of the attitude just described. For Bohr the dual aspect of light and matter is not the deplorable consequence of the absence of a satisfactory theory, but a fundamental feature of the microscopic level. For him the existence of this feature indicates that we have to revise, not only the classical physical *theories* of light and matter, but also the classical *ideal of explanation* (D 317).[8] "It would be a misconception to believe", he writes (A 16), "that the difficulties of the atomic theory may be evaded by eventually replacing the concepts of classical physics by new conceptual forms". We "must be prepared ... to accept the fact that a complete elucidation of one and the same object may require diverse points of view which defy a unique description" (A 96). Many physicists have expressed their belief that this feature of the microscopic level is final and irrevocable. How has this belief been defended?

---

[7] For an excellent description of the classical ideal and of its connection with realism cf. K. R. Popper, "The Aim of Science", *Ratio*, Vol. 1 (1957), pp. 24 ff.

[8] The following abbreviations will be used in the text: "A" for the collection of essays *Atomic Theory and the Description of Nature*, Cambridge, 1934. "E" for Bohr's essay in the Einstein-Volume of the *Library of Living Philosophers*. "D" for Bohr's article in *Dialectica*, 7/8 (1948), pp. 312 ff. And "R" for Bohr's article in *Phys. Rev.*, Vol. 48 (1936).

The argument most frequently met is based upon the adoption of a positivistic point of view. According to positivism "the conceptual schemes and the theories of science are not an attempt to go beyond experience and to understand the 'essence' of things; they are rather auxiliary systems for the registration and ordering of our sensory experiences".[9] It is evident that a physicist who subscribes to this doctrine will interpret duality as "something originally given" which "is not in need of explanation".[10] [80]He will refrain from the attempt to find a new *and unified* conceptual scheme not because it has been demonstrated that such a scheme contradicts physics, but because any such attempt would conflict with his philosophical conscience as a positivist. His opposition to what has above been called the classical ideal of explanation is prompted largely by a philosophical disagreement and by the acceptance of a different ideal of explanation (viz., explanation by incorporation into a predictive scheme). Many physicists, especially those of the Copenhagen school regard this transition to positivism as a matter of course and are completely unaware of the change of philosophical background involved. To them Popper's remark applies[11] that "the view of physical science founded by Cardinal Bellarmino and Bishop Berkeley has won the battle without a further shot being fired".

I do not believe that this remark can be applied to Niels Bohr himself. In spite of an occasional vagueness in his writings and in spite of some inconsistencies (which might easily be interpreted as the uncritical acceptance of a positivistic point of view), Niels Bohr is consciously concerned with the development of a new model of explanation which is supposed to lead to an understanding of microscopic phenomena as well as to theories for the description of such phenomena. He produces arguments why he thinks that in the case of duality the classical procedure is not likely to succeed. And he can point out, with some justification (see Section 6), that there exists a formalism (viz., elementary quantum mechanics) which supports his point of view, but which leads to great difficulties if we try to analyze it as a universal theory.

In the next two sections Bohr's point of view will be gradually developed. I shall try to show that it is consistent, that it has led to important results in physics and that it therefore cannot be dismissed easily. It will also turn out [81]that this point of view is closely related to

---

[9] P. Jordan, *Anschauliche Quantenmechanik* (1936), p. 277.   [10] *Loc. cit.*, 276.
[11] "Three Views Concerning Human Knowledge", *Contemporary British Philosophy* III (1956), p. 8.

the position of *positivism*: the issue between the classical model of explanation and complementarity is essentially an instance of the age-old issue between positivism and realism. In dealing with this issue I shall be mainly concerned with the form in which it arises in quantum mechanics; I shall not be concerned with its resolution. This I have attempted elsewhere.[12]

3. PHILOSOPHICAL BACKGROUND

In his analysis of physical conceptions Bohr is guided by two philosophical ideas which are so simple and at the same time so general that physicists either tend to regard them as obvious, or overlook them altogether. Yet the validity of Bohr's approach completely depends upon the validity of these two ideas.

The first idea is that "no content can be grasped without a form" (E 240) and that, more especially, "any *experience* makes its appearance within the frame of our customary points of view and forms of perception" (A 1). – This idea shows that Bohr's philosophy is different from positivism in an important respect. According to positivism proper the formal elements of knowledge are restricted to the laws which we use for the ordering of our experiences. The experiences themselves do not possess any formal property, they are unorganized simple elements, such as sensations of color, of touch, etc. According to Bohr, however, even our experiences (and, consequently, our ideas) are organized by "categories" (E 239), or "forms of perception" (A 1, 8, 17, 22, 94, etc.) and cannot exist without these forms.

Bohr's second idea is that "however far the new phenomena transcend the scope of classical physical *explanation*, the account of all evidence must be expressed in classical *terms*" (E 209; A 53, 77, 94, 114; D 313; R 702) which means that the "forms of perception" referred to above are, *and will always be*, those of classical physics.[82] "We can by no means dispense with those forms of perception which color our whole language and in terms of which all experience must ultimately be expressed" (A 5). – According to this second idea, the forms of perception postulated by the first idea, are imposed upon us and cannot be replaced by different forms, even if it were desirable. This shows that Bohr's point of view still retains an important element of positivism. Although he admits that scientific experience is more complicated than the immediate experience

---

[12] "An Attempt at a Realistic Interpretation of Experience", *Proc. Aristot. Soc.*, 1958 {reprinted in *PP1*, ch. 2}.

presupposed by the positivists, he still seems to assume that this more complicated experience is "given" and not capable of further analysis. One might therefore christen Bohr's point of view a *positivism of a higher order*.

A more concrete statement of the two ideas just explained may be obtained by connecting them with duality. As was stated in Section 2 all the experiences of light and matter divide into two groups. The experiences in the first group can be united by the wave theory. The experiences in the second group can be united by the particle picture. However, these two classical theories are not only important as devices which allow us to *summarize*, and to *unify* a host of facts in an economical way. Without the (key terms of the) wave theory the facts of the first group *could not even be stated*. Nor would it be possible to determine experimentally such physical magnitudes as the wave length of the first Balmer line without applying the wave theory for the construction of a suitable apparatus as well as for the interpretation of the results obtained by it. Similarly, the idea of light quanta is needed in order to account for the conservation of energy and momentum (A 17, 113). "Only with the help of classical ideas", writes Bohr, "is it possible to ascribe an unambiguous meaning to the results of observation" (A 17) and to "give tangible content to the question of the nature of light and matter" (A 16).

Now it is of course quite correct that at the present moment "every actual experiment we know is described with the help of classical terms and" that "we do not know how to do it differently".[13] It is also to be admitted that [83]without this terminology and without the methods of classical physics many facts could neither be ascertained, nor even stated. Abandoning the concepts and ideas of classical physics would lead, not to the "foundations of knowledge" as the positivists would have it, but to the end of knowledge. But does it follow that therefore our experiences can be obtained and reported "*only* with the help of classical concepts" (A 17)? i.e., does it follow that we can *never* go beyond the classical framework and the fact, characteristic for this framework, that a coherent account of all our experimental knowledge is impossible?

It is quite obvious that the use of classical concepts in present day physics can never justify such an assumption. For a theory may be found whose conceptual apparatus, when applied to the domain of validity of classical physics, would be just as comprehensive and useful as the

---

[13] C. F. v. Weizsäcker, *Zum Weltbild der Physik* (1954), p. 110.

classical apparatus, without coinciding with it. Such a situation is by no means uncommon. The behavior of the planets (with the exception of Mercury), of the sun and of the satellites can be accounted for by Newton's laws as well as by general relativity. The order introduced into our experiences by Newton's theory is *retained and improved upon* by relativity. This means that the concepts of relativity are sufficiently rich to allow us to state all the facts which were stated before with the help of Newtonian physics. Yet these two sets of concepts are completely different and bear no logical relation to each other.

An even more striking example is provided by the phenomena known as the "appearances of the devil". These phenomena are accounted for both by the assumption that the devil exists, and by some more recent psychological (and sociological) theories. The concepts used by these two schemes of explanation are in no way related to each other. Nevertheless the abandonment of the idea that the devil exists does not lead into experiential chaos, as the psychological scheme is rich enough to account for the order already [84]introduced. – From these and many similar examples the following lesson can be derived: although in reporting our experiences we make use, *and must make use*, of certain theoretical concepts, it does not follow that different concepts will not do the job equally well, or even better, because more coherently. After all, this is only an instance of the old truth that under certain conditions different things may look exactly alike. It follows that the permanence of the classical "forms of perception" cannot be justified by an analysis of the present "use" of the classical terms.[14]

Nor can it be justified by an analysis of "the ... idea of observation" (A 67). According to such an analysis, parts of which may again be found in Bohr's writings, the success and usefulness of an experiment presupposes (*a*) that the parts of the apparatus used interact in a well-determined way so as to give a certain result; and (*b*) that this result is intersubjectively valid, which means that its description allows for objective representation in space and time (after all, the process of communication is one which happens in ordinary space and time). Both conditions are satisfied by classical events. Hence, the demand to use classical terms for the report of experimental results is nothing but a "simple logical demand" (A 313). – Two objections must be raised against this argument. First, we can only say that the classical conceptual scheme *satisfies* the conditions of

---

[14] For a more detailed analysis of this point cf. my paper referred to in fn. 12.

causality and intersubjectivity. We cannot say that it is the *only* scheme satisfying these conditions. Secondly, as they are stated, the conditions are much too strict. It is sufficient to demand that any dependence which might exist (or might be implied by a new theory) between the observer and the situation observed, and any deviation from the normal space-time frame *be negligible on the level of experiment*, i.e., under the special conditions valid for macroscopic instruments and macroscopic organisms (human observers). Hence, any new and universal theory which, on the level of experiment, leads to a suitable approximation of observer-independence [85]and the normal space-time frame will satisfy the "logical demands" mentioned. It appears therefore that the permanence of the classical forms of perception is not guaranteed by either of the two arguments we have investigated so far.

Now Bohr seems to doubt that it would ever be possible to invent a universal theory transcending the classical frame. He emphasizes that there exist "general limits of man's capacity to create concepts" (A 96) and, more especially, that "it would be a misconception to believe that the difficulties of the atomic theory may be evaded by eventually replacing the concepts of classical physics by new conceptual forms" (A 16). This shows that we may have still overlooked what must be a decisive argument in favor of the permanence of the classical concepts.

So far as I can see there are only two further arguments left. The first argument proceeds from the assumption (we shall call it assumption P) that *we invent only such ideas, concepts, theories as are suggested by observation*. "Only by experience itself", writes Bohr, "do we come to recognize those laws which grant us a comprehensive view of the diversity of phenomena" (A 1). And it makes use of the (psycho-sociological) law, emphasized by Bohr and Heisenberg, that any conceptual scheme employed for the explanation and prediction of facts imprints itself upon our language, our experimental procedures, our expectations and thereby also upon our perceptions. The argument itself may be outlined as follows: if assumption P is correct, then *there will exist an upper limit* of our ability to invent new conceptual schemes and hence, of the usefulness of the classical approach. This upper limit will be reached as soon as the theories have become so general that no conceivable fact is outside the domain of their application. In this case they will, given sufficient time, imprint themselves upon *all* our experiences which means that experience will not any longer be able to suggest a *new* theory. It seems to me that Bohr's conviction that we cannot go beyond the classical concepts is due (1) to [86]his admirable insight into the effect, upon our language, our perceptions, our thoughts and our methods, of the use of a fairly general theory; (2) to the assumption that classical physics is general enough to have influenced all our perceptions in the way described; and (3) to the belief that experience is the only source of our ideas.

Neither the conviction, nor the arguments leading up to it, can be accepted without criticism. First, because the domain of application of classical concepts is not as universal as one might think. There are many situations which fall outside this domain such as, e.g. personal consciousness, the behavior of living organisms, sociological laws, etc. Secondly, even if it were the case that all facts suggest an interpretation in terms of a certain conceptual scheme, the invention of a different scheme need not be impossible as long as there exist abstract pictures of the world (metaphysical or otherwise) which may be turned into new physical theories. After all, a universal mythology was superseded by Aristotelian physics; and Aristotelian physics in its turn was superseded by the physics of Galileo, Newton, and Einstein. This shows convincingly that the scientific concepts which have been used since the Renaissance (and earlier) are not the offspring of experience; they are "free mental creations".[15] To sum up: neither (2), nor (3) can be accepted. And assumption P does not give a correct account of the relation between our ideas and our experiences.

However this assumption has also a second aspect which we have not yet analyzed. A radical empiricist who believes that "... in our description of nature the purpose is not to disclose the real essence of the phenomena, but only to track down, as far as it is possible, relations between the manifold aspects of our experience" (A 18) will be reluctant to use theories whose concepts have no obvious foundation in the [87]facts, even if he should be able to invent them.[16] For him the assertion that every theory has its foundation in experience will not play the rôle of a descriptive statement but of the *demand* not to accept any conceptual scheme which does not "mirror" the facts in a more or less obvious way. But in this case the "general limits of man's capacity to create concepts" (A 96) are due, not to a lack of imagination, nor to some natural limitation of human abilities which prevent the physicists from ever going beyond the classical ideas, but to the (more or less conscious) *decision* not to go beyond what is given in experience, or, to express it in different terms,

---

[15] A. Einstein, *Ideas and Opinions*, London, 1954, p. 291; Cf. also H. Butterfield, *The Origins of Modern Science*, London, 1957.

[16] But it is of course to be expected that this inductivistic bias will severely curtail his inventive powers.

by the *adoption of an inductivistic methodology*. It seems to me that the idea of complementarity is ambiguous in the sense that it may be interpreted either as a description of actual fact, or as a description of what would be the case if a certain procedure were adopted and certain demands satisfied.

Now if an inductivistic methodology is adopted ("consider only such theories and use only such concepts as have a basis in experience!") then *there will exist an upper limit to the classical ideal of explanation* (cf., what has been said above): a purely inductivistic physics sooner or later must either drop the idea of explanation by universal theories (of which it makes illegitimate use anyway) and consider new forms of explanation; or it will come to a standstill. This is precisely what has happened in quantum mechanics. Classical physics is so general that no conceivable *physical* event falls outside the domain of its applicability. No theory satisfying the inductivistic demand can therefore contain any terms but the classical terms. Hence, if the quantum of action is to be accounted for at all, this will have to be done on the basis of a completely new type of procedure. This challenge was realized, and met, by Niels Bohr who developed a consistent point of view which *retains inductivism*, and is yet able to get on with physics. Anticipating later results we may say that according to this point of view explanation does not consist [88] in relating facts to a universal theory, but in their incorporation into a predictive scheme none of whose concepts is universally applicable. The question arises whether we are *forced* to adopt this point of view.

It would seem that we are not. A look at the success of theories which did not stop at the experiences but tried to explain them in a deeper way[17] shows that the inductivistic ideal behind Bohr's standpoint possesses an alternative which has been very successful and which is not confronted with any upper limit of applicability. Is there any reason why this alternative should be given up? This leads to the last argument (in favor of the idea of the permanence of the classical forms of perception) I want to consider in this section.

This last argument consists in pointing out that the only coherent quantum theory we possess at the present moment, viz., elementary quantum mechanics, can be interpreted as an instrument for the prediction of classically describable situations, but does not easily lend itself to a realistic interpretation in terms of concepts which are (*a*) nonclassical and

---

[17] For the concept of the "depth" of a theory cf. K. R. Popper's paper referred to in fn. 7.

(*b*) universally applicable. Now it is quite obvious that this fact cannot force us to abandon the classical ideal. For from the fact that Bohr's model of explanation (incorporation into a predictive scheme) can be realized at a certain level it does not follow that the realistic ideal (subsumption under a general theory whose primitive concepts are universally applicable) cannot be realized at that level; and even less does it follow that this realistic ideal will have to be abandoned once for ever. This statement remains valid even if there should exist a proof (sometimes the Neumann proof is used for that purpose) to the effect that no realistic theory of the micro-phenomena would ever be compatible with elementary quantum mechanics in its present form. For as was pointed out above the adequacy of a realistic account of the micro-phenomena is established already by showing that it [89]contains elementary quantum mechanics as an approximation. It may therefore contradict quantum mechanics without violating the realistic demand for rational explanation. This means that the possibility of a realistic theory of microscopic phenomena does not depend on whether or not a realistic interpretation can be given of elementary quantum mechanics. *For a realist the solution of the problem of duality is not to be found in alternative interpretations of a theory which in all probability is nothing but a predictive scheme anyway, but in the attempt to devise a completely new universal theory.* What we have shown in this section is that neither physics, nor philosophy can provide any argument to the effect that such a theory is not possible.

It seems fitting to conclude this section with a quotation and the remark that what has been said above completely refutes it. The quotation is from a paper by L. Rosenfeld about the point of view of complementarity:[18] "We are here not presented with a point of view which we may adopt, or reject, according to whether it agrees, or does not agree with some philosophical criterion. It is the *unique*" (my italics) "result of an adaptation of our ideas to a new experimental situation in the domain of atomic physics. It is therefore completely on the plane of experience ... that we have to judge whether the new conceptions work in a satisfactory way".

4. GENERALIZATIONS OF CLASSICAL PHYSICS COMPLEMENTARITY

In this section a more detailed analysis will be given of the predictive devices which Bohr calls "natural generalizations of classical physics".

---

[18] *Louis de Broglie, Physicien et Penseur*, Paris, 1953, p. 44.

Special stress will be laid upon the difference between such generalizations on the one side and physical theories in the usual sense on the other.

We proceed from the problem how the idea (or the demand – cf., the last section) of the permanence of the classical concepts and the classical forms of perception can be reconciled with the fact that classical physics is contradicted by the quantum of action.

[90]According to Bohr a reconciliation is possible if we adopt a more "liberal attitude toward these concepts" (A 3), realizing that they are "*idealizations*" (A 5; italics in the original), or "abstractions" (A 63) whose suitability for description or explanation depends upon the smallness of the quantum of action and which therefore must "be handled with caution" (A 66) if applied to a new experimental domain. "Analysis of the elementary concepts" (A 66) has to reveal their limitations in these new fields (A 4, 5, 8, 13, 16, 53, 108) and new rules for their use have to be devised "in order to evade the quantum of action" (A 18). These rules must satisfy the following demands: (*a*) they must allow for the description of any conceivable experiment in classical terms; (*b*) they must "provide room for new laws" (R 701; A 3, 8, 19, 53), and especially for the quantum of action (A 18); (*c*) they must always lead to correct predictions. (*a*) is needed if we want to retain the idea that experience is to be described in classical terms; (*b*) is needed if we want to avoid any clash with the quantum of action; (*c*) is needed if this set of rules is to be as powerful as a physical theory in the usual sense. Any set of rules satisfying (*a*), (*b*), and (*c*) is called by Bohr a "natural generalization of the classical mode of description" (A 4, 56, 70, 92, 110; D 316; E 210, 239), or a "reinterpretation ... of the classical electron theory" (A 14). "The aim of regarding the quantum theory as a rational generalization of the classical theories led to the formulation of the ... correspondence principle" (A 70, 37, 110). The correspondence principle is the tool by means of which the needed generalizations may be, and have been accomplished. Bohr seems to assume that *any future quantum theory* can only be a rational generalization of classical physics in the sense just explained. Most physicists have followed him in this respect. We now proceed to a closer examination of the properties of such generalizations.

The most fundamental property of all natural generalizations and hence, according to Bohr, of any future quantum theory, is that they contain duality as a basic feature which [91]is accepted without explanation. Indeed, the fact that all the experiences of class (*a*) and class (*b*) (cf., sec. 2) can be ordered and correctly predicted by the wave theory and the particle theory respectively, indicates that the concepts of these two theories need not be abandoned if they are used without the realistic pretensions expressed in assertions A and B. Used in this more "liberal" way (A 3) they will enable us fully to utilize the tremendous predictive power they possess *within their respective fields*, without leading any longer to the paradoxes connected with the assumption of their universal (and objective) validity. Hence, the first step toward a "natural generalization" of classical physics will consist in regarding the classical concepts as nothing but "expedients which enable us to express in a consistent manner essential aspects of the phenomena" (A 12) or, to use different words, it will consist in eliminating their realistic bias. The second step consists in devising a set of rules for the handling of the classical concepts thus purified, which satisfy (*a*) and (*c*) ((*b*) is satisfied once the first step has been taken). It is clear that neither of these two steps leads, or will ever lead, to the elimination of duality in favor of a coherent explanation.

Secondly, one should note the difference between natural generalizations and physical theories in the "classical" sense of Galileo, Newton, and Einstein. A physical theory is a *universal statement* which may be used for the explanation of the facts of a certain domain. Such an explanation (in the classical sense) consists in showing that these facts are instances of a general law which can be expressed in terms of the theory. On the other hand, the concepts of a natural generalization do not allow for universal application, as they have been "purified" of their realistic bias. Hence, the derivation of a classical situation from a natural generalization does not amount to its explanation in the sense stated above. It is not even a derivation in the proper sense. For what is derived is not the description of a classical situation, but a statement which has no objective reference and whose concepts are only "expedients which enable us to express in [92]a consistent manner essential aspects of the phenomena" (A 12 – see above). This applies especially to classical mechanics itself. The terms of classical mechanics possess a surplus meaning not contained in the purified concepts of a natural generalization and never obtainable from them. It follows (*α*) that the classical theories cannot be "regarded simply as the limiting case of a vanishing quantum of action" (A 87); (*β*) that even after a certain natural generalization has been established, classical physics may still provide a guide for further development (A 88); and (*γ*) that quantum mechanics, interpreted as a natural generalization of classical physics, is in a sense poorer than classical physics. Therefore any future theory which attempts to give a universal account of the facts of the atomic level will have to utilize both the results of quantum mechanics, and the theories of classical physics. This novel situation (novel, because hitherto every theory contained as special cases *all* the general facts of its domain of applicability) has been duly emphasized by D. Bohm and plays an essential rôle in his philosophy. It also indicates that Bohr's procedure (transition from the use of universal theories to the use of predictive devices) only delays the breakdown of inductivism. For once the correspondence principle has exhausted the content of the classical concepts (and this is bound to happen sooner or later) no further progress will be possible for an inductivist; neither on the basis of universal theories (this has been shown by Bohr himself – cf., the last section), nor on the basis of predictive devices will he be able to promote physics. Only the elimination of the inductivistic doctrine itself and the return to the classical mode of explanation will then lead to further progress.[19]

[93]Thirdly, any set of rules satisfying (*a*) and (*c*) (cf., the beginning of the present section) will be incomplete without reference to the conditions under which the purified classical concepts may be applied. In classical physics such a reference does not occur as it is assumed that the classical concepts are applicable wherever classical physics is applicable. However, in order to circumvent duality and the quantum of action, these concepts have been "purified" and turned into predictive devices, and these predictive devices are successful only in certain parts of the classical level. Hence, any theory which wants to utilize their predictive power to the utmost without running into danger of being refuted will have to indicate their limits of applicability and this again in classical terms. A description, or prediction, produced by a natural generalization will therefore consist of two parts, viz., (1) of the application, for the purpose of prediction (or description), of a set of classical terms (e.g., of the terms of the wave picture); within classical physics this part is already logically complete. And (2) of a statement of the conditions under which these terms are being applied. The conditions may have been deliberately prepared in the course of a measurement, or they may be the result of independent development in some part of nature (e.g., inside the sun). It is impossible to separate (2) from (1) without making the description incomplete, or, if such a separation is intended to lead to an objective (and universal)

---

[19] Some physicists seem to assume that this point has already been reached. "... to-day", writes L. de Broglie (*La Physique quantique restera-t-elle indeterministe?* Paris, 1953, p. 22), "the explicative power of wave mechanics ... seems to be to a large extent exhausted ... and the champions of the probabilistic interpretations themselves try, without much success, so it seems, to introduce new, and even more abstract conceptions ...". The return to spatio-temporal pictures, advocated by de Broglie, does not seem to me to be an ideal solution, however.

interpretation of (1), without running into paradoxes. As within classical physics (1) was used to represent the behavior of objects (particles, waves) independently of their surroundings; and as (2) usually contains the description of an experimental arrangement, this situation may be expressed (in the "material mode of speech") by saying that "no sharp separation can be made between an independent behavior of the objects and their interaction with the measuring instruments which define the reference frame" (D 313; E 224).

The sum total of the "observations obtained in specific circumstances, including an account of the whole [94]experiment" has been called by Bohr a *phenomenon* (E 237 f; D 317). A phenomenon is what is described by a logically complete sentence. It is for this reason (and not because of the existence of an "indivisible quantum", linking the object to the measuring apparatus) that it must be regarded as an *indivisible block* in the sense that *even conceptually* it does not allow for an analysis into observed object and observing mechanism. Indeed, any such subdivision would be possible – *within the framework of a natural generalization* – only by again endowing (some of the) classical concepts used in parts (1) of the complete description with a realistic bias, a procedure which must at once lead back to the paradoxes outlined in Section 2.

In spite of these changes it is still possible to retain the useful notion of a physical object. But such an object is now characterized as a set of (classical) *appearances* only, without any indication being given as to its nature. The *principle of complementarity* consists in the assertion that this is the only possible way in which the concept of an object (if it is used at all) can be employed upon the microscopic level; as well as in a description of the way in which the various sides (or appearances) of the microscopic objects are related to each other. It rests, as will be clear from what has been said so far, upon the *fact*, explained in Section 2, of duality (and the quantum of action); and upon the *decision* (see Section 3) to overcome the difficulties connected with this fact by a "natural generalization" of classical physics rather than by a completely new theory. One may therefore say that the principle of complementarity is a natural consequence of the tendencies implicit in the principle of correspondence. To sum up: complementarity is a description, based upon duality, of the way in which the classical concepts appear within the predictive schemes which for Bohr replace classical physics on the atomic level (cf., also D 314).

It remains to develop the obvious consequences of complementarity. These consequences are (1) *indeterminism.* This indeterminism is not due to the absence of *causes* [95]which might be made responsible for changes in the behavior of microscopic systems; nor is it due to the alleged disturbance of the microscopic objects by the act of observation. Indeed, the impossibility, stated above, of dividing a phenomenon into an object and a measuring device disturbing the object would show that any such explanation is based upon an inadequate account of the situation we face in quantum mechanics (interpreted as a natural generalization of classical physics). This indeterminism is due to the fact that the different pictures utilized in the complementary mode of description do not allow for a unified account of motion and change *in the purely kinematic sense* (cf., A 57 ff.) which implies that *probabilities*, if they enter the scheme, must be regarded as irreducible. – (2) *Indivisibility*, as a consequence of the block character of the phenomena and the dependence, upon certain conditions, of any sentence aiming at a description of atomic objects. – (3) If these conditions are changed, a sudden *rupture* occurs in the mode of our description (e.g., from the wave picture to the particle picture, when an atom hits a scintillation screen) not because of a *physical* interaction between the screen and the matter wave of the atom (see the first point), but because of the *logical* dependence (within a natural generalization) of any description upon the experimental situation in which it has been obtained. – (4) Using the equations

$$E = h\gamma \qquad P = h/\lambda \tag{I}$$

together with the idea of complementarity we arrive at the result that a "separation" is necessary of "features which are united in the classical mode of description" (A 19) viz., of all the features connected with the configuration, in ordinary space, of a given system on the one hand; and of the features connected with the *motion* of this system on the other. By a careful analysis of the way in which the wave phenomena and the particle phenomena limit each other, Heisenberg[20] [96]succeeded in deriving, and interpreting his *uncertainty relation*. This analysis amounts to a proof of consistency of the point of view of complementarity.

5. THE COPENHAGEN INTERPRETATION

We obtain the Copenhagen interpretation of quantum mechanics when regarding this theory as a "natural generalization of the classical mode of

---

[20] Cf. his *Physical Principles of Quantum Mechanics*, Chicago, 1930.

description" (A 58) and applying to it all the results which have been shown to hold for such generalizations.

The most important consequence of this interpretation is that the characteristic symbols of the theory, such as the wave function (which must be clearly distinguished from the matter waves to which we refer when applying complementarity) or the Hermitian operators, are not interpreted as new concepts which, on the quantum level, replace the concepts of classical physics; they are interpreted as tools "for deriving predictions ... as regards information obtainable under experimental conditions described in classical terms" (D 314). It also follows that a complete account of the classical level in terms of the symbols of the quantum theory is impossible.[21] Hence, the dual aspect, characteristic for this level, of light and matter is still a basic feature of the theory and indeed of any future quantum theory. Any such theory amounts to a characterization, with the help of classical concepts, of the way in which atomic objects appear in different circumstances. These appearances are divided into classes which are complementary to each other. Indeterminism, indivisibility, the existence of sudden ruptures in the development of the atomic objects ("reduction of the wave packet") are further features characteristic of elementary quantum mechanics and, according to the opinion of many physicists, of any future theory of the microscopic level.

[97]At this point it seems to be advisable to discuss a certain unclarity in the Copenhagen point of view which is due to the persistence of some earlier ideas, especially of the idea that there exist "quantum jumps". The attempt has been made (by Bohr in his earlier writings, by Heisenberg and by some other followers of the Copenhagen point of view) to justify the sudden changes occurring during a measurement by reference to the fact that on the quantum level the disturbance of the object by the measuring device cannot be neglected (A5, 11, 15, 54, 68, 93; D 315). "Any observation", writes Niels Bohr (A 115), "necessitates an interference with the course of the phenomena which is of such a nature that it deprives us" of the possibility of going on utilizing the

---

[21] "In the first instance", writes Heisenberg (*Syllabus of the Gifford Lectures*, 1956, p. 8), "it would seem plausible that quantum theory has disproved and replaced classical mechanics. This view, however, cannot be maintained if one remembers that the concepts of classical physics are necessary for the definition of any experiment relevant for quantum theory".

concepts which were adequate before the measurement started. And Heisenberg points out[22] that "the most important difference between quantum theory and the classical theories consists in the fact that in the case of an observation we must carefully consider the disturbance, due to the experiment, of the system investigated". The inadequacy of this part of the Copenhagen interpretation was exposed by Einstein, Podolsky, and Rosen in a paper[23] which played a decisive rôle in the development of the interpretation of quantum mechanics. In that paper it is shown that (according to the formalism of elementary quantum mechanics) physical operations, such as measurements, may lead to sudden changes in the state of a system *which is in no physical interaction whatsoever* with the domain in which the measurement is being performed. These changes obviously cannot be the consequence of a "disturbance, due to the experiment, of the system investigated". In his answer (R 695) Bohr draws the distinction between a "mechanical disturbance", and "an influence upon the very conditions which define the possible types of prediction regarding the future behavior of the system". And in his later writings he "warned especially against phrases, often [98]found in the physical literature" (including his own earlier writings – cf. A 53 ff. and the further references quoted above) "such as 'disturbing the phenomenon by observation', or 'creating physical attributes to atomic objects by measurement'" and drew attention to the block character of physical phenomena which was discussed in the last section (E 237). It seems to me that this is the only reasonable way in which a "natural generalization" can account for the processes which happen during measurement; and that it is sufficient to introduce the quantum of action in a purely formal way, namely by using the equations (I).

These are some of the elements of the Copenhagen interpretation. It remains for us to enquire whether this interpretation provides an adequate picture of elementary quantum mechanics. Such an enquiry would seem to be desirable for two reasons. First, because, historically speaking, the Copenhagen interpretation was completed only *after* the invention of the formalism of elementary quantum mechanics. And secondly, because the invention of a certain mathematical representation

---

[22] *Naturwissenschaften* (1929), p. 495.
[23] *Phys. Rev.*, Vol. 47 (1935), 777. – For a different criticism cf. E. Schrödinger, *Naturwissenschaften*, Vol. 22, p. 341.

of this formalism, viz., wave mechanics, was stimulated by ideas completely different from those of the Copenhagen school.[24]

6. ELEMENTARY QUANTUM MECHANICS

In order to remain within the limits of a short paper I shall investigate only two results of the Copenhagen point of view, viz., (1) the assertion that elementary quantum mechanics does not admit of a realistic interpretation; and (2) the assertion that the classical concepts in the proper (i.e., nonpurified) sense are presupposed by quantum mechanics and cannot be derived from it. It will turn out that these two results are not entirely correct which seems to open the possibility of a realistic interpretation of quantum mechanics. Yet it should be repeated (cf., the end of Section 3) that the case of realism does by no means depend upon this possibility.

[99]Assertion (1) is usually defended by showing (a) that the classical concepts employed by quantum mechanics do not admit of universal application; and (b) that the ψ-function, one of the characteristic nonclassical symbols of the theory, cannot be interpreted realistically either. In the proof of (a) it is pointed out that atomic objects such as electrons, protons, etc., can be neither particles nor waves. The difficulties of both the particle- and the wave-interpretation have been discussed in Section 2. But as relying upon duality would already mean accepting the Copenhagen point of view, we have to consider objections which are based upon quantum mechanics (in the Born interpretation) and general physics only. Some of the objections in the case of the particle interpretation are (α) the difficulty of devising a particle model of interference. In order to appreciate this difficulty in more concrete terms we consider (β) the phenomenon known as the penetration of a potential barrier. Any particle model of this phenomenon would have to assume that at some time of the process the particle is dwelling *within* the barrier, which contradicts the principle of conservation of energy.

Another difficulty closely connected with the one just mentioned is (γ) the fact that the statistical ensembles corresponding to the pure states are irreducible in the sense that they do not contain any subensembles with statistical properties different from their own (this is one of the main

---

[24] It would therefore seem to be a little pretentious to assert, as Heisenberg has done (cf. his article in *Niels Bohr and the Development of Physics*, London, 1955), that this formalism is itself part of the Copenhagen interpretation.

results of von Neumann's proof). In order to appreciate this difficulty in more concrete terms we consider (δ) the following example, due to Weyl:[25] compare a measurement to a selection, by a filter, of a particle in a certain state. Call a filter which selects particles of the state C a C-filter, a filter which selects particles of state non-C and which rejects particles of state C, a C̄-filter, and so on. Then the particle interpretation implies that a C-A-C̄-filter must be impenetrable – which contradicts some very elementary quantum phenomena such as the polarization of light quanta. It is of [100]course true that each one of the objections stated so far may be overcome by some adjustment or another. But any such adjustment will be ad hoc and hence, unsatisfactory, unless we change the theory itself. The situation is similar with respect to the wave interpretation. We may therefore conclude that (1*a*) seems to be fairly well established. – The same applies to (1*b*): assume that the ψ-function of a system designates an objective property of this system. The difficulty of this suggestion lies (α) in the fact that the ψ-wave of a system consisting of two parts, S and S' (which may both be macroscopic and separated by a large distance) can be broken into two ψ-waves for S and S' respectively only under very special conditions. Furthermore any such account is faced (β) by the Schrödinger paradox as well as (γ) by the paradox of Einstein, Podolsky, and Rosen (cf., Section 5). And the interpretation of the ψ-field as a probability field leads back to the particle interpretation which has already been criticized.

Surveying these arguments we see that they are all right as far as they go: they prove that *the predictive apparatus of quantum mechanics* (which employs ψ-functions and classical concepts) *does not admit of a realistic interpretation.* However, assertion (1) depends upon the further assumption that quantum mechanics consists of this predictive apparatus only. That this further assumption cannot be correct has been shown by Schrödinger.[26] Schrödinger has pointed out that in quantum mechanics we are hardly ever concerned with the prediction of states, but rather with the determination of the nature of physical systems, such as the "eigen-values" of the Hamiltonian or of other important observables. These "eigen-values" obtain under all possible experimental conditions and can therefore be objectified (though not in ordinary space and time). This indicates that elementary quantum mechanics can be

[25] *Philosophy of Mathematics and the Natural Sciences*, Appendix.
[26] Schrödinger, *Naturwissenschaften*, Vol. 23 (1935), p. 810, as well as *Nuovo Cimento*, 1955, pl.

approached in two ways. It may be approached as a theory, similar to [101]celestial mechanics, which predicts *states* of systems with the help of dynamical equations. If we use this approach, then a realistic interpretation does not seem to be possible and the Copenhagen point of view will seem to be correct. But we may also approach it as a theory containing new knowledge about, and new concepts for the description of, the *nature* of physical systems. In this case a realistic interpretation of a rather unusual kind is definitely possible. In brief: assumption (1) is only partially correct.

So far as textbooks go, the relation between classical physics and quantum mechanics has been clarified long since either by Ehrenfest's theorem, or by certain approximation methods which only involve parameters of the physical systems concerned. That such a statement cannot be correct becomes evident from examples, like Einstein's,[27] which satisfy all the conditions of these approximations and which nevertheless exhibit some nonclassical features. Such cases apparently confirmed Bohr's contention that the classical theories cannot be "regarded simply as the limiting case of a vanishing quantum of action" (A 87), as well as the more general assertion (2) (see above). Yet some more recent results by Ludwig[28] seem to imply that this account can no longer be taken to be correct and that some decisive features of the classical level can be derived (under suitable conditions) from quantum mechanics. This again shows that a realistic interpretation of quantum mechanics, though excluded by the Copenhagen point of view, need not be impossible.

7. WIDER APPLICATION OF COMPLEMENTARITY SUMMARY

Bohr's point of view can be applied to domains different from physics. This application consists in the suggestion that progress might be achieved with the help of rational generalizations which utilize all the primitive concepts of the domain in question. In the case of psychology, e.g., such a [102]natural generalization would employ both the concepts of the physiological approach and the more vague (and familiar) concepts of the introspectionists. It would try to utilize their predictive power to the utmost without running into paradoxes (such as, e.g., the apparent paradox that thought is localized in the brain). In order to achieve this it

---

[27] Cf. Einstein's paper in the Volume of *Papers Presented to Max Born*, Edinburgh, 1953.
[28] *Zs. Physik*, Vol. 135 (1953), p. 483.

would purify them of their universalistic (and realistic) bias and use them as instruments for prediction only. The application of these instruments would be made to depend upon appropriate conditions, which means that an object, such as, e.g., a living being, could again be only characterized by various sets of appearances. It is obvious that the number of such sets need not be restricted to two (as Bohr seems to assume). In the case of psychology we are already confronted with at least three different possibilities: physiology, introspection, (molar) behaviorism. It must also be pointed out that Bohr's more particular arguments in favor of the application of his point of view to the nonphysical sciences are very weak and mostly invalid.[29] Yet the suggestion to employ natural generalizations instead of universal theories is not unreasonable in itself, *if* one is prepared to accept its consequences. In the present section I shall discuss only two such consequences.

First, the success of the correspondence principle was due to the fact that an adequate mathematical tool was found for a "general consistent utilization of the classical concepts ... through the transformation theory of Dirac and Jordan" (A 79). It is very unlikely that biology, psychology, sociology are advanced enough to lend themselves to a similar treatment along the lines of some generalized correspondence principle. Hence, the attempt to utilize all the ideas pertinent to such a field can only lead to a shallow [103]syncretism, or to a secularized revival of the old doctrine of the two kinds of truth. This movement toward a point of view which says that there is some truth in every idea and which leaves it at that, is well exemplified by the following quotation from Heisenberg:[30] "It has certainly been the pride of natural science since the beginning of rationalism to describe and to understand nature without using the concepts of god, and we do not want to give up any of the achievements of this period. But in modern atomic physics we have learned how cautious we should be in omitting essential concepts just because they lead to inconsistencies".

---

[29] In the case of biology he points out that "there is set a fundamental limit to the analysis of the phenomena of life in terms of physical concepts, since the interference necessitated by an observation which would be as complete as possible from the point of view of atomic theory, would cause the death of the organism" (A 22). An argument of this form could equally well be used to show that we cannot say that a bomb contains explosive powder as the test of the explosiveness of its content would destroy the whole bomb. – For a different criticism cf., E. Zilsel, *Erkenntnis*, Vol. 5 (1935), p. 56.

[30] *Syllabus*, p. 16.

Now we have tried to show in this paper (*esp.*, in Section 3) that the transition to natural generalizations and to the (restricted) use of all the concepts of a given level is a lesson which physics has in store *for the inductivist only*. Only if it is assumed that every concept must have a foundation in experience, only then the breakdown of classical physics forces us to give up the classical ideal of explanation with the help of universal theories. Realism, which is not based upon the inductivistic doctrine, need not give up this ideal. On the other hand, no theologian can possibly be delighted by the way in which the concept of god is here being reintroduced. For this concept, and, indeed, any concept used within a natural generalization of the ideas of some field, is introduced *without existential commitment*, as a purely predictive device for the description of some features of the universe. Or, to express it in different terms, it is introduced for the description of appearances (which may well be the appearances of some evil spirit).

Secondly, the adoption of the method of natural generalizations must sooner or later lead to stagnation. This consequence has already been discussed in Section 4 and it may be repeated here: every concept has only a finite content. This content is determined by the views which were first expressed with its help. The utilization of the various ideas [104]used within a special field will sooner or later exhaust this content – and then further progress is impossible. The stable picture of the world toward which this development must finally converge, and which, as Bohr himself has pointed out, will pervade all our beliefs, thoughts, perceptions, will be a kind of new myth, but without the coherence and inner unity of the myths known to us at the present moment. This will mean the end of science as a rational undertaking. Only the invention of a new set of ideas which boldly oppose appearances and common belief, and which attempt to explain both in a deeper way, can then lead to further progress and to the continuation of rational argument. This shows the close connection between what has been called the classical ideal, or realism on the one side, and scientific progress on the other.

To sum up: complementarity is based upon a new ideal of explanation. This ideal (which is similar to the positivistic ideal of explanation by incorporation into a predictive device) was introduced and elaborated by Bohr who realized that an inductivistic account of the quantum of action is incompatible with realism. It was also applied to domains outside physics. Whereas within physics this idea has led to some fruitful developments along the lines of the correspondence principle, such results are not available in other fields. But in all domains the application of the positivistic ideal must finally lead to stagnation. This can be avoided only by again adopting the realistic procedure. Neither physics nor philosophy possesses any argument to the effect that this procedure cannot succeed. But both physics and philosophy show (contrary to what is commonly believed) that it is the only procedure which makes scientific progress along rational lines possible.

6

Niels Bohr's Interpretation of the Quantum Theory
(1961)

I. INTRODUCTION

[371]Among the numerous interpretations of the quantum theory, the so-called Copenhagen interpretation has a fairly unique position.[1] It is closely connected with physical practice; for it may be regarded as a consequence of the "*Korrespondenzdenken*" whose fruitfulness within the older quantum theory is well known. It also seems to be the only interpretation that gives a more or less satisfactory account of the theory as it is formulated at the present moment (I am here referring to the elementary theory of Schrödinger and Heisenberg and not to the field theories). And it is rich enough to imply a point of view in epistemology, and in philosophy in general, that admits of application to domains outside physics. Yet in spite of all these apparent advantages it has been denounced as dogmatic and positivistic; it has been described as an unwanted and unwarranted intrusion of Berkeley's *esse est percipi* into the domain of physics; and it has even been attacked as a principle which, if consistently applied, would sooner or later constitute a grave danger for the further development of physics.

Now in order to become clear about the strong and weak points in these claims and counterclaims, it would seem to be appropriate to state as clearly as possible the basic assumptions of the Copenhagen interpretation. This attempt at once meets with the following difficulty: there is no such thing as the "Copenhagen interpretation". True, there are numerous

[1] The author is indebted to the National Science Foundation and the Minnesota Center for Philosophy of Science for support of research.

physicists who profess to follow either Heisenberg or Niels Bohr in their more general ideas, and who also call themselves adherents of the Copenhagen point of view. If one tries to discover [372]some common element in their beliefs, however, one is bound sooner or later to be disappointed. The uncertainty relations always play an important role, and so do Heisenberg's thought experiments (the γ-ray microscope, and the like). Yet the exact interpretation of these experiments as well as of Heisenberg's formulas is neither *clear*, nor is there a *single* such interpretation. On the contrary, we find all philosophical creeds, from extreme idealism (subjectivism) to dialectical materialism, superimposed upon these formulas. In this situation it seems advisable to repeat the simple qualitative arguments which have been developed by Bohr and which at a later time were incorporated into his idea of complementarity. It will then turn out that these arguments are much more plausible than the vague speculations which were later used in order to make them more acceptable. But it will also turn out that in spite of this plausibility none of the arguments is powerful enough to guarantee that "the new conceptions which we need will be obtained ... by a rational extension of the quantum theory";[2] and that new theories of the microscopic domain will, therefore, have to be increasingly indeterministic. Any such prediction is possible only if certain *philosophical* ideas are used as well. And while the physical arguments themselves are sound and ingenious, the philosophical ideas needed to confer absolute validity upon their conclusions are neither correct nor reasonable.

I shall now start with an outline of these physical elements in the idea of complementarity.

2. INDEFINITENESS OF STATE DESCRIPTIONS

As I see it the most important element of Bohr's ideas consists in the twin assertion that (a) the quantum theory will have to work with dynamical states which are only partly well defined and that (b) these dynamical states must be regarded as relations between the system and some appropriate measuring device. The present section will be devoted to the statement and the defence of the first part of this assertion. The second part will be discussed in the next section.

---

[2] L. Rosenfeld in *Observation and Interpretation*, ed. by S. Körner, New York: Academic Press, 1957, p. 45.

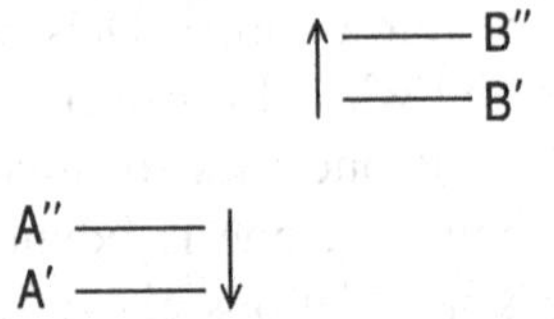

FIG. 6.1

In order to make plausible the need for indefinite state descriptions one should take into account the following two experimental facts: (1) a physical system is capable of discrete states of energy only, and it cannot dwell in intermediate states – this is also called the *quantum postulate*; (2) processes of interaction, emission, absorption do not take place instantaneously, but it takes some time before they are completed.

Now consider (Fig. 6.1) two systems, A and B, which interact in such a way that a certain amount of energy, ε, is transferred from A to B. [373] As long as the systems interact (see (2) above) A will not be in its initial state A'', nor will B be in its final state B''. However, according to (1) A and B cannot dwell in intermediate states either. This difficulty was resolved by Bohr on the basis of the assumption that during the interaction of A and B the dynamical states of these systems, and therefore also their energies, cease to be well defined so that it becomes *meaningless* (and not only *false*) to ascribe a definite energy to either of them.

This simple solution has been so often misrepresented that a few words of explanation seem to be needed. First of all it must be pointed out that in the above formulation the term "meaning" has not entered, as has been asserted by various critics, because of some connection with the now customary attitude of preferring semantical dissection to analysis of physical conditions. After all, there are well-known classical examples of terms which are meaningfully applicable only if certain physical conditions are first satisfied, and which become meaningless as soon as these conditions cease to hold. A good example is the term "scratchibility" (Mohs scale) which is applicable to rigid bodies only and which loses its significance as soon as these bodies start melting. Second, it should be noted that the proposed solution does not contain any reference to *knowledge*, or to the absence of knowledge. It is not asserted that during the time of transfer A and B *are* in some state, *unknown* to us. For the quantum postulate does not merely exclude the knowledge of intermediate states, it excludes these intermediate states themselves. Nor is it sufficient to assert that what has been shown is a lack of *predictability*. For also in this case it would be possible to assume that we could predict better if we only knew more, whereas Bohr's suggestion excludes the possibility that there is anything more to be known. The third remark concerns a suggestion to get around a kinematics of ill-defined states which has often been made in connection with wave mechanics. According to this suggestion, the difficulties which arise when we try to give a rational account of processes of interaction are due to the fact that the classical point [374]mechanics is not the correct theory for dealing with atomic systems and that the state descriptions of classical point mechanics are not adequate means for describing the states of systems upon the atomic level. According to this suggestion, we ought not to retain the classical notions, such as position and momentum, and make them less well defined. What we ought to do rather, is, to introduce completely new notions which are such that when *they* are used states and motions will again be well defined. Now if any such new system is to be adequate for the description of quantum phenomena, then it must contain means for expressing the quantum postulate, and it must, therefore, also contain adequate means for expressing the concept of energy. However, once this concept has been introduced, all our above considerations again apply with full force: when being part of $A + B$, neither A nor B can be said to possess a well-defined energy, from whence it follows that the new, and ingenious set of concepts will not lead to a well-defined and unambiguous kinematics. "It would", therefore, "be a misconception to believe", writes Niels Bohr,[3] "that the difficulties of the atomic theory may be evaded by eventually replacing the concepts of classical physics by new conceptual forms". I repeat that this last remark will be of great importance in connection with the interpretation of Schrödinger's wave mechanics.

The duality of light and matter leads to another argument for the need to replace the classical kinematics by a new set of assumptions. I shall try to explain this argument as carefully as possible in order to refute the idea, which is held by some writers,[4] that the interference properties of light and matter are an instance of a general type of statistical behavior which is best exemplified by such classical devices as pinboards and roulette games. Instead of a pinboard I shall use an arrangement containing two slits, and I shall compare the behavior of classical particles (dust particles

---

[3] Niels Bohr, *Atomic Theory and the Description of Nature*, Cambridge, Cambridge Univ. Press, 1934, p. 16.

[4] Cf. the paper by A. Landé in the present volume {"From Duality to Unity in Quantum Mechanics", in Herbert Feigl and Grover Maxwell (eds.), *Current Issues in the Philosophy of Science*, New York: Holt, Rinehart & Winston, 1961, pp. 350–370}.

falling through the slits – we shall neglect Brownian motion) and of nonclassical particles (electrons). There are certain *similarities* in both cases. For example, we may assume in both cases that the pattern on the receiving screen is independent of the density of the stream of particles (for a wide range of densities >0).[5] What is important for our discussion, however, are the *differences* between the two cases. The differences lie in the relation [375]between the two-slit pattern and the pattern that is obtained when only one slit is open. In the classical case the one-slit pattern is modified by the opening of the second slit only in the domain where both patterns overlap, and what we obtain here is a simple addition of the number of incoming particles. This shows that the behavior of each single particle depends only on the situation around the slit where it enters, and is not influenced by the situation in the neighborhood of the other slit. In the quantum-mechanical case, however, the opening of the second slit may lead to a minimum in a place (say *P*) in which the one-slit pattern shows a finite intensity. The last fact implies that when one slit is open, some electrons end up in *P* (we assume all the time, as is done by those who want to understand interference patterns along classical lines, that the electrons are particles which at any given time possess a well-defined position, a well-defined momentum, and a definite trajectory). As long as slit 2 is closed, some electrons will always arrive at slit 1 which, if 2 were to remain closed, would wander along towards *P*. Assume that *E* is such an electron and that it is about to enter 1. Now if we open 2 at this very moment, we have created a condition which is such that *E* must not arrive in *P*. Hence, the process of opening slit 2 must change the path of *E*. It is well known that the usual way of accounting for this change by referring to well-defined energies of interaction is not applicable here, since such energies cannot be introduced in a simple and satisfactory way. Let me point out, in this connection, that the more recent attempt to explain the change of the pattern as the individually indeterministic and yet statistically well-defined result of a change in the experimental conditions cannot be successful either – if it is also maintained that the electron always possesses a well-defined position and a well-defined momentum.[6]

[5] With respect to light this was shown, a. o. by Jánossy. Cf. the booklet edited by the *Hungarian Academy of Sciences*, 1957, where also previous experiments are reported, as well as Jánossy, *Acta Physica Hungarica*, Vol. IV, 1955, and *Nuovo Cimento*, Vol. VI, 1957.

[6] This suggestion is due to K. R. Popper. Cf. his "The Propensity Interpretation of Probability and the Quantum Theory", in *Observation and interpretation*, ed. by S. Körner, New York: Academic Press, 1957. I should like to point out that the argument

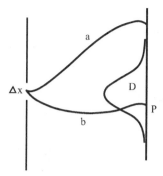

FIG. 6.2

For it is the peculiarity of atomic processes that they obey the conservation laws *individually* whereas the proposed solution could at most guarantee the *statistical* validity of these laws. The only reasonable assumption again seems to be the assumption that the behavior of the particles between the slits and the screen is no longer well defined.

In the present case this lack of definition can even be roughly calculated. Consider for that purpose (Fig. 6.2) a single slit of width Δx, and assume that, to start with, only an infinitely narrow part of it is being kept open. In this case the whole screen will be irradiated, and a [376]will be a possible path. Adding the remainder of the slit we obtain irradiation only within D, so that a will have to change into b, inside D. Now what *can* be explained, on the basis of the wave theory, is the change of the intensity distribution from the original irradiation of the whole screen to the irradiation of D only. Hence, if instead of referring to a definite trajectory b we let the state of the electron be undefined within D, then we shall be able to give an account of the changes that occurred. But this means that because of $\Delta x \sin \alpha \geq \lambda$ (α the direction of the first minimum; $\sin \alpha = \frac{\Delta p}{p}; x = \frac{h}{p}$)

$$\Delta x \Delta p \geq h \qquad (1)$$

What we have derived in this rather simple way can be shown to possess universal validity. Indeed, whenever a measurement is performed of either momentum or of position, the difficulties which we discussed in the last paragraphs can be avoided only if we relax in the definiteness of the dynamical state to the extent that is indicated by the wavelike character

in the text above can be repeated in a slightly different form even if the minimum should not be an absolute minimum.

of the elementary particles. It ought to be repeated that this relaxation corresponds to a real indefiniteness in the processes described. Only *after* this basic assumption has been adopted can Heisenberg's discussions and his attempt to derive various forms of (1) from an analysis of methods of measurement[7] lead to the proper interpretation of the relations thus obtained. Otherwise the objection could always be raised that what has been derived is a restriction of our ability to *know* the exact state of the dynamical system, or of our ability to *predict*, but not a restriction in the dynamical states themselves. Such a conclusion is the more likely to be drawn, since many dis[377]cussions of the matter are subjectivistic in the sense that they first interpret (1) as restrictive of knowledge,[8] which interpretation is then reconciled with the objective validity of the quantum postulate by identifying knowledge and physical reality, or by declaring that the idea of a real external world is "metaphysical".[9] It must also be pointed out, in accordance with what has been stressed above, that the objective interpretation of the uncertainties as restrictive of the meaningful applicability of such functors as "position" and "momentum" will have to be adopted by any physical theory that incorporates the quantum postulate and the dual nature of (light and) matter; more especially, it will have to be adopted by wave mechanics.

3. WAVE MECHANICS

The older quantum theory, although experimentally very successful, was yet regarded as unsatisfactory by many physicists. Its main fault was seen to lie in the fact that it combined classical and nonclassical assumptions in a way that made a coherent interpretation impossible. For many physicists it was, therefore, nothing more than a stepping stone on the way to a really satisfactory theory – that is, to a theory which could give us not only correct predictions but also some insight into the nature and the dynamics of microscopic entities. It is quite true that Bohr, Kramers, Heisenberg, and others worked along very different lines. Their main objective was not the construction of a new physical theory about a world that existed independently of measurement and observation; it was rather

---

[7] *Zs. Physik*, 43, 1927.

[8] Cf., for example, W. Heitler, "The Departure from Classical Thought in Modern Physics", in *Albert Einstein, Philosopher-Scientist*, edited by P. A. Schilpp, Evanston: Northwestern Univ. Press, 1949.

[9] For an analysis of these more general confusions cf. B. Fogarasi, *Kritik des Physikalischen Idealismus*, Budapest: 1953.

the construction of a logical machinery for the utilization of those parts of classical physics which could still be said to lead to correct predictions. However that may be – the philosophical spirit behind the *"Korrespondenzdenken"* was by no means shared by everybody. Thus de Broglie and Schrödinger tried to develop an entirely new theory for the description of the nature and the behavior of atoms, molecules, and their constituents. When this theory was finished it was hailed by many as the long-expected coherent account of the microscopic level. The hypothesis of the indefiniteness of state description, so it was thought, had only reflected the indefiniteness and incompleteness of the early theory, and it was, therefore, no longer necessary. More especially it was assumed either that states were now new, but well-defined entities (the ψ-waves); or it was assumed that whatever incompleteness occurred was due to the statistical character of the theory; that is, it was due to the fact that wave mechanics was "primarily a variety of statistical mechanics, similar to the classical statistical mechanics of [378]Gibbs.[10] I hope that our repeated arguments have made it clear that any such interpretation is bound to lead into inconsistencies. The only presupposition of the hypothesis of indefinite state-descriptions is the quantum postulate and the dual nature of light and matter. Both these facts are contained in the wave mechanics which will therefore be equally in need of the said hypothesis. A closer analysis of the two main alternatives to Bohr's hypothesis shows that this is indeed correct.

Let us first consider the suggestion that the quantum theory provides us with new types of complete and well-defined states – namely, the ψ-waves which are to replace the descriptions, in terms of trajectories, of the classical point mechanics. This suggestion has been made, and elaborated in some detail, by M. Planck. According to Planck[11] "the material waves constitute the primary elements of the new world picture.... In general the laws of the matter waves are basically different from those of the classical mechanics of the material points. However, the point of central importance is that the function characterizing matter waves, that is, the wave function ... is fully determined for all places and times by initial and boundary conditions".

---

[10] E. C. Kemble, *The Fundamental Principles of Quantum Mechanics*, New York: McGraw-Hill, 1937, p. 55.

[11] "The Concept of Causality in Physics", in *Scientific Autobiography and Other Papers*, New York: 1949, p. 135 ff. Similar sentiments have been expressed, much later, by E. C. S. Northrop (*The Logic of the Sciences and the Humanities*, New Haven: Yale Univ. Press, 1948. p. 27) and by E. Nagel (cf. his article in *Readings in the Philosophy of Science*, ed. Feigl and Brodbeck, New York: Appleton-Century, 1953).

Two objections must be raised against this point of view. The first objection is that it does not remove the indeterminateness of such magnitudes as the energy, or the momentum, or the position of a material particle, as the relation between the ψ-wave, and the values of these magnitudes is purely statistical.[12] It can of course be asserted that now the ψ-waves are the states. However, it must then also be admitted that these states do not uniquely determine the physical properties of the system concerned. But even if we were satisfied with this verbal manoeuvre, we would still not have solved the case from which the whole difficulty arose – that is, the case of the interaction of two systems. For it can be shown[13] that only in very special circumstances [379]it is possible to break up the ψ-wave for two systems into two ψ-waves, one for each of the components. In general the elements of a large system of interacting particles cannot be said to be in any state whatever.[14] This makes it quite obvious that wave mechanics does not provide any means for getting around the indeterminateness which we discussed in the previous section.

We now turn to the second suggestion, according to which "the description of the quantum theory" must be regarded as "an incomplete and indirect description of reality" such that "the ψ-function does not in any way describe a state which would be that of a single system; it rather relates to many systems, to an 'ensemble of systems' in the sense of statistical mechanics".[15] According to this interpretation the single system is always in a well-defined (classical) state, and any indeterminacy that is introduced is due to our lack of knowledge of this state. This interpretation has been dealt with and refuted in the last section. Let us not forget, however, that for anybody who does not know of the qualitative arguments developed there, there are strong reasons to assume that it is correct. These reasons consist above all in the fact that the connection

---

[12] Planck is, of course, aware of this fact, and he therefore distinguishes between the "world picture" of physics on the one hand, and "the sensory world" on the other; and he ascribes to the former a symbolic content only (*op. cit.*, 129). However, this cannot be reconciled with the claim that "determinism is ... strictly valid in the world picture of quantum mechanics" (136), as it is clear that the causal relation can hold between real events only. Still, the above-mentioned philosophers are not even aware of this difficulty.

[13] Von Neumann, *Mathematical Foundations of Quantum Mechanics*, Princeton: Univ. Press. 1955. VI/2 {Feyerabend's review of this book is reprinted as Chapter 17 of the present collection}.

[14] For an excellent presentation of this fact, cf. E. Schrödinger, *Naturwissenschaften*, 23, 1935.

[15] A. Einstein, "Physics and Reality", reprinted in *Ideas and Opinions*, London: 1954, p. 35.

between wave mechanics and "reality" is established by *statistical rules* (Born's rules): "It has not till now been taken sufficiently into account", writes Popper in his analysis of the uncertainty relation[16] "that to the mathematical derivation of the Heisenberg formulas from the fundamental equations there must correspond, precisely, a derivation of the *interpretation* of the Heisenberg formulas from the *interpretation* of these fundamental equations". And he points out that, given Born's rules, we must interpret Δx and Δp in (1) as the standard deviations, within large ensembles, of quantities *which are otherwise well defined*, and not as "statements imposing limitations upon the attainable precision of measurement".[17] This point of view seems to derive further support from von Neumann's systematic application of the frequency theory to wave mechanics, whose statistical ensembles "make again possible an objective description (which is independent of ... whether one measures, in a given state, the one or the other of two not simultaneously measurable quantities)".[18] Hence, "if we start from the assumption that the formulas which are peculiar to the quantum theory are ... statistical statements, then it is difficult to see how prohibitions of single events could be de[380]duced from a statistical theory of this character. ... The belief that single measurements can contradict the formulas of quantum physics seems logically untenable; just as untenable as the belief that a contradiction might one day be detected between a formally singular probability statement ... say "the probability that the throw k will be five equals $^1/_6$", and one of the following two statements: ... "the throw is in fact a five", or ... "the throw is in fact not a five".[19] This is, of course, completely correct reasoning *provided* the elements of the collectives with which we are dealing in the quantum theory are all in a state that is well defined from the classical point of view – that is, *provided we already know what kinds of entities are to be counted as the elements of the collectives*. Only then it is possible to derive the interpretation of the uncertainty relations which Popper wants to defend from the statistical character of the quantum theory. The arguments in the last section which deal with the *single case* show that such an interpretation is not admissible. Naturally this necessitates a reinterpretation of what is to be counted in the collectives. What is counted is now no longer the number of systems *possessing* a certain well-defined property (value of a classical magnitude); what *is* counted is the number of *transitions*, on measurement, from certain

---

[16] *Logic of Scientific Discovery*, New York: Basic Books, 1959, p. 227.   [17] *Op. cit.*, 224.
[18] *Op. cit.*, 300.   [19] Popper, *op. cit.*, 228f.

ill-defined states into other ill-defined states (namely, into such states in which that property possesses a well-defined value). Clearly the frequency approach, which deals with relations between *collectives* only, is neutral with respect to either the first or the second choice of the *elements*. Hence it follows that the occurrence of frequencies within the quantum theory cannot be used as an argument against the customary interpretation of formula (1).[20]

However, at this stage, the adherents of the second suggestion will point out that there exist further reasons for the correctness of their interpretation. They will refer to an argument by Einstein, Podolsky, and Rosen[21] according to which the formalism of the wave mechanics [381]is such that it demands the existence of exact simultaneous values of noncommuting variables. Clearly, if this should be the case, then Bohr's interpretation of formula (1) would have to be dropped, and it would have to be replaced by the interpretation of Einstein and Popper. Now a closer analysis of the argument will show two things. It will show (1) that it is conclusive only if it is assumed that dynamical states are *properties* of systems rather than *relations* between systems and measuring devices in action; and it will also show (2) that the conclusion of the argument is unacceptable and bound to lead to inconsistencies. In order to dispel the impression, felt by some writers, that the qualitative considerations of Section 2 count little when compared with the powerful utilization, by Einstein, Podolsky, and Rosen (EPR), of the apparatus of wave mechanics, I shall first deal with the second point.

Assume for that purpose that it has been demonstrated by EPR that an electron always possesses a well-defined position and a well-defined momentum. The *first difficulty* which arises concerns the *values* of these two variables at a particular time. If correct, EPR only demonstrates that

---

[20] Nor can it be used, as has been attempted by von Neumann in his famous "proof", as an argument for Bohr's interpretation of formula (1). This follows at once from Popper's analysis as it has been explained in the text. (For a more detailed investigation cf. my paper in *Zs. Physik.* 145, 421, 1956 {translated as Chapter 4 of the present collection}, as well as "On the Interpretation of the Uncertainty Relations", *Studia Filoz.*, 1960 {This paper, originally published in Polish as "O Interpretacij Relacyj Nieokreslonosci" (*Studia Filozoficzne*, 19, 1960, pp. 21–78), is the preliminary version of Chapter 7 of the present collection}). However, it ought to be pointed out that all these considerations refer to the general theory only and do not make any reference to the dynamical laws and the conservation laws. As soon as these laws are added to the principle of superposition (which is the core of the general theory), we obtain very powerful arguments against the possibility of a classical specification of state. This will be shown later in the text.

[21] *Phys. Rev.*, 47, 777, 1935.

electrons possess a well-defined position and a well-defined momentum. It does not allow us to *calculate* the simultaneous values of these two magnitudes. Now assume that we are dealing with an electron which is in a (quantum-mechanical) state of well-defined momentum. Considering the wavelength of its corresponding matter-wave we may obtain the value of its momentum, and this value has experimental significance; that is, it is such that a measurement of momentum will produce exactly this value. Now whatever position we ascribe to the electron, it will not be capable of independent test unless we *stipulate*[22] (rule R) that the state of a single particle be completely determined by the juxtaposition of the results of measurement of all the variables that can be measured in it. However, it turns out that the "superstate" thus defined contains what one might call descriptively redundant elements. What, after all, is the role of a dynamical state in physical theory? The reply is that the dynamical state contains part of the initial conditions which, if taken together with other initial conditions, such as mass and charge, as well as with some theories, will serve for the explanation, or the prediction, of the behavior of the system to which it belongs. It is obvious from this statement that an element of state will be superfluous if it does not play any role in any prediction or in any explanation. It will be even more superfluous if it [382]can be shown that this is true, not only for a finite span of time, but even for those measurements which are carried out *at the very time at which the occurrence of the element is asserted*. In this case there exists no possibility whatever of testing the assertion that the element has occurred, and this assertion may, therefore, be properly called descriptively redundant. It can easily be shown that any "superstate" that has been constructed in accordance with rule R is bound to contain descriptively redundant elements in exactly this sense. For if $(a, b, \ldots, g, \ldots, n)$ is the series of numbers obtained in accordance with rule R, and if $(A, B, \ldots, G, \ldots, N)$ is the corresponding series of observables, then the latter series will contain at least two elements which do not commute. Let G be such that $[G, N] \neq 0$, and that for any element a between G and $N, [a, N] = 0$. Then while $P[a'/(a, b, c, \ldots a', \ldots, n)] = 1$ (a' is the value obtained for a) and $P[a''/(a, b, c, \ldots a', \ldots, n)] = 0$ for any $a'' \neq a'$, $P[g'/(a, b, c, \ldots g \ldots, n)] = const$ for any $g' \neq g$ even if what is referred to is the exact moment at which $G = g$ is asserted. This means, of course, that what we shall find, on closer inquiry,

[22] Cf. the discussion of the role of such stipulations in H. Reichenbach, *Philosophic Foundations of Quantum Mechanics*, Berkeley: Univ. California Press, 1946, p. 118 ff. {Feyerabend's review of this book is reprinted as Chapter 23 of the present collection.}

in the system is completely independent of whatever value we have chosen for G. This part of the superstate is, therefore, descriptively redundant.

Now at this stage one might still be inclined to say that the use of superfluous information, although not very elegant, can at most be rejected on the basis of considerations of "taste".[23] That this is not so is shown by the *second difficulty* which I am now going to discuss. This second difficulty consists in the fact that the use of superstates is incompatible with the conservation laws.[24] In the case of the energy principle this becomes obvious from the fact that for the single free electron $E = p^2/2m$, so that after the measurement of the position of the particle any value of the energy may emerge, and, after a repetition of the measurement of q, any different value. One may, of course, try to escape this conclusion, as Popper apparently has been inclined to do,[25] by declaring that the energy principle be only statistically valid. It is very difficult however, to reconcile this hypothesis with the many independent experiments (spectral lines; experiment of Franck and Hertz; experiment of Bothe and Geiger) which show that energy is conserved also in the quantum theory, not only on the average, but for any single process of interaction. The difficulty of the point of view of Einstein and Popper becomes very obvious if we apply it to [383]the case which is known as the *penetration of the potential barrier*. In this case we may obtain rather drastic disagreement with the principle of the conservation of energy if we assert it in the form that for any superstate [Eqp] *as determined by three successive measurements*, E will satisfy the equation $E = p^2/2m + V(q)$.[26] These difficulties suffice to show that the conclusion of EPR cannot be correct. Our next step will, therefore, consist in an examination of the argument itself. [27]

The argument considers two systems which are separated to such an extent that no interaction occurs between them. Also, the state of the combined system [S'S''] is supposed to be such that the measurement of some observables in S' allows for the calculation of the values of some corresponding observables in S'' for at least two pairs of noncommuting

---

[23] This is Heisenberg's attitude. Cf. his *Physical Principles of Quantum Theory*, Chicago: Univ. of Chicago Press, 1930, p. 20.

[24] Cf. also the remark in the text to footnote 6.

[25] I am referring here to discussion I had with Professor Popper. The responsibility for the representation is, of course, entirely mine.

[26] For a numerical evaluation cf. D. I. Blochinzew, *Grundlagen der Quantenmechanik*, Berlin: 1953, p. 505.

[27] For a more detailed discussion, cf. my paper referred to in fn. 20.

observables. Now if S' and S'' no longer interact, then a disturbance of S' cannot influence the state of S''. More especially, a measurement within S' cannot influence the state of S'', which means, of course, that the state of S'' will be the same immediately before and immediately after the occurrence of the measurement. Hence, if the measurement is such that it results in information about S'', it follows at once that this information was valid even before the measurement was actually carried out. As the case is such that it allows us, by measurement, to obtain information about the values of two noncommuting observables we arrive at the result that these observables must always possess simultaneous sharp values.

As it stands the argument is unassailable. As a matter of fact, it turns out that most of the attempts at a refutation are quite unsatisfactory.[28] On the other hand, we have seen that the conclusion cannot be acceptable and that there exist strong independent reasons (such as the qualitative considerations in Section 2) to the effect that we must work with states which are only partly well defined. This being the case it is of paramount importance to make the argument compatible with Bohr's hypothesis of the indefiniteness of state description. This was achieved by Bohr with the help of the assumption (which we formulated at the very beginning of Section 2) that states are *relations* between systems and measuring devices in action, rather than properties of such systems. It is easily seen how this second basic postulate of Bohr's point of view makes indefiniteness of state description compatible with EPR. For while a property cannot be changed except by interference with the system that possesses this property, a relation can be changed without such interference. Thus the state A = being longer [384]than b, of a physical object a may change when we compress a – that is, when we physically interfere with a. But it may also change when we expand b without at all interfering with a. Hence, lack of physical interference excludes changes of state only if it has already been established that positions and momenta are *properties* of systems rather than *relations* between them and suitable measuring devices. "Of course", Bohr writes, referring to Einstein's example,[29] "there is in a case like the one ... considered no question of a mechanical disturbance of the system under investigation ... But even at this stage there is essentially the question of *an influence on the very conditions which define the possible types of prediction regarding the future behavior* of the system". He compares this influence with "the dependence on the reference system,

[28] For discussion of some such attempts cf. my paper referred to in fn. 20.
[29] *Phys. Rev.*, 48, 696, 1935.

in relativity theory, of all readings of scales and clocks".[30] I would like to repeat, at this stage, that Bohr's argument is not supposed to *prove* that quantum-mechanical states are indeterminate; it is only supposed to show under what conditions the indeterminacy of the quantum states can be made compatible with EPR. If this is overlooked, it is easy to get the impression that the argument is ad hoc.[31] One must, as it were, approach the argument with the realization that superstates cannot be incorporated into wave mechanics without leading to inconsistencies. Once this is admitted, there arises the need for a proper interpretation of the special case discussed by EPR which seems to lead to superstates. It is in this connection that Bohr's suggestion proves so extremely fruitful.

4. THE DOGMATIC ELEMENTS IN BOHR'S APPROACH

In the last two sections we have shown to what extent it is possible to justify Bohr's twin assertion that (a) the quantum theory will have to work with dynamical states which are only partly well defined and that (b) these states must be regarded as relations between systems and measuring devices rather than as properties of the systems. The first part of this assertion was justified by simple qualitative arguments as well as by reference to the fact that even within the quantum theory the conservation laws hold for the individual processes and not on the average only. The second part was introduced in order to make the first part compatible with the formalism of wave mechanics which, as was shown by EPR, allows for surprising correlations that are not easily understood unless it is assumed that states are relations. Now, as is well known, Bohr's point of view is not exhausted by the presentation of these two elements. Bohr's point of view results from an attempt to arrive at a coherent epistemology that contains these elements, but which also goes [385]beyond them. I have dealt with this epistemology in some other publications.[32] At the present moment I am only concerned with one of its consequences which is in radical disagreement with scientific practice from Galileo up to Einstein,

---

[30] *Loc. cit.*, p. 704.   [31] Popper, *op. cit.*, p. 446.
[32] "Complementarity", *Proc. Aristot. Soc.*, Suppl. Vol. XXXII, 1958; "Professor Bohm's Philosophy of Nature", *British Journal for the Philosophy of Science*, February, 1960; "On the Interpretation of the Uncertainty Relations", *Studia Filosoficzne*, 1960; "Explanation, Reduction and Empiricism", *Minnesota Studies in the Philosophy of Science*, Vol. III; "On the interpretation of Elementary Quantum Theory", ibid., Vol. IV (forthcoming) {see, respectively, Chapter 5 of the present collection; *PP1*, ch. 14; Chapter 7 of the present collection; and *PP1*, ch. 4. The latter paper was never published in that form}.

and which may have detrimental influence upon the further development of physics. For Bohr and his followers regard their point of view as based upon assumptions which are so general, and at the same time so firmly established, *that any future theory must conform to them.* It will be admitted that the quantum theory will have to undergo some decisive changes in order to be able to cope with new phenomena. But it will also be pointed out that, however large these changes may be, they will always leave untouched the two elements mentioned above – namely, the indefiniteness of state description and the complementary character of the main variables of the theory.[33] This conviction with respect to the future development of physics and the undisputable character of the point of view of complementarity is well expressed by Rosenfeld who writes:[34] "We are here" (that is, in the case of complementarity) "not presented with a point of view which we may adopt, or reject, according to whether it agrees or does not agree with some philosophical criterion. It is the *unique*" (my italics) "result of an adaptation of our ideas to a new experimental situation in the domain of atomic physics ..."; and who asserts that[35] "it is idle to 'hope' that the cure of our troubles will come from underpinning quantum theory with some deterministic substratum". All these sentiments are very much reminiscent of the belief, held in the nineteenth century, that the fundamental principles of classical physics are absolutely correct and that the only task for the physicist is to find the appropriate forces for the explanation, according to these principles, of the behavior of a mechanical system. But how have these sentiments been defended?

In order to give a reply to this question we need only repeat our discussion of the last two sections. The premises we used in this dis[386]-cussion were mainly three: (1) the quantum postulate; (2) the duality of light and matter; (3) the conservation principles. Given these premises the main elements of Bohr's point of view can be derived. However, now the question arises whether these premises can be regarded as unconditionally correct. To this question Bohr and the other members of the Copenhagen circle seem to have the following answer ready: (1)–(3) are experimental

---

[33] For this sentiment cf. W. Pauli, *Dialectica*, VIII, 124, 1954; L. Rosenfeld, *Louis de Broglie, Physicien et Penseur*, Paris: 1953, 41, 57; P. Jordan, *Anschauliche Quantentheorie*, Berlin: 1936, 1, 114f., 276; G. Ludwig. *Die Grundlagen der Quantenmechanik*, Berlin: 1954, 165ff.

[34] *Loc. cit.*, 44.

[35] *Observation and Interpretation*, ed. by S. Körner, New York: Academic Press, 1957, p. 41.

facts. Science is based upon facts. Therefore, in the course of any argument (1)–(3) must not be tampered with, but must be regarded as the unalterable basis for any further theoretical consideration. This answer, which is bound to be extremely popular with empiricists, is yet completely unsatisfactory. First of all, it is well known (though not well understood) that our theories frequently contradict our experiences and ascribe to them an approximate validity only. Consider for that purpose the relation between Newton's theory of gravitation and the law of the free fall. The latter law asserts that near the surface of the earth the acceleration of falling bodies is a constant, whereas Newton's theory implies that it must be a function of the distance. The difference between these two assertions is, of course, so small that it is undetectable by experiment. Yet from a strictly logical point of view, Newton's theory violates the requirement, implicit in Bohr's approach, that any theory should be compatible with the results of our observations. The same holds for Galileo's law of inertia which says, when applied to everyday experiences, that these experiences give us at most an approximate account of what is going on in reality. It is also very plausible that this should be so. After all, it would be very unreasonable to assume that while our theories may be mistaken, our experiences are yet a perfect indicator of what is going on in the world. It would be unreasonable because it would amount to the assumption that human individuals are better than any measuring instrument because they are not subjected to errors of measurement.

From all this it follows that the only condition to be satisfied by a future microscopic theory is *not* that it be *compatible* with 1–3 (which would indeed imply that it must work with indefinite states, complementary pairs of variables, and the like) but that it be compatible with these premises *to a certain degree of approximation* which will have to depend upon the precision of the experiments used for establishing (1), (2), and (3).

A completely analogous remark holds for the assertion, frequently made, that Planck's constant will have to enter *every* microscopic theory in an essential way. After all, it is quite possible that this constant has meaning only under certain well-defined conditions (just as the density of a fluid, or its viscosity, or its diffusion constant can be meaningfully [387]defined only for not-too-small volumes), and that all the experiments we have made so far explore only part of these conditions. Quite obviously the invariance of h in all these experiments cannot be used as an argument against such a possibility. But if neither the constancy of h nor duality can be guaranteed to hold in new domains of research, then the whole argument of Bohr and his followers is bound to break down: it does not guarantee the persistence of the familiar features of complementarity, of probabilistic laws, of quantum jumps, in future investigations.

Similar considerations apply to Bohr's insistence that it will be impossible to invent a new conceptual scheme which will replace the classical concepts in the quantum domain. According to Bohr, the classical concepts are the *only* concepts which we possess. As we cannot construct a theory, or a description of fact, out of concepts which we do not possess and as the classical concepts cannot anymore be applied in an unrestricted way, it follows that we are stuck with complementarity and indeterminateness of state description. Against this argument, which has been elaborated in some detail by Heisenberg[36] and von Weizsäcker,[37] it is sufficient to point out that introducing a set of concepts is not something that occurs independently of, and prior to, the construction of theories. And with respect to theories, it must be said that man is not only capable of *using* them for prediction and description, but that he is also capable of *inventing* them. How else could it have been possible, to mention only one example, to replace the Aristotelian physics, and the Aristotelian cosmology (which was much more general than our contemporary mechanics as it contained a *general* theory of change – not only a theory of *spatial* changes) by the new physics of Galileo and Newton? The only conceptual apparatus then available was the Aristotelian theory of change with its opposition of actual and potential properties, the four causes, and the like. Within this conceptual scheme, which was also used for the description of experimental results, Galileo's (or rather Descartes') law of inertia does not make sense, nor can it be formulated. Should, then, Galileo have followed Heisenberg's advice, and should he have tried to get on with the Aristotelian concepts as well as possible as[38] "his actual situation ... was such that the Aristotelian concepts *were* being used" and as "there is no use discussing what could have been done if the situation had been otherwise and his contemporaries different people from what they actually were"? By no means. What was needed was not improvement, or delimitation of the Aristotelian concepts; what was [388]needed was *an entirely new theory*. Are there (apart from pessimism with respect to the abilities of contemporary physicists) any reasons to assume that what was

[36] *Physics and Philosophy*, New York: 1958.
[37] *Zum Weltbild der Physik*, Leipzig: 1954.
[38] This is a paraphrase of a passage found on p. 56 of Heisenberg's book.

possible in the seventeenth century (and even earlier) will be impossible in the twentieth century? As far as I can understand, it is Bohr's contention that such reasons do indeed exist, that they are of a logical rather than of a sociological character, and that they are connected with the peculiar nature of classical physics.

Bohr's first argument in favor of this contention proceeds from the fact that we need our classical concepts not only if we want to give a *summary* of observational results, but that without these concepts the results to be summarized could not even be *stated*. As Kant before him, he observes that our experimental statements are always formulated with the help of certain theoretical terms and that the elimination of these terms must lead, not to the "foundations of knowledge" as the positivists would have it, but to complete chaos. "Any experience" he asserts,[39] "makes its appearance within the frame of our customary points of view and forms of perception" – and at the present moment the forms of perception are those of classical physics.

But does it follow, as is asserted by Bohr, that we can *never* go beyond this framework and that therefore *all* our future microscopic theories must have duality built into them?

It is quite obvious that the use of classical concepts for the description of experiments within contemporary physics can never justify any such assumption. For a theory may be found whose conceptual apparatus, when applied to the domain of validity of classical physics, would be just as comprehensive and useful as the classical apparatus, without yet coinciding with it. Such a situation is by no means uncommon. The behavior of the planets, of the sun, and of the satellites can be described both with the help of the Newtonian concepts, and with the concepts of relativity. The order introduced into our experiences by Newton's theory is retained, *and improved upon*, by relativity. This means that the concepts of relativity are sufficiently rich for the formulation of all the facts which were stated before with the help of Newtonian physics. Yet the two sets of concepts are completely different and bear no logical relation to each other.

An even more striking example is provided by the phenomena known as the "appearances of the devil". These phenomena are accounted for either by the assumption that the devil exists, or by more recent psychological (and sociological) theories. The concepts used by these two

---

[39] Bohr, *op. cit.*, p. 1.

schemes of explanation are in no way related to each other. Nevertheless, the abandonment of the idea that the devil exists does [389]not lead into experiential chaos, since the psychological frame is rich enough to account for the order already introduced.

To sum up: Although in reporting our experiences we make use, and must make use, of certain theoretical terms, it does not follow that different terms will not do the job equally well, or perhaps even better, because more coherently. And as our argument is quite general, it seems to apply to the classical concepts as well.

This is where Bohr's second argument comes in. According to this argument, which is quite ingenious, we shall have to stay with the classical concepts, because the human mind will never be able to invent a new, and different, conceptual scheme. The argument for this peculiar inability of the human mind rests upon the following *premises*: (a) we invent (or should use) only such ideas, concepts, theories, as are suggested by observation; (b) because of the formation of appropriate habits, any conceptual scheme employed for the explanation and prediction of facts will imprint itself upon our language, upon our experimental procedures, our expectations, and thereby also upon our perceptions; (c) classical physics is a universal conceptual scheme – that is, it is so general that no conceivable fact falls outside the domain of its application; (d) classical physics has been used long enough for the formation of habits, referred to under (b), to become operative. The *argument* itself runs as follows: If classical physics is universal (premise c), and has been used long enough (premise d), then all our experiences will be classical (premise b), and we shall be unable to conceive any concept which falls outside the classical scheme (premise a). Hence, the invention of a new conceptual scheme which might enable us to circumvent duality, is impossible.

Now the first remarkable thing about this argument is that all the premises apply to the Aristotelian physics, and premise (c) applies with even greater force to this theory than it applies to classical mechanics. For as is well known, Aristotelian physics contained a general theory of (qualitative and quantitative) changes, whereas classical physics deals with a special kind of changes only. Yet in spite of all this, the conclusion is false in the case of Aristotelian physics – it *was* superseded by the physics of Galileo and Newton. It is easily seen why this is so: it is quite clear that neither premise (a), nor (c), nor (d) is correct, and that premise (b) needs some qualification. The qualification of (b) derives from the demand that a scientist should always keep an open mind and that he should, therefore, always consider possible alternatives along with the theory he is favoring at a certain moment. If this demand is satisfied, then the habits cannot form, or at least they will not any longer completely determine the actions of the scien[390]tist. That premise (a) cannot possibly be true (is not a reasonable demand) is illustrated by the fact that transitions from universal points of view to other universal points of view *have* taken place. Finally, it cannot be admitted (premise c) that the classical scheme is universally valid. It is not applicable to such phenomena as the behavior of living organisms, personal consciousness, the formation and behavior of social groups, and the like. We have to conclude, therefore, that Bohr's arguments *against* the possibility of alternatives to complementarity and *for* the perennial character of uncertainties are all inconclusive. There is not the slightest reason to assume that "the new conceptions which we need will be obtained ... by a rational extension of the quantum theory"[40] and that the new theories will, therefore, be increasingly indeterministic. And there is, therefore, not the slightest need for any theoretician to restrict himself to theories of a certain type at the expense of simplicity, intuitive appeal, and universal validity. However ingenious and interesting Bohr's physical ideas may be, the attempt to regard them as absolute truths can only lead to stagnation and to a new kind of dogmatism. We have to realise that no element of our knowledge, physical or otherwise, can be held to be absolutely certain and that in our search for satisfactory explanations we are at liberty to change any part of our existing knowledge, however "fundamental" it may seem to be to those who are unable either to imagine or to comprehend alternatives.

[40] Rosenfeld in *Observation and Interpretation*, ed. by S. Körner, New York: Academic Press, 1957, p. 45.

Rejoinder to Hanson
(1961)

Editorial Note

Feyerabend's rejoinder to Norwood R. Hanson's paper – "Comments on Feyerabend's 'Niels Bohr's Interpretation of the Quantum Theory', or *die Feyerabendglocke für Copenhagen?*" (pp. 390–398) – is self-explicatory. We reproduce here only the final part of Hanson's text (pp. 397–398), which is explicitly referred to by Feyerabend:

The very *most* a Copenhagen theorist can ever be in a reasonable position to argue, is that:

(1) Any theory that incorporates the quantum postulate and the duality principle must also, if it would interpret the state uncertainties objectively, restrict the meaningful applicability of functors like "position" and "momentum".

(2) There are good contingent arguments in support of the expectation that any future microphysics *will* incorporate the quantum postulate and the duality principle – although no *conclusive* argument to this effect exists, or could exist.

(3) There exists now no genuine working alternative to the quantum theory we now have, notwithstanding all its awkward features.

Feyerabend justifiably jousts against Bohr's metaphysical excesses. But I do not anticipate that he would be deeply troubled with points (1), (2), and (3) above, although he might express himself differently concerning them. Bohr's metaphysics constitutes no indispensable part of the Copenhagen interpretation. Extended conversations with physicists convince me that most of those who feel an affinity with the "Copenhagen interpretation" in virtue of (1), (2), and (3) above feel no need whatever for Bohr's epistemology(ies).

My issue with Feyerabend here may be more verbal than real. If the Copenhagen interpretation is construed as Feyerabend construes it, then it *should* be fought at every turn. Philosophy of physics can ill afford to embrace dogmatism, even when endorsed by giants. But if the "Bohr Interpretation" is cut away, then the Copenhagen interpretation would seem to me (and perhaps even to Professor Feyerabend?) eminently defensible. Or does *die Feierabendglocke* continue to toll even for this "liberalized" version of the Copenhagen interpretation?

[398]There are two points in Professor Hanson's reply with which I cannot agree. The first point is his contention that the Copenhagen circle, like any other party, has its right wing as well as its liberals and that it is unfair

95

wholly to identify the two. This point would be well taken if the difference between them were a substantial one – that is, if the so-called "liberals" were to admit that a new approach is possible. Unfortunately I have not been able to detect any sign of such an admission. Quite on the contrary, all these alleged liberals to which Hanson refers share Bohr's attitude that his (Bohr's) interpretation of the quantum uncertainties has come to stay. Heisenberg, whose attitude seems to be acceptable to Hanson,[41] has given reasons for this permanence which are hardly different from the reasons given by Bohr. "It has sometimes been suggested" he remarks,[42] discussing Bohr's well-known contention, that all experimental results are, and will forever have to be, formulated in classical terms,[43] "that one should depart from the classical concepts altogether, and that a radical change in the concepts used for describing the experiments might possibly lead back to a nonstatistical and completely objective description of nature. This suggestion", he continues, "rests upon a misunderstanding. The con[399]-cepts of classical physics ... are an essential part of the language which forms the basis of all natural science. Our actual situation in science is such that we do use the classical concepts ... There is no use in discussing what could be done if we were other beings than we are". Now it may be agreed that here the impossibility of transcending the classical framework is ascribed (whether rightly or wrongly is not my concern at the present moment) to the *empirical* fact that human beings do think in certain ways and not in other ways. It is an *empirical* impossibility, whereas in Bohr's case one sometimes gets the impression that for him the permanence of the classical terminology on the experimental level is a matter of *logic*. But which physicist is going to be impressed by such a subtle distinction, or which physicist is going even to recognize such a distinction! Both Bohr and Heisenberg say that it is *impossible* to leave the classical framework in the domain of observation – and this is what determines their attitude. The remaining differences may comfort the philosophers, but they will not influence the course of theorizing. Moreover even this slight resemblance of a greater liberalism on Heisenberg's part disappears a few pages later where the possibility of discovering experimentally the hidden variables is likened to the possibility that two times two might be five – that is,

---

[41] Cf. his "Three Cautions for the Copenhagen Critics", *Philosophy of Science*, 1959; *Physics and Philosophy*, New York: 1958, p. 56.

[42] Cf. his "Three Cautions for the Copenhagen Critics", *Philosophy of Science*, 1959; *Physics and Philosophy*, New York: 1958, p. 56.

[43] "Discussions with Einstein", *Albert Einstein, Philosopher-Scientist*, Evanston: Row, Peterson, 1949, p. 209.

it is now declared to be a matter of logic. Is it perhaps this inconsistency on Heisenberg's part which has prompted Professor Hanson to count him among the liberals?

Now to turn to Hanson's own version of a point of view which he could accept. This is the second point on which I am in disagreement with him. I am referring to the three items appearing at the end of his paper which, so Hanson says, is "the very *most* a Copenhagen theorist can ever be in a reasonable position to argue". My contention is that this "very most" does not any more provide any material for polemics against suggestions such as those made by Bohm, de Broglie, Vigier (suggestions which Hanson himself has attacked in earlier papers) and, second, I contend that this "very most" is still too much to be accepted without criticism. My reasons are as follows:

Item one can be admitted *provided*, of course, that the quantum postulate and duality are taken to be strictly true in all domains of experimentation. However (*item two*), this *strict* correctness can never be established by experimentation. Experimentation at most shows that certain laws hold within a certain margin of error, and for a limited amount of instances (namely, all the experiments actually carried out). All we can therefore say is that the two principles mentioned have not yet been refuted by experience, and are to that extent in agreement with it, *confirmed* by it. *But so is any alternative theory* whose devia[400]tions from the accepted theory are small enough to fall within the domain of experimental error. Bohm's investigations have already indicated a path along which such alternatives may be constructed in such a manner that *item two applies to them as well.* Now concerning these attempts Hanson says (*item three*) that there does not yet exist such a fully developed alternative to the existing theory. If I read him correctly he seems to imply that the absence of such an alternative which is connected with Bohm's philosophy may be regarded as an argument against the latter – and he is not the only one to use this very fashionable pragmatic argument. However what argument is this? Being well versed in the history of the sciences and in the considerations which play a role in processes of discovery, Professor Hanson surely knows that elaborate theories are preceded by more or less general considerations which sometimes are inconsistent with the prevalent philosophy. But should Copernicus have abandoned the idea of Aristarch just because it was not yet worked out in as great detail as was the geocentric idea? By no means; he perceived, and justly so, that his idea was a possible one, and he had no reason to assume that the finished theory would be worse than Ptolemy's system. The same is true today – with a difference, however. For whereas the orthodox demand very little formal coherence of their theories and are content if they can cook up some predictions in a manner that would raise the hairs of any mathematician, the followers of Bohm demand a *coherent* theory which makes understandable different processes from a *single* point of view. Small wonder that their higher demands cannot be satisfied as easily and do not as easily lend themselves to theoretical elaboration, as do the demands of their opponents. Finally it ought to be pointed out that qua production of theories the point of view of complementarity is not better off than Bohm's point of view. For it was introduced *after* and not *before* the conception of wave mechanics, which was invented by Schrödinger on the basis of a completely different philosophy. Hence neither *item one*, nor *item two*, nor *item three*, can prove the superiority of the idea of complementarity over the new ideas of Bohm and his collaborators.

7

Problems of Microphysics
(1962)

[189]Descartes who seemed to me to be jealous of the fame of Galileo had the ambition to be regarded as the author of a new philosophy, to be taught in academies in place of Aristotelianism. He put forward his conjectures as verities, almost as if they could be proved by his affirming them on oath. He ought to have presented his system of physics as an attempt to show what might be anticipated as probable in this science, when no principles but those of mechanics were admitted: this would indeed have been praiseworthy; but he went further, and claimed to have revealed the precise truth thereby greatly impeding the discovery of genuine knowledge.

Huygens

ACKNOWLEDGMENTS

[190]For support of research the author is indebted to the National Science Foundation and to the Minnesota Center for the Philosophy of Science.

I. INTRODUCTION

[191]When the formalism of the elementary quantum theory was first conceived it was unclear how it was to be related to experience and what intuitive picture should be connected with its application. "The

---

[254]The passage from Huygens at the beginning of the paper appears in his annotations on Ballet's *Life of Descartes* (a book which was published in 1691) and is printed in V. Cousin's *Fragments Philosophiques* tome II, p. 155 Quoted from E. Whittaker, *A History of the Theories of Aether and Electricity*, Vol. I, London, 1951.

mathematical equipment of the ... theory", writes Heisenberg about this period,[1] "was ... complete in its most important parts by the middle of 1926, but the physical significance was still extremely unclear". There existed a variety of interpretations. However, in the course of time each of these interpretations turned out to be unsatisfactory. Only the suggestions of Niels Bohr and of his collaborators, which later on were presented in a more systematic manner and which then received the name "Copenhagen Interpretation"[2] seemed to succeed in solving most of the problems that had been fatal for its rivals. It was this interpretation which was finally accepted by the great majority of physicists, including some of those who had first objected to it on philosophical grounds.[3] From about 1930[4] (or rather from about 1935[5]) to 1950 the Copenhagen Interpretation was *the* microphilosophy, and the objections of a few opponents, notably of Einstein and Schrödinger[6] were taken less and less seriously. During these years the idea of complementarity was developed and proved to be fertile

---

[1] "The Development of the Interpretation of the Quantum Theory," in *Niels Bohr and the Development of Physics*, London, 1955, p. 13.

[2] Some very important elements of this interpretation were developed in connection with the older quantum theory and the correspondence principle. This is true of the assumption of the indefiniteness of state descriptions. However, it would be somewhat rash to assert, as has been done by Heisenberg and others (cf. the reference in the last footnote), that his interpretation was a natural outcome of the "Korrespondenzdenken" and that its growth was not at all influenced by considerations of an entirely different character. For this cf. fn. 4, 5, as well as Section 4.

[3] An example is L. de Broglie. Cf. *Une tentative d'interprétation causal et non-linéaire de la mécanique ondulatoire*, Paris, 1956, Introduction.

[4] To be more precise, the decisive date was the Fifth Solvay Conference which took place in Brussels in October, 1927 and which led to a complete victory for the point of view of Bohr and his collaborators. For this evaluation cf. Heisenberg *loc. cit.*, p. 16; de Broglie, *La Méchanique quantique, restera-t-elle indéterministe?*, Paris, 1953, Introduction; and Niels Bohr, "Discussions with Einstein of Epistemological Problems in Atomic Physics", originally published in *Albert Einstein, Philosopher-Scientist*, The Library of Living Philosophers, Inc. 1949, pp. 199ff, reprinted in *Atomic Physics and Human Knowledge*, New York, 1958, asp. pp. 41ff. The Proceedings of this very decisive Conference have been published; Institut International de Physique Solvay, *Rapport et discussions du 55 Conseil*, Paris, 1928. Cf. however the next footnote.

[5] For *it was not before 1935 that the idea of the relational character of the quantum mechanical states was added to the Copenhagen Interpretation*. Despite later assertions to the contrary this meant a tremendous change of point of view. For a more detailed discussion, cf. Section 6 and footnote 116.

[6] Schrödinger's attitude was not consistent. He has published papers which contain vigorous attacks upon the philosophical attitude of Bohr and his followers. However in private discussions he seemed to be much more impressed by the soundness of Bohr's point of view and of the positivistic theory of knowledge in general (by which I do not mean to imply that Bohr's ideas are positivistic).

in such fields as psychology[7] (Jung's included), biology,[8] ethics,[9] and theology.[10] The most decisive argument for the soundness of this idea of complementarity derived from its application (and its success[11]) in the domain of physics itself. Here it led to the development of a variety of theories such as Dirac's theory of the electron and the quantum theory of fields. The belief that these theories could not have been developed without support from the basic principles of the Copenhagen interpretation,[12] and their empirical fruitfulness, seemed to show the soundness of

[7] An example is Jordan's *Komplementarität und Verdrängung* where occult phenomena are also treated.

[255][8] An example of P. Jordan, *Die Physik und des Geheimnis des organischen Lebens*, Braunschweig 1943. For Bohr's own speculations cf. the relevant articles in *Atomic Physics and Human Knowledge*.

[9] In a talk in Askov (Denmark) in the year 1949 which I attended, Bohr pointed out that there may exist a complementary relationship between love and justice. References to this talk and to the assertion just mentioned may be found in one of the August numbers of *Berlingske Tidende*.

[10] In a draft of a paper "Complementarity in Quantum Mechanics: A Logical Analysis" (Draft, October, 1960) Messrs Hugo Bedau and Paul Oppenheim state their interest to investigate the possibility of applying Bohr's concept of complementarity to the relation between science and religion. For a discussion of already existing attempts in this direction and a criticism cf. P. Alexander, "Complementary Descriptions", *Mind*, Vol. LXV (1956), pp. 145–165. Cf. also MacKay, D. M., "Complementarity", *Aristotelian Society*, Suppl. Vol. XXXII (1958), pp. 105–122. Assertions involving gods are also found in W. Heisenberg, *Syllabus of the Gifford Lectures*, 1956, esp. p. 16: "It has certainly been the pride of natural science since the beginning of rationalism to describe and to understand nature without using the concept of God, and we do not want to give up any of the achievements of this period. But in modern atomic physics we have learned how cautious we should be in omitting essential concepts just because they lead to inconsistencies".

[11] It is necessary to point out that most of these successes have only been partial successes. The elementary quantum theory, Dirac's theory of the electron, the earlier field theories – all these theories have been found to be unsatisfactory in one way or another. That is they were successful in certain domains, completely unsuccessful in others. And the more recent theories are characterized by a much less close adherence to the principle of correspondence and the philosophy of complementarity connected with it. Cf. J. Schwinger, *Quantum Electrodynamics*, Dover, 1958, pp. xivff as well as Bogoliubov-Shirkov, *Introduction to the Theory of Quantized Fields*, New York, 1958, p. 16.

[12] Thus Hanson [*Am. Journal of Physics*, Vol. 27 (1959), pp. 4ff, reprinted in Danto-Morgenbesser, *Philosophy of Science*, New York, 1961, pp. 450ff; cf. esp. p. 454] points out that the development of the field theories and of Dirac's theory of the electron was strongly influenced by the point of view of complementarity. Despite Hanson's explicit assertion to the effect that Dirac himself has felt that way, I am somewhat doubtful as far as the second case is concerned: the paper where the theory is first developed uses some purely formal considerations (properties of the Hamiltonian) which seem in no way connected with the idea of complementarity; and the interpretation which this theory

the philosophy that had been developed by Bohr and his disciples. Here at last the practicing physicist was given a point of view which contained no arbitrary element, which allowed him correctly to interpret many applications of a fairly complicated theory, viz. the wave-mechanics, which was capable of providing a guide in research and all of whose basic assumptions were directly taken from experience. "We are here not presented", writes Professor Rosenfeld,[13] [192]"with a point of view which we may adopt, or reject, according to whether it agrees, or does not agree with some philosophical criterion. It is the unique result of an adaptation of our ideas to a new experimental situation in the domain of atomic physics. It is therefore completely on the plane of experience ... that we have to judge whether the new conceptions work in a satisfactory way". Considering the success of the theories which were built in accordance with the idea of complementarity, this judgment most certainly was believed to be a positive one.

And yet we are now witnessing the development of a counter-movement which demands that the basic assumptions of the Copenhagen Interpretation be given up and be replaced by a very different philosophy. The critics do not deny the *initial* success of the elementary theory, of Dirac's theory of the electron and of the field theories. They mainly point out two things. First, that at the present moment all these theories are in difficulty as far as empirical adequacy is concerned and that they are also not as simple and straightforward as they perhaps ought to be.[14] And secondly, they question the argument from empirical success itself. Complementarity is successful – let us take this for granted. This does not show that it is a reasonable point of view for its empirical success, the fact that it is confirmed by all actual, and perhaps even by all conceivable experimental results may well be due, not to its correctness, but to its being void

---

finally received (hole-theory) even runs counter to the assumption of [256]the relational character of quantum mechanical states. Still, it is quite possible that in the *preparation* of this paper Dirac has made use of some ideas which are essential to Bohr's point of view. Which is only one more argument in support of my demand which I have voiced since 1954 that the history of the quantum theory be based upon *live interviews* (to be carried out as speedily as possible) and not on papers only. It should be realized that very little of the thought that has led to the invention of the Copenhagen Interpretation has found *immediate* expression in papers. And it should also be realized that a history of the quantum theory will therefore be of value only if it is based upon what is to be found in print as well as upon carefully prepared interviews of its main participants.

[13] "L'évidence de la complémentarité", in *Louis de Broglie, Physicien et Penseur*, Paris, 1953, p. 44.

[14] Cf. fn. 11.

of empirical content.[15] Nor would its empirical success prove the non-existence of valid *alternatives* to complementarity. Indeed, one of the most valuable contributions of the revolutionaries consists in their having shown that there is not a single sound argument, empirical or mathematical, establishing that complementarity is the last word in matters microphysical and that a theoretician who intends to improve quantum theory will be successful only if he works with theories that possess inbuilt uncertainties. At least to this writer it is now clear that future research need not (and should not[16]) be intimidated by the restrictions which some high priests of complementarity want to impose upon it.

It is unfortunate that the majority of physicists is still opposed to these very clear results. There are various reasons for this situation. The most prominent reason is the soporific effect which partial empirical success seems to have upon many physicists. Also many physicists are very practical people and not very fond of philosophy. This being the case, they will take for granted and not further investigate[17] those philosophical ideas which they have learned in their youth and which by now seem to them, and indeed to the whole community of practicing [193]scientists, to be the expression of physical common sense. In most cases these ideas are part of the Copenhagen Interpretation.

A second reason for the persistence of the creed of complementarity in the face of decisive objections is to be found in the *vagueness* of the main principles of this creed.[18] This vagueness allows the defendants to take

---

[15] This point has been made by D. Bohm; cf. his "Quantum Theory in Terms of 'Hidden Variables'", *Phys. Rev.*, Vol. 85 (1951), pp. 166ff.

[16] As I have shown in "Explanation, Reduction, and Empiricism", *Minnesota Studies in the Philosophy of Science*, Vol. III {reprinted in *PP1*, ch. 4}, the empirical content of a theory of the generality of the present quantum theory depends to a decisive degree on the number of alternative theories which, although in agreement with all the relevant facts, are yet inconsistent with the theory in question. The smaller this number, the smaller the empirical content of the theory. The invention of alternatives which are inconsistent with the present quantum theory is therefore a necessary demand of empiricism.

[17] For a more detailed description of this undesirable state of affairs cf. section 7 of my paper referred to in footnote 16.

[18] von Weizsäcker ["Komplementarität und Logik", in *Die Naturwissenschaften*, Vol. 17 (1955), pp. 521ff] and Groenewold [private communication] have asserted that the fruitfulness of the Copenhagen interpretation was to a large extent due to the vague and indefinite manner in which it has been formulated, and discussed. I completely agree that precision may (and often does) go hand in hand with sterility (which is one of the reasons why I cannot embrace a great deal of contemporary philosophy of science). However, it cannot be allowed that vagueness is made the handmaid of dogmatism. That is it cannot be allowed that a theory is made vague in such a manner that it cannot any more be reached by criticism.

care of objections by *development* rather than by *reformulation*, a procedure which will of course create the impression that the correct answer has been there all the time and that it was overlooked by the critic. Bohr's followers, and also Bohr himself, have made full use of this possibility even in cases where the necessity of a reformulation was clearly indicated. Their attitude has very often been one of people who have the task to clear up the misunderstandings of their opponents rather than to admit their own mistakes. A very important instance of this kind of behavior which will be discussed later in the paper is Bohr's answer to Einstein's very decisive objection of 1935.[19] Here I want to mention another example where this attitude is exhibited most clearly. Before Bohm published his well-known paper on hidden variables[20] it was commonly assumed that the quantum theory was incompatible with any interpretation that employed hidden parameters and that this character of the theory was guaranteed by experience.[21] "There is *only one* interpretation", writes Pascual Jordan,[22] "which is capable of conceptually ordering the ... totality of experimental results in the field of atomic physics". In defending his interpretation against Einstein's remarks Bohr points out that, in his opinion, "there could be no other way to deem a logically consistent mathematical formalism as inadequate than by demonstrating the departure of its consequences from experience, or by proving that its predictions did not exhaust the possibilities of observation, and Einstein's argumentation could be directed to neither of these ends".[23] Considering the idea of hidden parameters Pauli writes:[24] "In this connection I may also refer to von Neumann's well-known proof that the consequences of quantum mechanics cannot be amended by additional statements on the distribution of values of observables, based on the fixing of values of some hidden parameters, without changing some consequences of present quantum mechanics". Von Neumann went even further. Thus summarizing his often quoted but otherwise little known "proof" he writes:[25] "Not only is the measurement impossible [of the hidden parameters] *but so is any reasonable theoretical definition*, i.e.,

[19] Cf. section 4 and footnote 116.   [20] Cf. the reference in footnote 15.
[21] Cf. the quotation from Rosenfeld in text to footnote 13.
[257][22] *Die Physik und des Geheimnis des organischen Lebens*, Braunschweig, 1943, p. 114.
[23] "Discussions with Einstein", in *Albert Einstein, Philosopher–Scientist*, Evanston, 1948, p. 229.
[24] *Dialectica*, Vol. 2 (1948), p. 309 (editorial of a special issue on the interpretation of the quantum theory).
[25] *Mathematical Foundations of Quantum Mechanics*, Princeton, 1955, p. 326.

any definition which, although incapable of experimental proof, would also be incapable of experimental refutation". The gist of all these statements is that the Copenhagen Interpretation is the only interpretation that is empirically adequate and [193]formally satisfactory. This contention was refuted in Bohm's paper. Has this fact been admitted by the Copenhagen school? It has not been admitted. On the contrary, Bohm's paper was attacked in a manner that suggested he had overlooked some essential elements of the idea of complementarity. Thus Heisenberg remarked that the model "says nothing about physics that is different from what the Copenhagen language says",[26] and he also insinuated that it was *ad hoc*.[27] It has also been said that "given the ψ-function, the values of the parameters cannot manifest themselves, neither directly, nor indirectly";[28] and that they are for this reason to be regarded as "purely metaphysical".[29] As regards the first assertion we need only return to Bohr's reply to Einstein which we have quoted above. For if Bohr's interpretation is unassailable by virtue of the fact that it satisfies the criteria outlined in this quotation (full empirical adequacy, and formal satisfactoriness), then so is Bohm's which, as Heisenberg explicitly admits, is empirically equivalent with it and formally without reproach. This refutes the idea, expressed, among others, by Jordan, that there exists only one interpretation that possesses these desirable characteristics.[30] And as regards the second assertion, we must point out that, as is evident from the quotation, von Neumann asserted that even metaphysical hidden parameters could not be made to agree with the formalism of

---

[26] "The Development of the Interpretation of the Quantum Theory", *op. cit.*, p. 19.

[27] *Op. cit.*, p. 18.

[28] W. Pauli, "Remarques sur le problème des paramètres caches dans la mécanique quantique et sur la théorie de l'onde pilote", in *Louis de Broglie*, etc., p. 40.

[29] Pauli, *op. cit.*, p. 41.

[30] Cf. the above quotation from Jordan. The irrelevance of all these replies to Bohm's investigations has been stated very clearly by J. Agassi [*British Journal for the Philosophy of Science*, Vol. IX (1958), p. 63]: "That Bohm's theory is *factually* false, or at least, very much *ad hoc* is, of course, entirely irrelevant, since the question is whether von Neumann's argument consists of a proof that a theory like Bohm's must be *logically* false".

That the idea of complementarity asserts the logical impossibility of a model like Bohm's has been asserted by P. Jordan. According to him [*Anschauliche Quantentheorie*, Leipzig, 1936, p. 116; my italics] a model like Bohm's which works with well-defined trajectories "would be inconsistent, not only with the (changeable) notions of classical physics, *it would even be inconsistent with the laws of logic*". Similar sentiments have been expressed, in a less definite form, by Bohr and Heisenberg. Cf. the latter's *Physics and Philosophy*, New York, 1958, p. 132 as well as *Niels Bohr and the Development of Physics*, p. 18.

the theory. All this is a good example of the way in which complementarity has been reinterpreted in the light of objections, thereby giving the impression that these objections had really missed the point.[31]

A third reason for the persistence of the creed of complementarity is to be sought in the fact that many objections against it are irrelevant. Most critics interpret the two main principles of the Copenhagen Interpretation, viz., the principle of the indeterminateness of state descriptions and the principle of the relational character of quantum mechanical states not as *physical* assumptions which describe objective features of physical systems; they interpret them as the direct result of a positivistic *epistemology* and reject them together with the latter.[32] Now it is of course quite correct that this is the way in which some members of the Copenhagen Circle have introduced these principles, but it should also be clear that this is not the only way to justify them and that much better arguments are available, arguments which are directly derived from physical practice. This situation, by the way, accounts for the strangely unreal character of many discussions on the foundations of the present theory. The members of the Copenhagen school are confident that their point of view with whose fruitfulness they are well acquainted is satisfactory and [195]superior to a good many alternatives. But when writing about it, they do not draw sufficient attention to its physical merits, but wander off into philosophy and especially into positivism. Here they become an easy prey to all sorts of philosophical realists who quickly (or not so quickly) exhibit the mistakes in their arguments without thereby convincing them of the incorrectness of their point of view – and quite justly so, for this point of view can stand upon its own feet and does not need any support from philosophy. So the discussion between physicists and philosophers goes back and forth without ever getting anywhere.

In the paper to follow I shall try to get out of this vicious circle. I shall try to give a purely physical explanation of the main ideas behind the

---

[31] There are also other, and more technical objections against Bohm's model. Thus Pauli has pointed out in 1927 that the theory of the pilot wave requires an electron in an s-state to be at rest whereas electrons are always found in a well-defined state of motion. This objection which has been repeated, more recently, by Einstein (cf. his paper in *Scientific Papers Presented to Max Born*, Edinburgh, 1953) can however be answered by the Bohm-model (cf. Bohm's comments in the same volume).

[32] An excellent example is the discussion of the quantum theory in Chs. VIII and IX of M. Bunge's *Metascientific Queries*, Springfield, 1959. (Cf. my review of that book in *Phil. Rev.*, Vol. LXX (1961).) Similar remarks apply to some of Popper's criticisms of the quantum theory. For details cf. fn. 123.

Copenhagen Interpretation. It will turn out that these ideas and the physical arguments leading up to them are much more plausible than the vague speculations which were later used in order to make them more acceptable. But it will also turn out that despite this plausibility none of the arguments is powerful enough to guarantee the absolute validity of these ideas and to justify the demand that the theories of the microscopic domain will forever have to conform to a certain pattern.[33] Such restrictions are possible only if certain *philosophical* ideas are used as well. *These* philosophical ideas will be investigated in the second part of the present paper. The result of our investigation will be that while the physical arguments dealt with in the first part are sound and ingenious, the philosophical ideas needed to confer absolute validity upon them are neither correct nor reasonable. The result of the criticism is, of course, that now as ever we are free to consider whatever theories we want when attempting to explain the structure of matter.

2. THE EARLY QUANTUM THEORY: WAVE-PARTICLE DUALITY

Not long after Planck had introduced the quantum of action[34] it was realized that this innovation was bound to lead to a complete recasting of the principles of motion of material systems. It was Poincaré[35] who first pointed out that the idea of a continuous motion along a well-defined path could no longer be upheld and that what was needed was not only a new *dynamics*, i.e., a new set of assumptions about the acting forces, but also a new *kinematics*, i.e., a new set of assumptions about the kind of motion initiated by these forces. Both Bohr's older theory and the dual nature of light and matter further accentuated this need. One of the problems arising in the older quantum theory was the treatment of the interaction between two mechanical systems.[36] Assume that two systems, A and B, interact in such a manner that a certain amount of energy, ε, is [196]transferred from A to B. During the interaction the system A+B possesses a well-defined energy. Experience shows that the transfer of ε does not occur immediately, but that it takes a finite amount of time. This seems to suggest that

---

[33] L. Rosenfeld in *Observation and Interpretation*, London, 1957, p. 45.

[258][34] I am here referring to what is known as Planck's *First Theory* in which both absorption and emission were regarded as discontinuous processes [*Verh. d. phys. Gesellschaft*, Vol. II (1900), p. 237], and which also implies discontinuities in space (Cf. Whittaker, *History of the Theories of Aether and Electricity*, Vol. II, Edinburgh, 1953, p. 103).

[35] *Journal de Physique*, V. II (1912), p. 1.

[36] N. Bohr, *Atomic Theory and the Description of Nature*, Cambridge, 1932, p. 65.

both A and B change their state gradually, i.e., A gradually falls from 2 to 1, while B gradually rises from 1 to 2. However, such a mode of description would be inconsistent with the *quantum postulate* according to which a mechanical system can be only either in state 1, or in state 2 (we shall assume that there are no admissible states between 1 and 2), and it is incapable of being in an intermediate state. How shall we reconcile the fact that the transfer takes a finite amount of time with the non-existence of intermediate states between 1 and 2?

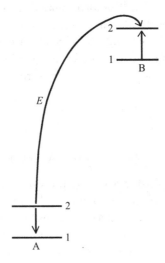

This difficulty was resolved by Bohr[37] on the basis of the assumption that during the interaction of A and B the dynamical states of both A and B cease to be well defined so that it becomes *meaningless* (rather than *false*) to ascribe a definite energy to either of them.[38]

This simple and ingenious physical hypothesis has so often been misrepresented that a few words of explanation seem to be needed. First of all it must be pointed out that in the above formulation the term "meaning" has not entered, as has been asserted by various critics,

[37] I shall not contend that this is the only possible way of getting around the difficulty, but it is a very reasonable physical hypothesis which has not yet been refuted by any of the arguments aimed against it.

[38] By the expression "dynamical state" we refer to "quantities which are characteristic of the motion" of the system concerned (such as the positions and the momenta of its components) rather than those quantities which, like mass and charge, serve as a characteristic of what kind of system it is. For this explanation cf. Landau-Lifshitz, *Quantum Mechanics*, London, 1958, p. 2 as well as N. Bohr, *Atomic Physics and Human Knowledge*, p. 90; cf. also H. A. Kramers, *Quantum Mechanics*, New York, 1957, p. 62.

because of [197]some connection with the now customary attitude of preferring semantical analysis to an investigation of physical conditions.[39] After all, there are well-known classical examples of terms which are meaningfully applicable only if certain physical conditions are first satisfied and which become inapplicable and therefore meaningless as soon as these conditions cease to hold. A good example is the term "scratchibility" (Mohs scale), which is applicable to rigid bodies only and which loses its significance as soon as the bodies start melting.

---

[39] It is to be admitted, however, that most derivations of the uncertainties, and especially those based upon Heisenberg's famous thought-experiments do make use of philosophical theories of meaning. Usually these arguments (and other arguments which proceed from the commutation relations of the elementary theory) only establish that inside a certain interval *measurements cannot be carried out*, or that the products of the mean deviations of certain magnitudes *cannot be ascertained* below Planck's constant *h*. The transition from this stage of the argument to the assertion that it would be *meaningless* to ascribe definite value to the magnitudes in this interval is then achieved on the basis of the principle that what cannot be ascertained by measurement cannot be meaningfully asserted to exist. This premature use of untenable philosophical theories of meaning has led to a very curious situation. It has led to a situation where physical principles, such as the principle of the indefiniteness of state descriptions are attacked, *and defended* for the wrong reasons. The principle "that it is impossible, in the description of the state of a mechanical system, to attach definite values to both of two canonically conjugate variables" (N. Bohr in *Phys. Rev.*, Vol. 48 (1935), p. 696) has been attacked because it was believed to be the result of positivistic considerations. And it has also been *defended* by such considerations. (Cf. M. Schlick, *Naturwissenschaften*, Vol. 19 (1931), pp. 159ff as well as Heisenberg, *The Physical Principles of the Quantum Theory*, University of Chicago Press, 1930, esp. p. 15. In the case of Heisenberg the result obtained is not even the *universal* indefiniteness[259] of state descriptions as it is admitted that "the uncertainty relation does not refer to the past" – p. 20. It is also interesting to note the difference between the terminology of the German and the English version. In the German version assertions concerning the situation inside the domain of indefiniteness are called "inhaltsleer" (cf. p. 11 of the German edition, *Die Physikalischen Prinzipien der Quantenmechanik*, Leipzig, 1930) – "devoid of content", whereas the English version used the term "meaningless".) The positivistic defense of the principle seems to have been adopted even by some physicists, and this from the very beginning. And yet this principle can – and, so we should add, *must* – be interpreted as a physical hypothesis according to which a relaxation is necessary in the classical description which, being an "idealization" (Bohr, *Atomic Description*, etc., p. 5; cf. also p. 631), goes beyond the evidence available at the time when it was first introduced and has now been found in need of modification. It seems that this was also Bohr's own point of view. Despite occasional lapses into positivistic jargon and argumentation, he has always claimed to be a realist and he has therefore been somewhat critical of Heisenberg's positivism (Bohm and Groenewold, private communication). The actual historical development is still far from clear however and this because of the fact that little has been published of the early discussions which later on led to the development of the Copenhagen point of view. Here is a challenging task for contemporary historians of science. Get the information, by letter, on tape, in discussions, before the main actors of this fascinating intellectual drama have died!

Secondly, it should be noted that the proposed solution does not contain any reference to *knowledge*, or *observability*. It is not asserted that during the time of transfer A and B may be in some state which is unknown to us, or which cannot be observed. For the quantum postulate does not merely exclude the knowledge or the observability of the intermediate states; it excludes these intermediate states themselves. Nor must the argument be read as asserting, as is implied in many presentations by physicists, that the intermediate states do not exist *because* they cannot be observed. For it refers to a postulate (the quantum postulate) which deals with existence, and not with observability. The emphasis upon the absence of *predictability* is not satisfactory either. For this way of speaking would again suggest that we could predict better if we only knew more about the things that exist in the universe, whereas Bohr's suggestion denies that there are such things whose detection would make our knowledge more definite. The third remark concerns a suggestion for getting around the kinematics of ill-defined states which has often been made in connection with wave mechanics and which will be discussed in detail later in the present paper. According to this suggestion, the difficulties which arise when we try to give a rational account of processes of interaction are due to the fact that the classical point mechanics is not the correct theory for dealing with atomic systems, and the state descriptions of classical point mechanics are not the adequate means for describing the state of systems upon the atomic level. According to this suggestion, we ought not to retain the classical notions, such as position and momentum, and make them less specific. What we ought to do is rather to introduce completely new notions which are such that when *they* are used states and motions will again be well defined. Now if any such new system is to be adequate for the description of quantum phenomena, then it must contain means for expressing the quantum postulate which is one of the most fundamental microlaws, and it must therefore also contain adequate means for ex pressing the concept of energy. However, once this concept has been introduced, in the very same moment all our above considerations apply [198]again with full force: while being part of A+B, neither A nor B can be said to possess a well-defined energy, from whence it follows at once that also the new and ingenious set of concepts will not lead to a well-defined and unambiguous kinematics. "It would", therefore, "be a misconception to believe", writes Bohr,[40] "that the difficulties

---

[40] *Atomic Theory*, etc., p. 16.

of the atomic theory may be evaded by eventually replacing the concepts of classical physics by new conceptual forms". This last remark will be of great importance in connection with the interpretation of Schrödinger's wave mechanics.

The empirical adequacy of the proposed solution is shown by such phenomena as the natural line breadth which in some cases (such as in the absorption leading to states preceding Auger effect) may be quite considerable.

Its consequence is, of course, the *renunciation of the kinematics of classical physics*. For if during the interaction of A and B neither A nor B can be said to be in a well-defined state, then the change of these states, i.e., the *motion* of both A and B will not be well defined either. More particularly, it will no longer be possible to ascribe a definite trajectory to any one of the elements of either A or B. One may now attempt to retain the idea of a well-defined motion and merely make indefinite the relation between the energy and the parameters characterizing this motion. The considerations in the next paragraph show that this attempt encounters considerable difficulties.

The reason is that the *duality of light and matter* provides an even more decisive argument for the need to replace the classical kinematics by a new set of assumptions. It ought to be pointed out, in this connection, that dealing with both light and matter on the basis of a single general principle, such as the principle of duality, may be somewhat misleading. For example, whereas the idea of the position of a light quantum has no definite meaning,[41] such meaning can be given to the position of an electron. Also, no account is given in this picture of the coherence length of light. It is therefore advisable to restrict the discussion of duality to elementary particles with a rest mass different from zero, for example to electrons.

Now it has been asserted[42] that the interference properties of light and matter, and the duality resulting from them, are but an instance of statistical behavior in general, which latter is then thought to be best explainable by reference to such classical devices as pin boards and

---

[41] Cf. E. Heitler, *Quantum Theory of Radiation*, Oxford, 1957, p. 65; D. Bohm, *Quantum Theory*, Princeton, 1951, pp. 97f.

[42] Cf. for example A Landé, "The Logic of Quanta", *British Journal for the Philosophy of Science*, Vol. VI (1956), p. 300 as well as "From Duality to Unity in Quantum Mechanics," *Current Issues in the Philosophy of Science*, New York, 1961, pp. 350ff.

roulette games. According to this assertion the elementary particles move along well-defined trajectories and possess a well-defined momentum at any instant of their motion. It is sometimes admitted that their energy [199]may occasionally undergo sudden and perhaps individually unexplainable changes. But it is still maintained that this will not lead to any *indefiniteness* of the state that experiences these sudden changes.

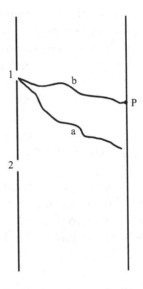

I shall now try to show that this assumption cannot give a coherent account of the wave properties of matter and of the conservation laws. It is sufficient, for this purpose, to consider the following two facts of interference: (1) interference patterns are independent of the number of particles which at a given moment are dwelling in the apparatus. To use an example: We obtain the same pattern on a photographic plate whether we now use strong light and short time of exposure or whether we use very weak light and a long time of exposure.[43] (2) The two-slit interference pattern is not simply the arithmetical sum of the patterns created by

---

[43] With respect to light this was shown by Jánossi. Cf. the booklet edited by the *Hungarian Academy of Sciences*, 1957, where also previous experiments are reported, as well as Jánossi, *Acta Physica Hungarica*, Vol. IV (1955) and *Nuovo Cimento*, Vol. VI (1957).

each single slit. It is quite possible for the two-slit pattern to possess a minimum in a place, say *P*, in which the one-slit pattern shows a finite intensity. (1) allows us to neglect the mutual interaction (if any) between the particles. Considering (2) we may reason as follows: If it is correct that each particle always possesses a well-defined trajectory, then the [200]finite intensity in *P* is due to the fact that some particles arrived there along the path *b*. As long as slit two remains closed there will always be some particles which, having passed slit one, will travel along *b*. Assume that *E* is such a particle, for example an electron, and that it is about to enter slit one. Now if we open slit two at this very moment we have thereby (by virtue of fact (2) above) created conditions which are such that *E* must not arrive at *P*. Hence, the process of opening slit two must lead to a change in the former path of *E*. How can this change be accounted for?

It cannot be accounted for by assuming action at a distance. There is no room in the conservation laws (which are valid also in the quantum theory) for energies deriving from such an action. Furthermore,[44] the alleged action works not everywhere in space, but only along those surfaces which in the wave picture are surfaces of equal phase and is therefore nothing but a misleading way of bringing in wave-notions.

According to Popper[45] and Landé[46] the change of path of the individual particle is not in need of explanation. What can be explained, by reference to the change of physical conditions (opening of slit two), is the emergence of a new stochastic process that leads to a new interference pattern. This position is *indeterministic*, as it admits

---

[44] The idea of action at a distance has been discussed, and regarded as a possible explanation by Hans Reichenbach. Cf. his *Philosophic Foundations of Quantum Mechanics*, Berkeley and Los Angeles, 1945, sec. 7 {Feyerabend's review of this book is reprinted as Chapter 23 of the present collection}. Action at a distance is not the solution which Reichenbach himself adopts. For an evaluation of Reichenbach's analysis and of his own solution (three-valued logic) cf. my paper "Reichenbach's Interpretation of the Quantum Theory", *Philosophical* [260]*Studies*, Vol. IX (1958), pp. 47ff. {Reprinted in *PP1*, ch. 15.}

[45] *Observation and Interpretation*, ed. Körner, London, 1957, pp. 65ff. Landé's suggestions are in many respects similar to those of Popper. I should like to point out that the arguments in the text can be repeated in a slightly different fashion even if the minimum should not be an absolute minimum. (An attempt to invalidate the argument by reference to the fact that the minima may not be absolute is due to Professor A. Landé.)

[46] A. Landé, *Quantum Theory, A Study of Continuity and Symmetry*, New Haven, 1955.

the existence of uncaused individual changes and its indeterminism is about as radical as that of the Copenhagen point of view. It also shares with that point of view its emphasis on the importance of the *experimental situation*: predictions are valid only for certain experimental conditions, they are not universally valid (this point will be explained in greater detail in Section 6). However, it *differs* from the Copenhagen point of view insofar as it works with well-defined states and well-defined trajectories. This being the case it must and does admit[47] that the conservation laws are valid only for large ensembles of particles in a certain situation and that they may be violated in the individual case. It is here that the difficulties arise. For, as is well known,[48] energy and momentum are conserved in each individual case of interaction also for elementary particles. This point of view must therefore be rejected *unless* it is developed in such detail that an alternative account can be given of all those experiments which have convinced the physicists of the individual validity of the conservation laws. *Until* such a more detailed account has been given (and nobody can say in advance that it is impossible!) we must again regard the assumption of the indefiniteness of state descriptions as the most satisfactory explanation. Let it be noted, by the way, that this restriction applies to all the arguments we are going to develop in favor of the assumption of the [201]indeterminateness of state descriptions. All these arguments presuppose the validity of certain experimental findings such as the quantum postulate, the laws of interference, the individual validity of the conservation laws; and the arguments show that *given* these experimental findings we are forced to reject various interpretations which do not work with intrinsically indefinite state descriptions. It follows that an alternative to the idea of complementarity is likely to be successful only if it implies that at least some of these experimental results are not strictly valid without, of course, contradicting them inside the domain (of error, or experimentation) where they were first found to apply. The issue concerning the foundations of the quantum theory can therefore be solved only by the construction of a *new theory* as well as by the demonstration that this

---

[47] Professor K. R. Popper, private communication.

[48] I am referring here to the experiments of Bothe and Geiger, *Zs. Physik*, Vol. 35 (1926) pp. 639ff as well as to the results of Compton and Simon, *Phys. Rev.*, Vol. 25 (1925) pp. 306ff.

new theory is experimentally at least as valuable as the theory that is being used at the present time; it cannot be solved by alternative *interpretations* of the present theory.[49]

---

[49] For this point cf. p. 89 of my "Complementarity", *Proceedings of The Aristotelian Society*, Suppl. Vol. XXXII (1958) {reprinted as Chapter 5 of the present collection}. If I am correct in this, then all those philosophers who try to solve the quantum riddle by trying to provide an alternative interpretation of the *current theory* which leaves all the laws of this theory unchanged are wasting their time. Those who are not satisfied with the Copenhagen point of view must realize that only a new theory will be capable of satisfying their demands. Of course, they may try in advance to consider the effects which the success of such a new theory might have upon the interpretation of the current theory. However, an essential part of any such interpretation will have to be the admission that the current theory and the empirical laws upon which it is based *are not entirely correct.*

It is interesting to note the similarity between the present situation in the quantum theory and the problems which confronted the followers of Copernicus. As is well known, the issue was then not the predictive correctness of the Copernican theory (which was admitted by the opponents) but the extent to which the Copernican hypothesis could be regarded to mirror the actual structure of the universe. That is the church did not contest the predictive value of the Copernican hypothesis. What it objected to was the realistic interpretation of this hypothesis, i.e., the assumption that the hypothesis could be regarded as a description of the world. Now it is most important to realize that this move, on the part of the church, was neither wholly due to a philosophical conservatism, nor was it wholly due to the fear that a realistic interpretation of Copernicus' hypothesis might do considerable damage to theological dogma. There were weighty *physical* arguments against the assumption of a real motion of the earth. These physical arguments were an immediate consequence of the then popular Aristotelian physics. In short the situation was as follows: the conjunction of Aristotelian physics and a realistically interpreted Coperni[261]can theory was inconsistent with some widely known empirical results. Aristotelian physics was highly confirmed by experiment. On the other hand the Copernican theory led to the correct celestial predictions and had therefore to be regarded as empirically adequate. Was it not natural, in these circumstances, to assume that no realistic importance should be ascribed to Copernicus and that the merit of this theory consisted merely in having found a coordinate system in which the problem of the planets assumed an especially simple form? This move could not be countered by a purely *philosophical* criticism and the demand that every theory be interpreted in a realistic fashion. A philosophical criticism of the instrumentalistic interpretation of the Copernican theory quite obviously could not remove the inconsistency between a Copernican universe and the Aristotelian dynamics. *Nothing less than a new theory of motion would do*, a new theory of motion, moreover, which would not be as strictly empirical and commonsensical as the Aristotelian theory and which could therefore count on strong opposition. This was clearly realized by Galileo. "Against the physical principles of conventional cosmology, which were always brought out against him, he needed an equally solid set of principles – indeed, more solid – because he did not appeal to ordinary experience and common sense as his opponents did". (G. di Santillana, *The Crime of Galileo*, Chicago, 1955, p. 31; cf. also the pages following the quotation as well as my article on the philosophy of nature in *Fischer Lexikon, Band Philosophie*, Frankfurt/ Main, 1958 {reprinted as Chapter 24 of the present collection}.) It seems to me that this feature of the situation has not always been realized by philosophers (as an example we may take K. R. Popper's "Three Views Concerning Human Knowledge", *Contemporary*

Now it is the belief of many followers of the "orthodox" point of view that such a new theory is impossible. To be more specific, they believe that such a theory will either be internally inconsistent, or inconsistent with some very important experimental results. They therefore not only suggest an interpretation of the known experimental results in terms of indefinite state descriptions. They also suggest that this interpretation *be retained forever* and that it be the foundation of any future theory of the microlevel. It is at this point that we shall have to part company. I am prepared to defend the Copenhagen Interpretation as a physical hypothesis and I am also prepared to admit that it is superior to a host of alternative interpretations. The whole first part of the present paper will be devoted to showing its superiority as a physical hypothesis. But I shall also show that any argument that wants to establish this interpretation more firmly is doomed to failure. Now, as ever, the future development of physics is a completely open matter.

British Philosophy, Vol. III (1956), pp. 2ff). Nor has it been realized that the present situation in the quantum theory is very similar. "The view of physical science founded by Cardinal Bellarmino and Bishop Berkeley has won the battle without a further shot being fired" writes Popper (*loc. cit.*, p. 8). This is simply not true. First of all there is a tremendous difference between the instrumentalism of Bellarmino and the instrumentalism of Berkeley. The instrumentalism of Bellarmino *could* have been supported by physical arguments drawn from contemporary physical theory. The instrumentalism of Berkeley could not have been so supported and was of a purely philosophical nature. Secondly the arguments in the text above should have shown that there exist weighty *physical* reasons why at the present moment a realistic interpretation of the wave mechanics does not seem to be feasible (see also the arguments in the next section). A philosophical crusade for realism alone will not be able to eliminate these arguments. At best, it can ignore them. What is needed is a new theory. Nothing less will do.

[262]I have to admit, however, in view of a criticism of the above passage by J. W. N. Watkins, that philosophical arguments for realism, though not sufficient, are therefore not unnecessary. It has been shown that given the laws of wave mechanics, it is impossible to construct a realistic interpretation of this very same theory. That is, it has been shown that the usual philosophical arguments in favor of a realistic interpretation of theoretical terms do not work in the case of wave mechanics (for such arguments cf. my paper "Das Problem der Existenz Theoretischer Entitäten" in *Probleme der Erkenntnis Theorie, Festschrift für Viktor Kraft*, Vienna, 1960 {translated in *PP3*, ch. 1}). However, there still remains the fact that theories which do admit of a realistic interpretation are definitely preferable to theories which do not. It was this belief which has inspired Einstein, Schrödinger, Bohm, Vigier and others to look for a modification of the present theory that makes realism again possible. The main aim of the present article is to show that there are no valid reasons to assume that this valiant attempt is bound to be unsuccessful. For some reasons why realistic theories are preferable to instrumentalistic ones, cf. section 4 of my paper "Professor Bohm's Philosophy of Nature", *British Journal for the Philosophy of Science*, Vol. X (1960), pp. 326ff. {Reprinted in *PP1*, ch. 14.}

However, if we are not prepared, at the present stage at least, to doubt the validity of the conservation laws and of the laws of interference then the only possible account of the properties of interference patterns seems again to be in terms of the assumption that the behavior of the particles between the slits and the screen is no longer well defined. In the case of a single slit this lack of definiteness can even be roughly calculated. As is well known[50] the result is that the indefiniteness Δx of position and the indefiniteness Δp_x of momentum required for circumventing the described difficulties are related by the formula

$$\Delta x \cdot \Delta p_x \geq h \tag{1}$$

[202]This result can be shown to possess universal validity. Indeed, whenever the conditions of experimentation are such that interference effects are to be expected, the definiteness of state descriptions will have to be restricted just to the extent to which the wave picture remains applicable. The exact amount of this restriction can be calculated if we combine de Broglie's ideas with the mathematics of the Fourier analysis.[51] The result is again given by formula (1)[51] which thereby has been shown to possess universal validity: *inherent indefiniteness is a universal and objective property of matter.*[52] Only after this basic assumption has been adopted, can Heisenberg's discussions, and his attempt to derive various forms of (1)[53] from an analysis of methods of measurement,[54] lead to a proper interpretation of the relations thus obtained as well as to the realization that the reference, made in this analysis, to the process of *measurement* is purely

---

[50] Cf. i.e., Max Born, *Atomic Physics*, London, 1957, p. 76.

[51] For the mathematics cf. Heisenberg, *The Physical Principles of the Quantum Theory*, Chapter II, section 1, or L. Schiff, *Quantum Mechanics*, New York, 1955, pp. 54–56. For a more careful discussion cf. E. L. Hill, *Lecture Notes on Quantum Mechanics 1958–1959*, University of Minnesota, sections 4.9 and 4.10.

[52] There is a difference in the constants, but otherwise the formula is the same.

[53] In his essay *Zur Metatheorie der Quantenmechanik*, Helsinki, 1950 E. Kaila has asserted that this manner of derivation exhibits a limitation in the validity of the uncertainty relations. Every case of motion is regarded as a case of diffraction and therefore assumes that certain boundary conditions are first satisfied. According to Kaila this shows that the uncertainties are a result of the fact that boundary conditions have been imposed. A particle whose motion is not restricted by any boundaries may therefore possess a well-defined position and a well-defined momentum. This assumption can of course be made. But it is not very much more reasonable than the assumption that a glass-pane alone in the universe is as hard as iron but ceases to be that hard immediately after the creation of the first stone.

[54] *Zs. Physik*, Vol. 43 (1927) as well as Chapter II of *The Physical Principles of the Quantum Theory*.

accidental.[55] It is of course true that arrangements such as systems with slits, or Geiger counters, can be used for measuring properties of physical objects. However the idea of the indefiniteness of state description and, more specifically, the uncertainty relations emerge, not because the arrangements can be thus used, but because of the peculiarity of the physical processes going on in them. If this is not taken into account, then it is always possible to raise the objection that what has been derived is a restriction of our ability to know the exact state of the dynamical system, or of our ability to *predict*, but not a restriction of the dynamical states themselves. Such a conclusion is the more likely to be drawn, as many discussions of the matter are subjectivistic in the sense that they first interpret (1) as restrictive of *knowledge*,[56] which interpretation is then reconciled with the objective validity of the quantum postulate and of other empirical laws by identifying knowledge with physical reality[57] or by declaring that the idea of a real external world is "metaphysical".[58] This, by the way, explains the strange unreality of so many discussions on

---

[55] I.e., it can lead to a physical interpretation of (1) rather than to a situation where physical principles and a rather doubtful epistemology are superimposed upon each other. For this cf. also fn. 39.

[262]The fact that the customary interpretation of the uncertainty relations cannot be derived from wave mechanics in connection with Born's rules alone and that some further assumptions must be added, this has been felt by many thinkers. (An example is K. R. Popper, *Logic of Scientific Discovery*, New York, 1959, Ch. IX.) However, they were mistaken when assuming that the missing link would necessarily have to be a positivistic theory of knowledge. It is quite correct that "Heisenberg's rejection of the concept of path, and his talk of 'non-observable magnitudes' clearly show the influence of philosophical and especially of positivistic ideas" (Popper, *op. cit.*, p. 232). But this is not the only possible way to arrive at the assumption of the indefiniteness of state descriptions, nor is it the best one. And an attack upon Heisenberg's procedure will therefore at most reveal his arguments as unsatisfactory. It does not amount to a refutation of indefiniteness. Nor does it remove the difficulties which led to the assumption of indefiniteness in the first place.

[56] Cf. i.e., W. Heitler in *Albert Einstein, Philosopher-Scientist*, ed. Schilpp, Evanston, 1949, pp. 181–198.

[57] Thus Heisenberg writes in *The Physicist's Conception of Nature*, London, 1958, p. 25: "... the new mathematical formulae no longer describe nature itself, but *our knowledge* of nature". The similar position of Eddington, Jeans and of others is too well known to be repeated here. All this sounds, of course, very pleasant in the ears of obscurantists as is confirmed by numerous so called "philosophical" articles: in physics as in real life the principle is valid that when the cat's away, the mice will play.

[58] For an analysis of the more general confusions in the philosophy of quantum mechanics, cf. B. Fogarasi, *Kritik des Physikalischen Idealismus*, Budapest, 1953. For a comparative evaluation of the various interpretations of (1) (and there exists quite a lot of them!) cf. M. Bunge, *Metascientific Queries*, Springfield, Illinois, 1959, Ch. IX/4 as well as the literature quoted therein. A most valuable and detailed attempt to separate physical and philosophical assumptions in the interpretation of the quantum theory has been made by

the foundations of the quantum theory: the defenders use bad and irrelevant arguments for a point of view with whose physical fertility they are intimately acquainted and of whose value they have therefore a very high opinion. The opponents, ignorant of the features of physical practice but well acquainted with the irrelevant descriptions of it, set out to destroy these irrelevant arguments and believe that they have thereby destroyed the point of view these arguments were supposed to support. This, of course, will not be admitted by the physicists, and so the battle goes on and on without any hope of being resolved in a satisfactory manner. It is mainly this feature of which I [203]think when hearing the fashionable bon mot that nowadays physics and philosophy are much closer than they ever were in the 19th century.[59]

3. WAVE MECHANICS

The older quantum theory, although experimentally quite successful and also extremely fruitful in its power to unite a host of otherwise disconnected facts,[60] was yet regarded as unsatisfactory by many physicists. Its main fault was seen to lie in the manner in which it combined classical and non-classical assumptions and which made a coherent interpretation impossible.[61] For many physicists it was therefore nothing more than a stepping stone on the way to a really satisfactory theory, i.e., to a theory which could give us not only correct predictions, but also some insight into the nature and dynamics of microscopic entities. It is quite true that Bohr, Kramers, Heisenberg, and others worked along very different lines. Their main objective was not the construction of a new physical theory about a world that existed independently of measurement and

---

Prof. A. Grünbaum in his paper "Complementarity in Quantum Physics and its Philosophical Generalizations", *The Journal of Philosophy*, Vol. LIV (1957), pp. 713–727.

[59] For this cf. also the introduction.

[60] In order to get an idea about the amount of these facts the reader should consult an early edition (for example the third edition) of Vol. I of Sommerfeld's *Atombau und Spektrallinien*. Cf. also Geiger-Scheel, *Handbuch der Physik*, first edition, Vols. 4 (1929), 20 (1928), 21 (1929), 23 (1926), as well as Wien-Harms, *Handbuch de Physik*, Vols. 21 (1927), 22 (1929).

[264][61] Some physicists felt, with some justification, that the theory was close to being *ad hoc*. "... it was so direct a transcription of the Balmer formulae that there could be little credit in such a performance as it stood". B. Hoffmann, *The Strange Story of the Quantum*, Dover, 1959, p. 58. The reader should not stop, however, after this quotation but he should go on reading and learn in this manner to what extent the theory was more than just a transcription of Balmer's formulae.

observation; their main objective was rather the construction of a logical machinery for the utilization of those parts of classical physics which could still be said to lead to correct predictions.[62] The inspiration for this

---

[62] For a very clear presentation of this idea behind the correspondence principle cf. G. Ludwig, *Die Grundlagen der Quantenmechanik*, Berlin, 1954, Ch. I. As Aage Petersen has pointed out to me, Bohr's ideas may be compared with Hankel's principle of the permanence of rules of calculation in new domains (for this principle cf. chapter 4 of F. Waismann's *Einführung in des Mathematische Denken*, Vienna, 1947). According to Hankel's principle the transition from a domain of mathematical entities to a more embracing domain should be carried out in such a manner that as many rules of calculation as possible are taken over from the old domain to the new one. For example, the transition from natural numbers to rational numbers should be carried out in such a manner as to leave unchanged as many rules of calculation as possible. In the case of mathematics, this principle has very fruitful applications. Its application to microphysics is suggested by the fact that some important classical laws remain *strictly valid* in the quantum domain. A complete replacement of the classical formalism seems therefore to be unnecessary. All that is needed is a modification of that formalism which retains the laws that have found to be valid and makes room for those new laws which express the specific behavior of the quantum mechanical entities. According to Bohr the modification must be based upon a more "liberal attitude towards" the classical concepts (*Atomic Description*, etc., p. 3). We must realize that these concepts are *"idealizations"* (p. 5; italics in the original), or "abstractions" (p. 63) whose suitability for description or explanation depends upon the relative smallness of the quantum of action and which must therefore be "handled with caution" (p. 66) in new experimental domains. "Analysis of the elementary concepts" (66) has to reveal their limitations in these new fields (4, 5, 8, 13, 15, 53, 108) and new rules for their use have to be devised "in order to evade the quantum of action" (18). These rules must satisfy the following demands: (a) they must allow for the description of any conceivable experiment in classical terms – for it is, in classical terms that results of measurement and experimentation are expressed; (b) they must "provide room for new laws" ("Can Quantum Mechanical Description of Physical Reality Be Considered Complete?", *Phys. Rev.*, Vol. 48 (1935), p. 701; *Atomic Theory*, etc. pp. 3, 8, 19, 53), and especially for the quantum of action (18); (c) they must always lead to correct [265]predictions. (a) is needed if we want to retain the idea, to be discussed later on in Section 7, that experience must be described in classical terms; (b) is needed if we want to avoid any clash with the quantum of action; (c) is needed if this set of rules is to be as powerful as a physical theory in the usual sense. Any set of rules satisfying (a), (b), and (c) is called by Bohr a "natural generalization of the classical mode of description" (4, 56, 70, 92, 110; "Causality and Complementarity", in *Dialiectica* 7/8 (1948), p. 316; "Discussions with Einstein", in *Albert Einstein, Philosopher-Scientist*, pp. 210, 239), or a "reinterpretation … of the classical electron theory" (*Atomic Theory*, etc., p. 14). "The aim of regarding the quantum theory as a rational generalization of the classical theories" writes Bohr (*Atomic Theory*, etc., pp. 70; 37; 110) "has led to the formulation of the … correspondence principle". The correspondence principle is the tool by means of which the generalizations may be, and have been obtained.

Now it is very important to realize that a "rational generalization" in the sense just explained does not admit of a realistic interpretation of any one of its terms. The classical terms cannot be interpreted in a realistic manner as their application is restricted to a description of experimental results. The remaining terms cannot be interpreted realistically either as they have been introduced for the explicit purpose of enabling the physicist

lay no doubt in the surprising fact that many classical laws remained *strictly valid* even on the quantum level. This suggested that what was needed was not the elimination, and complete replacement, of classical physics but rather a modification of it. However that may be – the philosophical spirit behind the "Korrespondenzdenken" was not shared by everybody. Thus de Broglie and Schrödinger tried to develop an entirely new theory for the description of the nature and the behavior of atoms, molecules, and their constituents. When this theory was finished it was hailed by many as the long expected coherent account of the microcosmic level. The hypothesis of the indefiniteness of state descriptions, so it was thought, had only reflected the indefiniteness and incompleteness of the early theory, and it was now no longer necessary. More especially, it was assumed either that the states were now new, but well-defined entities (the ψ-waves), or it was assumed that whatever incompleteness occurred was due to the statistical character of the theory, i.e., it was due to the fact that wave mechanics was "primarily a variety of statistical mechanics, similar to the classical statistical mechanics of Gibbs".[63] These two interpretations still survive. I hope that our arguments in the last section have made it clear that any such interpretation of wave mechanics is bound to lead into inconsistencies. The only presupposition of the hypothesis of

to handle the classical terms properly. The instrumentalism of the quantum theory is therefore *not a philosophical manoeuvre that has been wilfully superimposed upon a theory which would have looked much better when interpreted in a realistic fashion. It is a demand for theory construction that was imposed from the very beginning and in accordance with which, part of the quantum theory was actually obtained.* Now at this point the historical situation becomes complicated for the following reason: the full quantum theory (and by this we mean the full elementary theory) was created by Schrödinger who was a realist and who hoped to have found a theory that was more than a "rational generalization of classical mechanics" in the sense just explained. That is, the full quantum theory we owe historically to a metaphysics that was diametrically opposed to the philosophical point of view of Niels Bohr and his disciples. This is quite an important historical fact as the adherents of the Copenhagen picture very often criticize the metaphysics of Bohm and Vigier by pointing out that no *physical theory* has as yet been developed on that basis. (For such a criticism, cf. N. R. Hanson, "Five Cautions for the Copenhagen Critics", *Philosophy of Science*, Vol. XXVI (1959), pp. 325–337, esp. pp. 334–337.) They forget that the Copenhagen way of thinking *has not produced a theory either.* What it has produced is the proper interpretation of Schrödinger's wave mechanics *after* this theory had been invented. For it turned out that Schrödinger's wave [266]mechanics was just that complete rational generalization of the classical theory that Bohr, Heisenberg and their collaborators had been looking for and parts of which they had already succeeded in developing.

[63] E. C. Kemble, *The Fundamental Principles of Quantum Mechanics*, New York, 1937, p. 55.

indefinite state descriptions is the quantum postulate and the [204]dual nature of light and matter (taken together with the individual conservation of energy and momentum). Both these facts are contained in the wave mechanics which will therefore be equally in need of the said hypothesis.[64] A closer analysis of the two main alternatives shows that this is indeed correct.

Let us first consider the suggestion that the quantum theory provides new types of complete and well-defined states, viz., the ψ-waves which are to replace the descriptions, in terms of trajectories, of the classical point mechanics. This assumption has been made, and elaborated in some detail, by Planck. It is also identical with Schrödinger's original interpretation of his theory. According to Planck[65] (and Schrödinger) "the material waves constitute the primary elements of the new world picture. ... In general the laws of the matter waves are basically different from those of the classical mechanics of material points. However the point of central importance is that the function characterizing matter waves, i.e., the wave function ... is fully determined for all places and times by initial and boundary conditions".

Two objections must be raised against this point of view. The first objection is that it does not remove the indeterminateness of such magnitudes as the energy, or the momentum, or the position of a material particle; the reason is that the relation between the ψ-wave and the values of these magnitudes is purely statistical.[66] It can of course be asserted that now the ψ-waves are the states. However it must then be also admitted that these states do not uniquely determine the physical properties of the system concerned. But even if we were satisfied with the verbal

---

[64] "... the paradoxical aspect of the quantum theory" writes Niels Bohr ("Discussions with Einstein", quoted from *Atomic Physics and Human Knowledge*, p. 37) "were in no way ameliorated, but even emphasized, by the apparent contradiction between the exigencies of the general superposition principle of the wave description and the feature of individuality in the elementary atomic processes".

[65] Cf. *Scientific Autobiography and Other Papers*, New York, 1949, pp. 135f. Similar sentiments have been expressed, much later, by F. C. S. Northrop (*Logic of the Sciences and Humanities*, New Haven, 1948, p. 27) and by E. Nagel ("The Causal Character of Modern Physical Theory," *Readings in the Philosophy of Science*, ed. Feigl-Brodbeck, New York, 1953, pp. 419–437).

[66] Planck is, of course, aware of this fact, and he therefore distinguishes between the "world picture" of physics on the one hand, and "the sensory world" on the other, and he ascribes to the former a merely symbolic content (*op. cit.*, p. 129). However, this cannot be reconciled with the claim that "determinism is ... strictly valid in the world picture of quantum mechanics" (136) as it is clear that the causal relation can hold between real events only.

manoeuvre implicit in this move we would still not have solved the case from which the whole difficulty arose, i.e., the case of the interaction of two systems. For it can be shown[67] that only in very special circumstances it is possible to break up the ψ-function for two systems into two ψ-functions, one for each of the components. In general, the elements of a large system of interacting particles cannot be said to be in any state whatever.[68] This makes it obvious that the wave mechanics does not provide any means for getting around the indeterminateness which we discussed in the previous section. The most decisive objection against Planck's proposal is however this. Planck assumes that the only changes of the wave function are those which occur in accordance with Schrödinger's equation. This omits the reduction of the wave packet, which does not obey Schrödinger's equation[69] but which is necessary if we want to utilize the results of measurement for future calculation. A [295]closer analysis of this process of reduction shows that it cannot be interpreted in a simple, realistic manner.[70]

We now turn to the second suggestion according to which "the description of the quantum theory" must be regarded as "an incomplete and indirect description of reality", such that the ψ-function "does not in any way describe a state which would be that of a single system; it rather relates to many systems, to an 'ensemble of systems' in the sense of statistical mechanics".[71] According to this interpretation the single system is always in a well-defined (classical) state, and any indeterminacy that is

---

[67] Von Neumann, *Mathematical Foundations of Quantum Mechanics*, Princeton, 1955, VI/2.

[68] For an excellent and intuitive presentation of this fact cf. E. Schrödinger, "Die gegenwärtige Lage in der Quantenmechanik", *Naturwissenschaften*, Vol. 23 (1935).

[69] For details cf. Section 11.

[70] The reason is as follows: the development, according to the Schrödinger equation, of the combined state of the system plus the measuring apparatus results in a state that contains the possible outcomes of measurement as interfering parts. If we now assume that states of quantum mechanical systems exist in reality and that therefore any change of state corresponds to a real change, then we shall have to admit that a look at a macroscopic object leads to an immediate elimination of interference terms at distant places.

[71] A. Einstein, "Physics and Reality", reprinted in *Ideas and Opinions*, London, 1954, p. 35. The point of view expressed by Einstein and Popper (see the next footnote) is held also by J. Slater (*J. Frankl. Inst.*, Vol. 207 (1929), p. 449; cf. also Van Vleck on pp. 475ff of the same volume), and by E. C. Kemble (*op. cit.*) who says (p. 55) that "we are ... led to conceive of quantum mechan[267]ics as primarily a variety of statistical mechanics similar to the classical statistical mechanics of Gibbs". However, Kemble is aware of the fact, to be explained later on in the present paper, that the elements of the collectives are no longer systems which are existing in a well-defined classical state (note on page 55). It ought to be pointed out that despite appearances the point of view of Blochinzev is not identical with that of Einstein either. It will be discussed later in the present paper.

introduced is due to our lack of knowledge of this state. Quite obviously this assumption is inconsistent with Bohr's hypothesis, and it is also not easy to see how it could be made compatible with the quantum postulate. However, let us not forget that for anybody who does not know of the qualitative arguments presented in the last sections, and this includes a good many philosophers, there are strong reasons to assume that it is correct. These reasons consist above all in the fact that within wave mechanics the connection between the theory and "reality" is established by *statistical* rules (Born's rules): "It has not till now been taken sufficiently into account", writes Popper in his analysis of the uncertainty relations,[72] "that to the mathematical derivation of the Heisenberg formulae there must correspond, precisely, a derivation of the *interpretation* of these fundamental equations." And he points out that, given Born's rules *and nothing else*, we must interpret Δx and Δp_x in formula (1) as the standard deviations, within large ensembles, of quantities, *which are otherwise well defined*, and *not* as "statements imposing limitations upon the attainable precision of measurement".[73] This point of view seems to derive further support from von Neumann's systematic application of the frequency theory to wave mechanics whose statistical ensembles "make again possible an objective interpretation (which is independent of ... whether one measures, in a given state, the one or the other of two not simultaneously measurable quantities)".[74] Hence, "if we start from the assumption that the formulae which are peculiar to the quantum theory are ... statistical statements, then it is difficult to see how prohibitions of single events could be deduced from a statistical theory of this character.... The belief that single measurements can contradict the formulae of quantum physics seems logically untenable; just as untenable as the belief that a contradiction might one day be deduced between a formally singular probability statement ... say 'the probability that the throw k will be five equals $^1/_6$', and one of the following two statements: ... 'the throw is in fact a five', or ... 'the throw is in fact not a five'".[75]

[206]This is of course completely correct reasoning *provided* the elements of the collectives with which we are dealing in the quantum theory are all in a state that is well defined from the classical point of view, i.e., provided we already know what kinds of entities are to be counted as the

---

[72] *Logic of Scientific Discovery*, New York, 1959, p. 227. The book is a translation of the *Logik der Forschung* of 1935.
[73] *Op. cit.*, p. 224.   [74] Von Neumann, *op. cit.*, p. 300.   [75] Popper, *op. cit.*, pp. 228f.

elements of the collectives. Only if it is assumed that these elements are systems which are in classically well-defined states, is it possible to derive from the statistical character of the quantum theory the interpretation which Popper wants to defend. However it is also evident that the statistical character of the theory *taken by itself* is never sufficient for deriving such an assumption. For from the fact that a theory is statistical we can only infer that it works with collectives of objects, events, processes, rather than with these objects, events, processes themselves. We cannot draw any inference about the *individual properties* of these objects, events, processes. Any such information would have to be given *in addition* to the laws characterizing the relation between the frequencies in the collectives considered. Does Born's interpretation which, after all, establishes a connection between the formalism and results of measurement provide such information? It does not, or at least it should not if proper care is taken. For Born's rules, properly interpreted, do not make any assertion about the character of the elements of the quantum mechanical collectives. They only make assertions concerning the expectation values which these elements *exhibit on measurement*. That is, they leave still open at least two alternatives: (1) the elements possess their values before the measurement takes place and retain them throughout the measurement; (2) the elements do not possess their values before the measurement but are transformed, by the measurement, into a state containing these values in a well-defined manner. However great the empirical success of the statistical interpretation, it does not provide any means for deciding between (1) and (2). And this is a well-known characteristic of all statistical theories: even death statistics do not allow us to draw any conclusion concerning the manner in which death occurred, nor do they allow us to infer that human beings are entities whose traits are independent of observation, i.e., which can be assumed to be either alive or dead *independently* of the occasions on which they were *found* to be either alive or dead. In the case of human beings we possess of course independent evidence about their permanence, in a well-defined state between the moments of observation. The point is that this information is independent of the fact that in death statistics we are dealing with a statistical theory.

[207]The same is true for the quantum theory: Popper's idea that an elementary particle always possesses a well-defined value of all the magnitudes that can be measured in it does not follow from Born's interpretation. It is an additional idea that must be justified, or at least discussed. It is precisely this idea that has been found to be inconsistent with the dual nature of light and matter and the individual validity of the conservation

laws and that therefore had to be replaced, long ago, by the assumption of the indefiniteness of state descriptions.[76] To sum up: Popper's argument is invalid and its conclusion is false.

This being the case we must now look for a new interpretation of the elements of the quantum mechanical collectives. We must admit that what is being counted is not the number of systems *possessing* a certain well-defined property (for example a precise value of a classical magnitude): what is counted is rather the number of *transitions*, on measurement, from certain ill-defined states into other ill-defined states (viz. states in which the mentioned property possesses a well-defined value). And neither the frequency approach which deals with relations between collectives only and is neutral with respect to either the first or the second choice of elements; nor Born's rules which deal with results of measurement without allowing any inferences as regards the behavior that led to these results, can therefore be used as an argument against the customary (i.e., Bohr's) interpretation of formula (1).

It is worth pointing out, in this connection, that Born's rules cannot be used as an argument for this interpretation, either. More specifically, it is impossible to derive Bohr's hypothesis of indefinite state descriptions from the formalism of the wave mechanics plus the Born interpretation. Any such attempt would again involve an illegitimate transition from the properties of collectives of systems and their appearance at the end of measurements to the individual properties of these systems themselves and their behavior before and after the process of measurement. It is for this reason that von Neumann's famous "proof"[77] cannot succeed. A brief look at this proof shows at once where it fails.[78] The essence of the proof consists in the derivation of two theorems, the first of which asserts that no ensemble of quantum mechanical systems is dispersion free, whereas the second points out that the ensembles of minimum dispersion, the so-called homogeneous, or pure ensembles, are such that

---

[76] Cf. our argument in the last section and footnote 55. Popper quite correctly observes that the Copenhagen Interpretation goes beyond the assertion of the validity of Born's rules. He is also correct when pointing out that physicists like Heisenberg attempted to base their stronger assertion upon a positivistic theory of meaning that was far from satisfactory. However, he seems to have overlooked the fact that the stronger interpretation of the uncertainty relations, stronger, that is, than the interpretation he would have been prepared to accept, can be defended by purely physical arguments.

[77] Von Neumann, *op. cit.*, IV/2.

[78] Cf. also my paper "Eine Bemerkung zum Neumannschen Beweis", *Zeitschrift für Physik*, Vol. 145 (1956), pp. 421–423 {translated as Chapter 4 of the present collection}, where I used suggestions first made by Popper and Agassi.

any one of its (finite) subensembles possesses the same statistical properties as it itself. Now it must be realized that the subensembles considered are restricted to those that can be described with the help of a statistical operator and which therefore, in the case of a pure state, have been prepared in [208]exactly the same manner as the ensemble of which they all form a part. Applied to pure ensembles the second theorem therefore asserts no more than the principle of the excluded gambling system which by itself does not allow any inference as to the individual events that constitute the game. Conclusion: it is neither possible, as is attempted by von Neumann, to *replace* the qualitative considerations of Section 2 by arguments drawn from the formalism of wave mechanics plus the Born interpretation; nor is it possible, as is attempted by Popper, to *reject* them on the basis of such arguments.[79] Indeed these qualitative considerations are needed in addition to Born's interpretations if a full understanding of the theory is to be achieved.

Now at this stage the adherents of the second suggestion will point out that there exist further reasons for the correctness of their interpretation. They will refer to an argument by Einstein, Podolsky, and Rosen[80] (EPR for short) according to which the formalism of wave mechanics is such that it demands the existence of exact simultaneous values of non-commuting variables. Clearly, if this should be the case, then Bohr's interpretation of formula (1) would have to be dropped and it would have to be replaced by the interpretation of Einstein and Popper. At the same time all those difficulties which prompted Bohr to invent the hypothesis of the indefiniteness of state descriptions would reappear. Even worse – it would seem that an inconsistency has been discovered in the very foundations of the quantum theory. For if it should indeed be the

---

[79] It ought to be pointed out that our discussion so far was based upon the general part of wave mechanics only, i.e., upon those laws which describe the properties of the space of the ψ-functions. It is this general theory (which alone is presupposed in Born's interpretation) with which we were concerned when discussing the views of Popper and von Neumann. However, if we add the dynamical laws (the Schrödinger equation; the conservation laws) to the general theory, then we obtain very strong arguments for the necessity of an interpretation as it was outlined in Section 2. This formal feature of the theory mirrors the fact to which we have repeatedly pointed that the indefiniteness of state descriptions is necessary in order to account for the individual conservation of energy and momentum. Von Neumann's arguments are not completely useless, though. They may be regarded as proof to the effect that the general theory is *consistent* with Bohr's interpretation (this has been pointed out to me by H. J. Groenewold).

[80] "Can quantum mechanical description of Physical Reality be Considered Complete?", *Phys. Review*, Vol. 47 (1935), pp. 777ff. In the next section the [268]argument will be presented in a generalized form.

case that the only way of combining duality, the quantum postulate, and the conservation laws consists in assuming indefiniteness of state descriptions, then the case of Einstein, Podolsky, and Rosen would show that wave mechanics is intrinsically unable to allow for a coherent account embracing these three experimental facts. We now turn to a closer analysis of the argument of Einstein, Podolsky, and Rosen.

4. THE ARGUMENT OF EINSTEIN, PODOLSKY, AND ROSEN

Assume that a system S (coordinates q, q', q'', ..., q''; r, r', r'', ..., r'''; or (qr) for short) which is in the state Φ (qr) has been (either mentally or by physical separation) divided into two subsystems, S'(q) and S''(r). It can be shown[81] that it is always possible to select a pair of observables α(q), and β(r) (corresponding to sets of mutually orthogonal situations in S' and S'' respectively) with sets of eigenstates $[|a_i(q)>]$ such that

$$|\Phi(qr)>=\sum_i c_i|a_i>|\beta_i> \qquad (2)$$

[209]A pair of observables with the property just mentioned will be called *correlative* with respect to Φ, S' and S''.

The special case discussed by Einstein, Podolsky, and Rosen is characterized by the following three conditions:[82]

[i]: there exists more than one pair of observables which are correlative with respect to Φ, S' and S''.

[ii] : assume that $(\alpha\beta)$; $(\gamma\delta)$; $(\varepsilon\xi)$; ... are pairs of observables which are correlative with respect to Φ, S' and S''; then there is at least one pair of pairs, say $(\alpha\beta)$, and $(\gamma\delta)$, such that

$$\alpha\gamma \neq \gamma\alpha \cdot \beta\delta \neq \delta\beta \qquad (3)$$

[i] and [ii] are satisfied if and only if the constants c_i in (2) satisfy the conditions

$$|c_i|^2 = \text{const.} \qquad (4)$$

[81] Von Neumann, *op. cit.*, pp. 429ff.

[82] For proof cf. Schrödinger, *Proc. Camb. Soc.*, Vol. 31 (1935), pp. 555ff; Vol. 32 (1936), pp. 446ff as well as H. J. Groenewold, *Physica*, Vol. 12 (1948).

[iii] : the systems S'(q) and S''(r), i.e., the spatial regions defined by the (q) and (r), are separated such that no physical interaction is possible between them.[83]

From [i] it follows that, if and are the eigenfunctions of γ and δ in S' and S'' respectively, then Φ may also be represented in the form:

$$|\Phi(qr)>= \sum_i c'_i|\gamma_i>|\delta_i > \qquad (5)$$

Now assume that the magnitude corresponding to α is measured in S' with the result α_k. From 2 as well as from the assumption

that any measurement leaves the system in an eigenstate of the variable measured
$$(6)$$

it follows that the state of S'' after the measurement will be the state $|\beta_k >$, i.e., it will be an eigenstate of β.

Assume, on the other hand, that γ is measured in S' with the result $|\gamma_l >$. It then follows from (5) and (6) that the state of S'' after the measurement will be $|\delta_l >$, i.e., it will be an eigenstate of S.

Adding [iii] we may now derive the following result. First step: if S' I and S'' are separated in space such that no physical interaction is possible, then it is impossible that a physical change of S' will influence the state of S''. More especially, it is impossible that a measurement in S' (which leads to a physical change of S') will influence the state of S''. More especially still: if there is a measurement in S' whose result allows us to

---

[83] This condition suggests choosing the α, β, γ, δ in such a manner that they all commute both with the (q) and with the (r). An example which possesses this property has been described by D. Bohm in his *Quantum Theory*, Princeton, 1951, Ch. 22. If such a choice is not made then it is always possible to assail the argument with reasons that have to do only with the specific form in which it is presented. For such an unintentionally irrelevant attack, cf. de Broglie, *Une Tentative d'Interprétation Causale et non Linaire de la Mécanique Ondulatoire*, Paris, 1956, pp. 76ff. de Broglie discusses Bohr's example (*Phys. Rev.*, Vol. 48 (1935), pp. 696ff) of a pair of particles with known $q_1 - q_2$ and $p_1 + p_2$ respectively. Now the state just described (which satisfies [i] and [ii] above, q_1 and p_1 being the relevant variables) is realized only as long as the pair dwells near the two-slit-screen whose slit-distance defines $q_1 - q_2$. However, in this case [iii] will be violated. The argument is of course irrelevant as Einstein's point is independent of the way in which the state satisfying [i] and [ii] has been created. It can also be shown (cf. Schrödinger, "Die gegenwärtige Lage in der Quantenmechanik", *Naturwissenschaften*, 1935) that t seconds after the pair has left the slit Einstein's argument can be raised with respect to the variables p_1 and $q_1 - (P_1/m)t$. Still, it is an advantage to possess an example that cannot be criticized for the reasons just mentioned. Also Bohm's example is not beset by the difficulties of the continuous case (for these difficulties cf. Bohm and Aharonov, *Phys. Rev.*, Vol. 108 (1957), pp. 1070ff, Appendix).

infer that S″ is in a particular physical state, then, as no interaction can have taken place, we have to assume that S″ was in the inferred state already *before* the performance of the measurement, and would have been in this state even if the measurement had never been carried out. Second step: applying the above result to the measurement of α and γ we may [210]reason as follows: immediately after the measurement of α which yields the result α_k, S″ is in the state $|\beta_k>$, and this is, according to what has been said above, also the state of S″ immediately before the measurement. Hence, S″ must be in some eigenstate of β whether or not a measurement of α has been performed (we assume, of course, that all observables are complete commuting sets). By the same argument it must also be in some eigenstate of δ, whether or not a measurement of γ has been performed. In short, it must always be in a classically well-defined state. From (3) and Bohr's hypothesis it follows that this is impossible. This contradiction between Bohr's assumption of the indefiniteness of state descriptions and the above argument has been called the paradox of Einstein, Podolsky, and Rosen. It is preferable, however, not to talk of a paradox, but rather of argument.[84]

For Einstein (and Popper) this argument refutes Bohr's hypothesis. Having presented it, Einstein therefore feels justified to regard the quantum theory "as an incomplete and indirect description of reality" such that "the ψ-function ... relates ... to an 'ensemble of systems' in the sense of statistical mechanics".[85] I shall soon deal with the merits of this argument. But before doing so I wish to discuss some of the objections

---

[84] The paradoxical aspect arises as soon as the argument is combined with the completeness assumption. In this case we obtain the result that changes of state may occur which are (1) well predictable (although their exact outcome is not predictable); and which (2) occur in places which are very far from the reach of physical forces. It is this feature of the argument that among other things has led to the assumption of a sub-quantum mechanical level involving laws different from the laws of the quantum theory which allow for the occurrence of coordinated fluctuations. Such an assumption would of course lead to predictions that in some respect are different from the predictions of the present theory. For example, it would imply a disturbance of the correlations if the measuring apparatus interacting with the first system is turned around very rapidly. Such a difference of predicted results is far from undesirable, however. It is a guide to new experiments which then will be able to decide as to which point of view should be adopted in the end. Cf. D. Bohm, *Causality* [269]*and Chance in Modern Physics*, Routlege and Kegan Paul, London, Chs. III, IV as well as *Observation and Interpretation*, ed. Körner, London, 1957, pp. 33–40, 86–87. For a more general discussion concerning the relation of theories cf. my essay "Explanation, Reduction, and Empiricism" in the *Minnesota Studies for the Philosophy of Science*, Vol. III, Minneapolis, 1962 {reprinted in *PP1*, ch. 4}.

[85] Cf. footnote 71 as well as *Albert Einstein, Philosopher-Scientist*, ed. Schilpp, Evanston, Illinois, 1948, pp. 666ff.

which have been raised against it and all of which to me seem to be unsatisfactory.

Thus A. D. Alexandrov[86] and J. L. R. Cooper[87] have argued that the paradox cannot arise; for as long as [S'S''] is described by a wave function which cannot be broken up into a product, [iii] does not hold. To this it must be replied that within [iii] "separated", and "interaction" refer to *classical fields,* or at least to fields which contribute to the energy present in a certain space-time domain. Alexandrov and Cooper seem to assume that the ψ-field can be interpreted as such a field which is not borne out by the facts.

A much more typical counter argument is contained in a paper by Furry.[88] This paper has the advantage of being clear and straightforward and it also seems to reflect the attitude of a good many physicists. The physical process (interaction between S' and S'') which terminates in the paradoxical combined state is construed by Furry in two ways, A and B. According to assumption A this process leads to transitions in the quantum mechanical states of S' and S''. The transitions occur according to the laws of probability, but they terminate in well-defined states for both S' and S'', which are related to each other in such a way that a measurement in S' will indeed lead to *certain* information about the state of S'' (and vice versa). The correlation is between *states which are* al[211]*ready there* or, to put it formally, it is due to the fact that the state of S' + S'' after the interaction is given by the mixture

$$P_{\Phi'} = \sum_i |c_i|^2 |\alpha_i\rangle \langle \alpha_i | \beta_i\rangle \langle \beta_i| \tag{7}$$

We see that in this mixture the pair $(\alpha\beta)$ is such that if α is measured in S' with the result α_k, then β will be certain to exhibit, on measurement, the value β_k.

$$[\text{Proof}: Tr\{P_{|\alpha_k\rangle} P_{\Phi'} P_{|\alpha_k\rangle} P_{|\beta_l\rangle}\}/Tr\{P_{|\alpha_k\rangle} P_{\Phi'}\} = \delta_{kl}]$$

Another feature of this case is that it leads to the violation of some conservation laws.[89] Furry alleges that A is the assumption made by Einstein.

---

[86] *Proc. Acad. UdSSR,* Vol. 84 (1952), p. 253.

[87] *Proc. Cambr. Phil. Soc.,* Vol. 46 pt. 4 (1951), p. 620.

[88] *Phys. Rev.,* Vol. 49 (1936), pp. 397ff; cf. also Groenewold *loc. cit.*; M.H.L. Pryce, too, has used the argument in private discussions.

[89] Thus in Bohm's example of footnote 83 angular momentum will not be conserved.

Method B which, he says, is the method adopted by the quantum theory and which also preserves the conservation laws, implies that an interaction between two systems, S' and S'', will in general lead to a pure state

$$|\Phi\rangle = \sum_{ik} c_{ik}|\alpha_i\rangle|\beta_k\rangle \tag{8}$$

and that correlations between the values of α and β will occur if and only if $c_{ik} = c_i\delta_{ik}$, or if Φ has form (2). The difference between method A and method B, therefore, consists in the fact that method A represents a state with the above properties by (7) whereas method B represents it by (2). It can easily be shown that the two methods of representation do not lead to the same observational results.

$$\left[\text{Proof} : \text{ for } [\alpha\gamma]; [\beta\delta] \neq 0, \text{ in general} \right.$$

$$\left. Tr\{P_{|\gamma_l\rangle}P_\Phi P_{|\gamma_l\rangle}\delta\} \neq Tr\{P_{|\gamma_l\rangle}P_{\Phi'|\gamma_l\rangle}\delta\} \right]$$

Method A is therefore inconsistent with the method adopted by the quantum theory, which latter is a direct consequence of the superposition principle. This many physicists regard as a refutation of Einstein's idea (which was silently applied in the first step of the derivation of the paradox) that "if without in any way disturbing a system we can predict with certainty the value of a physical quantity, then there exists an *element of physical reality* corresponding to this physical quantity",[90] i.e., then this value belongs to the system independently of whether or not a measurement has been, or can be, carried out.

Now, concerning this argument we must consider two points. To start with, the fact that A differs from the method which is adopted by the quantum theory can be regarded as detrimental only if it has first been shown that in the special case dealt with by EPR the predictions of the quantum theory are empirically successful whereas the predictions implied by method A are not. The fact that the turn quantum theory has been confirmed by a good many experimental results is of no avail in this case, [212]for we are now asking for the behavior of systems under conditions which have not yet been tested.[91] And it was indeed Einstein's

[90] *Phys. Rev.*, Vol. 47 (1935), p. 777.

[91] For a more detailed account of the problems of confirmation that arise in this case and for a criticism of the (implied) attitude of the orthodox cf. section 7 of my paper referred to in fn. 84. {The original text lacks a reference to this footnote.}

guess that the current formulation of the many-body problem in quantum mechanics might break down when particles are far enough apart.[92] The second point is that even if B were to give the correct statistical predictions, even then Einstein's argument would stand unassailed. For it is not an argument for a particular description, *in terms of statistical operators*, of the state of systems which are far apart. It is rather an argument against the assumption that any such description in terms of statistical operators can be regarded as complete, or against the assumption that "the ψ-function is ... unambiguously coordinated to the physical state",[93] or against the assumption that under all circumstances a physical system will be completely and exhaustively described by its statistical operator.

In the literature this last assumption has become known as the *completeness assumption*. As has been indicated in the preceding paragraph the completeness assumption implies Bohr's hypothesis. One may therefore interpret the argument of EPR both as an argument against the completeness assumption and as an argument against Bohr's hypothesis. In what follows both these interpretations will be used.

As regards the first point we may additionally remark that it has been attempted to decide experimentally between assumption A and assumption B. The experiment was carried out by C. S. Wu[94] and it consisted in studying the polarization properties of correlated photons. Such photons are produced in the annihilation radiation of positron-electron pairs. In this case each photon is always emitted in a state of polarization orthogonal to that of the other, which is similar to Bohm's example of EPR.[95] The result of the experiment was a definite refutation of the prediction made on the basis of method A and a confirmation of B, the method that is in accordance with the general validity of the superposition principle. However, as we have shown above, this result cannot be used for refuting the contention, which is the core of Einstein's argument, that what is realized in *both* cases is an ensemble of classically well-defined systems rather than a single case.[96]

---

[92] Cf. Bohm-Aharonov, *loc. cit.*, p. 1071 as well as *Observation and Interpretation*, pp. 86ff.

[93] Einstein, "Physics and Reality", *loc. cit.*, p. 317.

[94] *Phys. Rev.*, Vol. 77 (1950), pp. 136ff. That the case is equivalent to the one discussed by EPR is shown in Heitler, *Quantum Theory of Radiation*, p. 269. Cf. also the analysis by Bohm-Aharonov, *loc. cit.*

[95] For proof of equivalence with the case of EPR cf. the last footnote.

[96] A similar position is held by D. R. Inglis, *Revs. of Modern Physics*, Vol. 33 (1961), pp. 1–7, especially the last section.

Similar remarks apply to the analysis given by Blochinzev.[97] It is true, Blochinzev emphasizes that "within quantum theory we do not describe the 'state in itself' of the particle, but rather its relation to the one, or the other (mixed or pure) collective".[98] He points out that this relation is of a completely objective character and that any measurement is to be regarded as a process which separates certain subcollectives from the collective in which they were originally embedded. Now, the assertion that [213]the quantum theory describes the elementary particles only insofar as they are elements of a collective would seem to make Blochinzev's interpretation coincide with the interpretation of Einstein, Slater, and Popper which also asserts that the present quantum theory is a theory of collectives. However, the important difference is that for Blochinzev the individual particle does not possess any state-property over and above its membership in a particular collective. "Our experiments", he writes,[99] "are precise enough to show us that the pair (p,q) of a single particle does not exist in nature". This means, of course, that Blochinzev, too, accepts the completeness assumption which is just the point at issue. Again, he has not shown that Einstein's argument contradicts the quantum theory; or that it contradicts "experience"; he has only shown that it contradicts the completeness assumption, i.e., he has shown that it serves the purpose for which it was constructed.

We see that all the arguments against EPR which we have discussed so far fail to refute it.[100] Apart from those suggestions which consist in introducing a sub-quantum mechanical level, and which thereby deny the *correctness* and not only the completeness of the present theory, the only argument left seems to be the one that led to Bohr's hypothesis in the first place and which is not dependent on the more detailed formal features of either the elementary quantum theory, or of any one of its improved alternatives (Dirac's theory of the electron; field theories). However with respect to these qualitative considerations many writers seem to be of the opinion that they count little when compared with the impressive utilization, by EPR, of the powerful apparatus of wave mechanics. In order to dispel this impression (which seems to reflect an overconfident and uncritical attitude with respect to mathematical formalisms) I shall now discuss the difficulties which arise when the

---

[97] *Sowjetwissenschaft*, Naturwissenschaftliche Reihe, Vol. VI (1954), pp. 545ff. *Grundlagen der Quantenmechanik*, Berlin, 1953, pp. 497–505.
[98] *Sowjetwissenschaft*, etc., p. 564.   [99] *Grundlagen*, p. 50.
[100] This is admitted, implicitly, by D. R. Inglis, *loc. cit.*

conclusion of EPR is assumed to be correct, i.e., when it is assumed both that the elementary theory is correct and that a more detailed description of state can be given than is admitted by the completeness assumption. A state referred to by such a more detailed description will be called a *superstate*.[101]

5. SUPERSTATES

Given the laws of the quantum theory,[102] are superstates possible? In the present section I shall develop some arguments (i.e., additional to the qualitative arguments of Section 2) against the possibility of superstates. The *first argument* will show that, dependent on the way in which they have been introduced, superstates either contain redundant elements, or are empirically inaccessible. To show that they contain redundant parts [214]a few explanations must be given concerning the role of a state in physical theory.

What is the role of a dynamical state in a physical theory? The reply is that it contains part of the initial conditions which, if taken together with other initial conditions such as mass and charge, with boundary conditions (properties of the acting fields), as well as with some theories, will serve for the explanation, or the prediction, of the behavior of the systems which it is supposed to describe.[103] According to this definition an element of state will be superfluous, if it does not play any role in any prediction and explanation. It will be even more superfluous if this applies not only to the *future* properties and behavior of the system, but also to the properties and the behavior it possesses at the very moment at which the occurrence of this element is being asserted. In this case there exists no possibility whatever to test the assertion that the element has occurred and this element may properly be called *descriptively redundant*. To show that a superstate will contain descriptively redundant elements let us consider the case of a particle with total spin σ. Assume that σ and σ_x have been measured and the values σ' and σ'_x obtained. Then, according to the theory an immediate repetition of the measurement will again give these values. On the other hand, assume that o is measured when the

---

[101] The first to use this expression seems to have been H. J. Groenewold, *loc. cit.*

[102] I would like to repeat that in this section and in the following sections it will be assumed that the elementary theory is essentially correct. Interpretations which dispute the absolute correctness of the elementary theory will be discussed later.

[103] Cf. Popper, *op. cit.*, sec. 12.

measurement of σ and σ_x has been completed. Then the formalism of the theory will inform us that *any* value of σ_y may be obtained which shows quite clearly that adding a specific value of σ_y (and, for that matter of σ_x) to the set $[\sigma'\ \sigma'_x]$ does not in the least change the informative content of our assertion. By generalization we obtain the following result: a set of magnitudes specifying the outcome of the measurement of a complete set of commuting observables has maximum informative content. Any addition to this set is descriptively redundant, *whatever the method* (EPR or other) by means of which it has been obtained, provided, of course, that this method did not involve a disturbance of the state already realized.

The superstates we have been discussing so far had the property that only part of them could be used for deriving information about the actual state of the physical system: i.e., assuming P and Q to be two different complete commuting sets pertaining to the same system, the superstate [PQ] was chosen in such a manner that and $\mathrm{Prob}(Q/[PQ]) = \mathrm{const}. \neq 1$, or the other way round. We may now want to define pair [PQ] of new variables in the following manner.

$$\mathrm{Prob}(P/[PQ]) = \mathrm{Prob}(Q/[PQ]) = \mathrm{Prob}(P/[QP]) = \mathrm{Prob}(Q/[QP]) = 1$$
$$(9)$$

[215]Let us now investigate what is the consequence of such a definition (which has been adopted, implicitly, by D. Bohm[104]).

To start with it should be noted that the first and the last equation of (9) are part of the *definition* of a superstate of this new kind, whereas in the case of the P, Q these equations follow from the quantum theory. Note further that we are here dealing with a minimum condition which is trivially satisfied in the classical case; the condition is that a series of statements describing a superstate of the kind discussed here should be such that it allows for the derivation of the value of any one of the elements of the superstate. We have not yet considered any dynamical law, such as perhaps a law governing the temporal development of the superstates or of functions of the superstate. Nor has the existence of dynamical laws (and of this specific case of dynamical laws, the conservation laws) been postulated. But it is clear that if (9) is not satisfied, then deterministic laws will not be possible. Thus the conditions (9) are a necessary presupposition of determinism in the quantum theory (and in any other theory). Let us now see where these conditions are going to lead us.

---

[104] Cf. the reference in footnote 15.

From (9) we can at once derive that

$$\text{Prob}(P/[QP]) = \text{Prob}(P \cdot [QP])/\text{Prob}\,[QP] = \text{Prob}([QP]/P)/\text{Prob}([QP]) \cdot \text{Prob}\,(P)$$
$$\text{Prob}(P/[PQ]) = \text{Prob}(P \cdot [PQ])/\text{Prob}\,[PQ] = \text{Prob}([PQ]/P)/\text{Prob}([PQ]) \cdot \text{Prob}\,(P)$$

hence $\text{Prob}([QP]/P)/\text{Prob}([PQ]/P) = \text{Prob}([QP])/\text{Prob}\,([PQ])$.

If we now postulate that the absolute probabilities of the superstates be independent of the order of their elements (and we indicate adherence to this postulate by now writing "PQ" instead of "[PQ]"), then we obtain [from (9) for any pair PQ satisfying (9) above, i.e.,

the elements of the newly introduced superstates are statistically independent

$$\tag{10}$$

Now let us assume, in order to provide this abstract scheme with some empirical content, that

$$P \leftrightarrow p \tag{11}$$

where p is a complete set of commuting observables in the sense of the quantum theory. We shall also assume that the systems discussed are fully described by the complementary sets p and q.

On the basis of Bayes' theorem we obtain,

$\text{Prob}\,(q/PQ) = \text{Prob}\,(P/qQ) \cdot \text{Prob}\,(q/Q)/\text{Prob}\,(P/Q)$, which leads to the following value for $\text{Prob}\,(q/Q)$

$\text{Prob}\,(q/Q) = \text{Prob}\,(P/Q) \cdot \text{Prob}\,(q/PQ)/\text{Prob}\,(P/q)$ which is, by virtue of (10) equal to $\text{Prob}\,(P) \cdot \text{Prob}\,(q/PQ)/\text{Prob}\,(P/q) =$ by virtue of (11) $=$ $\text{Prob}\,(P) \cdot \text{Prob}\,(q/pQ)/\text{Prob}\,(p/q)$

Now we have from the quantum theory that
$\text{Prob}\,(p/q) = \text{Prob}\,(p/q) = |< p|q|^2$, therefore

$$\text{Prob}\,(q/Q) = \text{Prob}\,(P). \tag{12}$$

In a completely analogous manner
$\text{Prob}\,(p/PQ) = \text{Prob}\,(P/pQ) \cdot \text{Prob}\,(q/Q)/\text{Prob}\,(P/Q)$ leads to

$$\text{Prob}\,(p/Q) = \text{Prob}\,(P). \tag{13}$$

Now $\text{Prob}\,(qQ) = \text{Prob}\,(P) \cdot \text{Prob}\,(Q) = \text{Prob}\,(PQ)$
$\text{Prob}\,(pQ) = \text{Prob}\,(P) \cdot \text{Prob}\,(Q) = \text{Prob}\,(PQ)$, hence

$$\text{Prob}\,(p) = \text{Prob}\,(q). \tag{14}$$

Finally, $\text{Prob}\,(p) = [\text{by virtue of (11)}] = \text{Prob}\,(P) = \text{Prob}\,(q/Q) = [\text{using}(14)] = \text{Prob}\,(q)$, i.e., *q and Q* (and, as can be easily shown, also p and Q) *are statistically independent.* Result: if we assume that there exist superstates which satisfy conditions (9); and if we also assume that one of the elements of these superstates is accessible to experimental investigation as it is provided by quantum mechanical measurement, *then the rest of the superstate will be statistically independent of any physical magnitude that can be measured in the system under consideration and cannot therefore be said to possess any empirical or even any ontological content.* Again it emerges that the maximum of information producible about a quantum mechanical system is given by the assertion that one of its complete sets of commuting observables possesses a certain value. Any additional assertion is arbitrary and not accessible to independent experimental test. Adopting Bohr's hypothesis of indefinite state descriptions we can easily explain this fact by pointing out that this inaccessibility is not due to the intricacies of the measuring process which forbid us to obtain more detailed information about nature, but rather to the *absence of more detailed features of nature itself.* That this is so has been explained in detail in the Sections 2 and 3.

Now at this stage one might still be inclined to say, in opposition to Bohr's hypothesis, that the use of superfluous information, although not very elegant, and certainly metaphysical, can at most be rejected on the basis of considerations of "taste".[105] That this is not so is shown by the *second argument* against the admissibility of superstates which I am now going to discuss. According to this second argument, which is but a more detailed repetition of the arguments used in Section 2 of the present paper, the use of superstates is incompatible with the conservation laws [217]and with the dynamical laws in general. In the case of the energy principle this becomes evident from the fact that for the single electron $E = p^2/2m$, so that after a measurement of position any value of the energy may emerge and, after a repetition of the measurement, any different value. One may try to escape this conclusion, as Popper apparently has been inclined to,[106] by declaring that the energy principle is only statistically valid. However, it is very difficult to reconcile this hypothesis with the many independent experiments (spectral lines; experiment of Franck and Hertz; experiment of Bothe and Geiger) which show that energy is conserved also in the quantum theory and this not only on the average, *but for any*

---

[105] This is Heisenberg's attitude. Cf. his *Physical Principles*, etc., p. 15.

[270][106] I am here referring to discussions I had with Professor Popper. The responsibility for the presentation is, however, entirely mine.

single process of interaction. The difficulty of the point of view of Einstein and Popper, which works with superstates of the first kind, i.e., with superstates which do not satisfy (9) and whose elements are determined by successive observations, becomes very obvious if we apply it to the case which is known as the *penetration of the potential barrier.* In this case we may obtain rather drastic disagreement with the principle of the conservation of energy[107] if we assert it in the form that for any superstate [Eqp] as determined by three successive measurements E will satisfy the equation $E = p^2/2m + V(q)$.[108] If we add these difficulties to the arguments leading up to Bohr's hypothesis of indefinite state description in the first place we obtain very powerful reasons indeed to the effect that the conclusion of EPR cannot possibly be correct.[109] This makes it imperative to show how the argument can be made compatible with that hypothesis. An attempt in this direction and, to my mind, a quite satisfactory attempt, has been made by Bohr.[110]

6. THE RELATIONAL CHARACTER OF THE QUANTUM MECHANICAL STATES

If I understand Bohr correctly,[111] he asserts that the logic of a quantum mechanical state is not as is supposed by EPR. EPR seems to assume that what we determine when all interference has been eliminated is a *property* of the system investigated. As opposed to this Bohr maintains that all state descriptions of quantum mechanical systems are *relations* between the

---

[107] In conversation, Landé has expressed the hope that further development of the point of view discussed here will lead to a satisfactory account of the case of the penetration of a potential barrier. This is quite possible. However what I am concerned with here is to show the strength of the Copenhagen Interpretation to those who are of the opinion that the transition to a different interpretation is more or less a matter of philosophical taste rather than of physical inquiry.

[108] For a numerical evaluation cf. Blochinzev, *Grundlagen*, p. 505. Cf. also Heisenberg, *op. cit.*, pp. 30ff.

[109] It ought to be mentioned that Bohm (*loc. cit.*, in footnote 15) has shown how superstates which obey conditions (9) can be made compatible with the dynamical laws. However the unsatisfactory feature remains that these superstates violate the principle of independent testability and that their introduction must therefore be regarded as a purely verbal manoeuvre. Yet it is important to repeat that von Neumann (*op. cit.*, p. 326) thought that his "proof" would be strong enough to exclude even such verbal manoeuvres.

[110] Einstein, too, regards Bohr's attempt as coming "nearest to doing justice to the problem". *Albert Einstein, Philosopher-Scientist*, p. 681.

[111] *Phys. Rev.*, Vol. 48 (1936), pp. 696ff. Cf. also D. R. Inglis, *loc. cit.*

systems and measuring devices in action and are therefore dependent upon the existence of other systems suitable for carrying out the measurement. It is easily seen how this second basic postulate of Bohr's point of view makes indefiniteness of state description compatible with EPR. For while a property cannot be changed except by *interference* with the system that possessed that property, a relation can be changed without such interference. Thus the state "being longer than *b*" of a rubber band [218]may change when we compress the rubber band, i.e., when we physically interfere with it. But it may also change when we change *b* without at all interfering with the rubber band. Hence, lack of physical interference excludes changes of state only if it has already been established that positions and momenta and other magnitudes are properties of systems, rather than relations between them and suitable measuring devices. "Of course", writes Bohr, referring to Einstein's example[112] "there is in a case like the one ... considered no question of a mechanical disturbance of the system under investigation.... But even at this stage there is essentially the question of *an influence on the very conditions which define the possible types of prediction regarding the future behavior* of the system", and he compares this influence with "the dependence on the reference system, in relativity theory, of all readings of scales and clocks".[113] I would like to repeat, at this stage, that Bohr's argument is not supposed to prove that quantum mechanical states are relational and indeterminate; it is only supposed to show under what conditions the indefiniteness assumption *which is assumed to have been established by independent arguments* can be made compatible with the case of Einstein, Podolsky, and Rosen. If this is overlooked one may easily get the impression that the argument is either circular, or *ad hoc*. That the argument is circular has been asserted by Professor H. Putnam.[114] In the case of relativity, says Putnam, we may set up two different reference systems and obtain *simultaneously* two different readings for the *same* physical system. This is not possible in the quantum theory, for it would presuppose, what is denied by Bohr, that we can make simultaneous measurements of position and momentum in S′ (see the discussion in Section 4), or that we can even *imagine* that position and momentum both possess definite values in S′. However this appearance of circularity disappears as soon as we realize that the hypothesis of indefiniteness of state descriptions is *presupposed* and that a way is sought to make it compatible with

---

[112] *Loc. cit.*   [113] *Loc. cit.*, p. 704.   [114] Private communication.

EPR. A similar remark applies to Popper's criticism[115] that the argument is *ad hoc*. One must as it were approach the argument from the realization that superstates cannot be incorporated into wave mechanics without leading to inconsistencies. Once this is admitted there arises the need for a proper interpretation of the very surprising case discussed by EPR. It is in this connection that Bohr's suggestion proves so extremely helpful.[116] Finally, we ought to discuss briefly the assumption which is silently made by almost all opponents of the Copenhagen point of view, that EPR creates trouble for this point of view but not for the quantum theory (the elementary theory, that is) itself. This overlooks that there is no [218]interpretation available that gives as satisfactory an account of all the facts united by the theory as does the idea of the indefiniteness of state descriptions. If we therefore interpret EPR as fatal for this idea, then we are forced to the conclusion *that the theory itself is in trouble.* (This conclusion has been drawn, a.o., by Bohm and by Schrödinger).[117]

It is very important to realize the far-reaching consequences of Bohr's hypothesis. Within classical physics the interaction between a measuring instrument and an investigated system can be described in terms of the appropriate theory. Such a description allows for an evaluation of the effect, upon the system investigated, of the measurement, and it thereby allows us to select the best possible instrument for the purpose at hand.

---

[115] *Op. cit.*, pp. 445ff.

[116] It ought to be pointed out, however, that there is one assumption in the earlier speculations about the nature of microscopic objects which has been definitely refuted by EPR. It is the assumption that "the most important difference between quantum theory and the classical theories consists in the fact that in the case of an observation we must carefully consider the disturbance, due to experiment, of the system investigated." (Heisenberg, *Naturwissenschaften* (1929), p. 495; cf. also Bohr, *Atomic Theory*, etc., pp. 5, 11, 15, 54, 68, 93, 115; also *Dialectica* 7/8 (1948), p. 315). And it is the corresponding assumption that the indeterminacy of the state of quantum mechanical systems is essentially due to *this* disturbance (cf. Bohm-Aharonov, *Phys. Rev.*, Vol. 108 (1957), pp. 1070ff as well as K. R. Popper, *op. cit.*, pp. 445ff). What is shown by EPR is that physical operations, such as measurements, may lead to sudden changes in the state of systems which are *in no physical connection whatever* with the domain in which the measurement is being performed. Unfortunately the attitude of the adherents of the Copenhagen point of view with respect to this [271]argument had very often been that the reply which was given by Bohr (and which cost him, as is reported, some headaches) was already implicit in the earlier ideas which would mean that these ideas were much more vague than one would at first have been inclined to believe.

[117] For Bohm cf. footnote 84. According to Schrödinger the paradox is an indication of the fact that the elementary quantum theory is a non-relativistic theory. Cf. his essay "Die gegenwärtige Lage in der Quantentheorie", *Naturwissenschaften*, 1935, especially the last section.

Hence, within classical physics, the classification of the measuring instruments is achieved, at least partly, by the theory that is being investigated. Now according to Bohr a quantum mechanical state is a relation between (microscopic) systems and (macroscopic) devices. Also a system does not possess any properties over and above those that are derivable from its state description (this is the completeness assumption). This being the case it is not possible, even conceptually, to speak of an *interaction* between the measuring instrument and the system investigated. The logical error committed by such a manner of speaking would be similar to the error committed by a person who wanted to explain changes of velocity of an object created by the transition to a different reference system as the result of an interaction between the object and the reference system. This has been made very clear by Bohr ever since the publication of EPR which refutes the earlier picture[118] where a measurement glues together, with the help of an indivisible quantum of action, *two different* entities, viz. the apparatus on the one hand and the investigated object on the other.[119] But if we cannot separate the microsystem from its relation to a classical apparatus, then the evaluation of a measuring instrument will have to be very different from the way in which such evaluation took place in classical physics; that is, it will no longer be possible to refer to the type of *interaction* occurring as a means of classifying measurements.[120]

---

[118] Cf. footnote 116.

[119] This is sometimes obscured by the fact that Bohr's account of measurement is not the only one. Very often physicists rely on a simplified version of von Neumann's theory where the relation between the measuring instrument and the system under investigation is indeed treated as an interaction (this theory will be discussed later in the present paper, especially in Sections 10 and 11), or else they use a theory of measurement similar to the one explained by Bohm (*Quantum Theory*, Princeton, 1951, Ch. XXII) which is also a theory of interaction. Heisenberg himself had treated measurement as interactions from the very beginning and he had also pointed to the fact, which is proved in von Neumann's theory, that the "cut" ("Schnitt") between the object and the measuring device can be shifted in an arbitrary manner. Very often such more formal accounts have been regarded as elaborations of Bohr's own point of view. This is not the case. Bohr's theory of measurement and von Neumann's theory (or any other theory that treats measurement as an interaction) are *two entirely different theories*. As will be shown later von Neumann's theory encounters difficulties which do not appear in Bohr's account. Bohr himself does not agree with von Neumann's account (private communication, Ascov 1949). A formal theory which is very close to Bohr's own point of view has been developed by Groenewold. Cf. his essay in *Observation and Interpretation*, pp. 196–203.

[120] Cf. fn. 84 as well as Section 4 of my paper "Complementarity", *Proc. Arist. Soc.*, Suppl. Vol. XXXII (1958). {Reprinted as Chapter 5 of the present collection.}

This has led to the assertion[121] that the classification of measuring instruments that is used by the quantum theory can at most consist in giving a list without being able to justify the presence of any member in the list. Such an assertion does not seem to be correct. First of all a proper application of the correspondence principle will at once provide means of measurement for position and momentum. Speaking more abstractly we may also say that now a measurement in a system whose ψ-function is element of a Hilbert space H leads to a destruction of [220]coherence between certain subspaces H', H'', H''' – of H and can be characterized by operators P′, P″, P‴, effecting projection into exactly these subspaces. It is, of course, required to give an interpretation of the P′s – but this problem is identical with the corresponding problem in classical physics which is the interpretation of the primitive descriptive terms of the theory.[122]

We may sum up the results of the foregoing investigation in the following manner. We first presented a physical hypothesis which was introduced by Bohr in order to explain certain features of microscopic systems (for example, their wave properties). It was pointed out that this physical hypothesis is of a purely objective character and that it is also needed, *in addition to Born's rules*, for a satisfactory interpretation of the formalism of wave mechanics.[123] The argument of EPR then showed that

---

[121] This assertion has been made by Hilary Putnam.

[122] For the specific difficulties of the quantum mechanical case, cf. Section 11 of the present paper.

[123] This means, of course, that formula (1) can be derived in two entirely different manners. The first derivation is of a fairly qualitative character. It makes use of the considerations which we put forth in Section 2 and introduces the quantum of action with the help of de Broglie's formula . This derivation makes it very clear to what extent the existence of duality and the quantum of action forces us to restrict the application of such classical terms as [272]position, momentum, time, energy, and so on, and it thereby transfers some intuitive content to formula (1). The second derivation which is completely formal in character makes use of the commutation relations of the elementary theory (cf. for example H. Weyl, *Gruppentheorie und Quantenmechanik*, Berlin, 1931, pp. 68 and 345; English translation pp. 77 and 393ff). Now it is very important to realize that the result of the first and intuitive derivation *may be regarded as a test of the adequacy* of the wave mechanics and indeed of any future quantum theory. For assume that the wave mechanics would give an uncertainty that is much smaller than the one derived with the help of duality (which is a highly confirmed empirical fact), de Broglie's relation (which is also a highly confirmed empirical fact) and Bohr's assumption of the indefiniteness of state descriptions (which is the only reasonable hypothesis that allows for the incorporation of the quantum postulate and the conservation laws). This would amount *to a refutation of wave mechanics*, i.e., it would amount to the proof that the wave mechanics is not capable of giving an adequate account of duality, the quantum postulate, and the

a further assumption must be introduced in order to make Bohr's hypothesis compatible with this formalism. According to this further assumption, the state of a physical system is a relation rather than a property and it presupposes that an adequate measurement is being performed, or has been performed immediately before the statement that the state obtains. By a "measurement" is meant, in this connection, a certain type of macroscopic process – a terminological peculiarity which is rather unfortunate and which must be blamed for the many subjectivistic conclusions that have been drawn from Bohr's ideas.

Now it would be incorrect to say that the presentation of the point of view of Bohr and of his followers is completed with the presentation of the two ideas we have just explained. For as is well known it has been attempted, both by Bohr, and by some other members of the Copenhagen circle, to give greater credibility to these ideas by incorporating them into a whole philosophical (ontological) system that comprises physics, biology, psychology, sociology, and perhaps even ethics. Now the attempt to relate physical ideas to a more general background and the correlated attempt to make them intuitively plausible is by no means to be underestimated. Quite on the contrary, it is to be welcomed that these physicists undertook the arduous task to adapt also the more general notions of philosophy to two physical ideas which, as has been pointed out in Section 2, possess some very radical implications. However this philosophical backing has led to a situation that is by no means desirable.

---

conservation laws. On the other hand, the agreement between the qualitative result and the quantitative result now transfers an intuitive content to the formalism.

The fact that formula (1) can be derived in two entirely different manners and that the quantum theory combines both derivations has been realized by various thinkers. Thus Popper (*op. cit.*, p. 224) points out "that Heisenberg's formulae ... result as logical conclusions from the theory; but the *interpretation* of these formulae as rules limiting attainable precision of measurement in Heisenberg's sense, does not follow from the theory". And Kaila (*Zur Metatheorie der Quantenmechanik*, Helsinki 1950) has made the existence of various interpretations of (1) the basis of an attack against the quantum theory. Now as against Popper it must be pointed out that the interpretation in question does follow from the theory provided the theory has been interpreted in accordance with the intentions of Bohr and Heisenberg. For in this case the interpretation uses, in addition to Born's rules, also the hypothesis of the indefiniteness of state descriptions. Popper regards such an addition to Born's rules as illegitimate and as a result of positivistic inclinations. We have already shown (footnote 76) that this is incorrect and that there are physical reasons which demand indefiniteness. Unfortunately these physical reasons are almost always presented in positivistic language which creates the impression that the peculiarity of the quantum theory, i.e., the features which are ascribed to it over and above the Born interpretation are indeed due to an epistemological manoeuvre.

Above all, this philosophical "backing", like so many philosophical arguments before, has led to the belief in the uniqueness and the absolute
validity of both of Bohr's assumptions. It will, of course, be admitted that
the quantum theory will have to undergo some very decisive changes
[221]in order to be able to cope with new phenomena (the first step of
these changes is indicated by the transition from the elementary theory to
Dirac's theory of the electron; the second step by the transition to the field
theories). But it will also be pointed out that, however large these changes
may be, they will always leave untouched the two elements mentioned,
viz. the indefiniteness of state descriptions and the relational character of
the quantum mechanical states, which, so it will be added, cannot be
replaced by different ideas without leading either into formal inconsistencies, or into inconsistencies with experiment.[124] "The new conceptions" asserts L. Rosenfeld[125] "which we need" in order to cope with
new phenomena "will be obtained ... by a rational extension of the
quantum theory"[126] which *preserves* the indeterminacies; and the new
theories of the microcosm will therefore be increasingly indeterministic.
Today this dogmatic *philosophical* attitude with respect to fundamentals
seems to be fairly widespread.[127] In the remaining sections of the present

---

[124] For this sentiment cf. W. Pauli, *Dialectica*, Vol. VIII (1954), p. 124; L. Rosenfeld, *Louis
de Broglie, Physicien et Penseur*, Paris, 1953, pp. 41, 57; P. Jordan, [273]*Anschauliche
Quantentheorie*, Berlin, 1936, pp. 1, 114f, 276; G. Ludwig, *Die Grundlagen der Quantenmechanik*, Berlin, 1954, pp. 165ff.
    In the last footnote we have shown how the adequacy of the formal uncertainties, i.e.,
of the uncertainties that follow from the commutation relation can be tested by qualitative considerations concerning the dual character of elementary particles. As has been
shown by Bohr and Rosenfeld (*Dan. Mat.-Phys. Medd.*, Vol. XII (1933) Nr. 8 as well as
Bohr-Rosenfeld, *Phys. Rev.*, Vol. 78 (1950), pp. 794ff) the adequacy of the field theories
and their consistency with the required restriction of the applicability of the classical
terms can be shown in a similar manner. Cf. also L. Rosenfeld, "On Quantum Electrodynamics", in *Niels Bohr and the Development of Physics*, London, 1955, pp. 70–95 as
well as Heitler, *Quantum Theory of Radiation*, Oxford, 1957, pp. 79ff.
[125] *Observation and Interpretation*, p. 45. For the idea of a "rational extension", or a
"rational generalization" cf. footnote 62. A "rational extension" of the quantum theory
would be any formalism that is consistent with the qualitatively derived uncertainties.
[126] Cf. also my paper "Complementarity" {reprinted as Chapter 5 of the present collection}.
[127] It is interesting to note that we are here presented with a dogmatic *empiricism*. Which
shows that empiricism is no better antidote against dogmatism than is, say, Platonism. It
is easily seen why this must be so: both empiricism and Platonism (to mention only one
philosophical alternative) make use of the idea of sources of knowledge; and sources, be
they now intuitive ideas, or experiences, are assumed to be infallible, or at least very
nearly so. Only a little consideration will show, however, that neither can give us an
undistorted picture of reality as neither our brains, nor our senses, can be regarded as
faithful mirrors. For the similarities between empiricism and Platonism cf. my paper

paper I shall try to give my reasons why I believe it to be completely unfounded and why I moreover regard it as a very unfortunate feature of part of contemporary science.[128]

However, before going into details, the following remarks seem to be in order: the particular interpretation of the microscopic theories (and especially of the quantum theory of Schrödinger and Heisenberg) which results from the combination of these theories with Bohr's two hypotheses and with the more general philosophical background referred to above, this interpretation has been called the *Copenhagen Interpretation*. A close look at this interpretation at once shows that it is not *one* interpretation, but a variety of them. True, the indefiniteness assumption and, to a lesser extent, the assumption of the relational character of the quantum mechanical states always play an important role, and so do the uncertainty relations. Yet the exact interpretation of these assumptions and of Heisenberg's formulae is neither *clear*, nor is there a *single* such interpretation. Quite the contrary – what we find is that all philosophical creeds, from extreme idealism (positivism, subjectivism) to dialectical materialism, have been imposed upon these physical elements. Heisenberg[129] and

---

"Explanation, Reduction, and Empiricism", in *Minnesota Studies in the Philosophy of Science*, Vol. III {reprinted in *PP1*, ch. 4}. For the idea of sources of knowledge behind both these philosophies cf. K. R. Popper, "On the Sources of Knowledge and Ignorance", read to the *British Academy* on 20 January 1960.

[128] For a more detailed account cf. my papers "Complementarity", *loc. cit.*; "Professor Bohm's Philosophy of Nature", *British Journal for the Philosophy of Science*, Feb. 1960; "Niels Bohr's Interpretation of the Quantum Theory", in *Current Issues in the Philosophy of Science*, New York, 1960, as well as "Explanation, Reduction, and Empiricism," *loc. cit.* {Reprinted, respectively, as Chapter 5 of the present collection; in *PP1*, ch. 14; Chapter 6 of the present collection; and in *PP1*, ch. 4.}

[129] Cf. *Physics and Philosophy*, New York, 1958. In their physics, too, Heisenberg and Bohr went different ways. "Bohr tried to make the dualism between the wave picture and the particle picture the starting point of a physical interpretation" writes Heisenberg (*Theoretical Physics in the Twentieth Century, A Memorial Volume to Wolfgang Pauli*, ad. Fierz and Weisskopf, New York, [274]1960, p. 45) "whereas I attempted to continue on the way of the quantum theory and Dirac's transformation theory without trying to get any help from the wave mechanics". "Bohr" writes Heisenberg at a different place (*Niels Bohr and the Development of Physics*, p. 15) "intended to work the new simple pictures, obtained by wave mechanics, into the interpretation of the theory, while I for my part attempted to extend the physical significance of the transformation matrices in such a way that a complete interpretation was obtained which would take account of all possible experiments". On the whole Bohr's approach was more intuitive, whereas Heisenberg's approach was more formalistic, indeed so much so that Pauli felt called upon to demand that "it must be attempted to free ... Heisenberg's mechanics a little more from the flood of formalism characteristic for the Göttinger savants [vom

von Weizsäcker[130] present a more Kantian version; Rosenfeld[131] has injected dialectics into his account of the matter; whereas Bohr himself is reported[132] to have criticized all these versions as not being in agreement with his own point of view. Quite obviously the fictitious unity conveyed by the term "Copenhagen Interpretation" must be given up. Instead we shall try to discuss only those philosophical ideas which Bohr himself has provided, and we shall refer to other authors only [222]if their contributions can be regarded as an elaboration of such ideas. The outline of the general background will be started with a discussion of the idea of complementarity.

7. COMPLEMENTARITY

Bohr's hypothesis of indefinite state descriptions referred to description in terms of *classical concepts*, i.e., it referred to description in terms of either Newtonian mechanics (including the different formulations which were provided later by Lagrange and Hamilton), or of theories which employ contact action, or field theories. The hypothesis amounted to the assertion that description in terms of these concepts must be made "more liberal"[133] if agreement with experiment is to be obtained. The principle of complementarity expresses in more general terms this peculiar restriction, forced upon us by experiment, in the handling of the classical concepts. In the form in which this principle is applied it is based mainly upon two empirical premises as well as upon some further premises which are neither empirical, nor mathematical, and which may therefore be properly called "metaphysical".[134] The

Göttinger formalen Gelehrsamkeitsschwall]"; (letter from Pauli to Kronig of October 9th, 1925; quoted from *Theoretical Physics in the Twentieth Century*, p. 26).

[130] Von Weizsäcker's point of view is most clearly explained in his book *Zum Weltbild der Physik*, Leipzig, 1954.

[131] Cf. his article in *Louis de Broglie*, etc. As opposed to Rosenfeld P. Jordan (*op. cit.*) and Pauli seem to represent a purely positivistic position.

[132] D. Bohm and H. J. Groenewold, private communication.

[133] N. Bohr, *Atomic Theory*, etc., p. 3. Cf. also footnote 62.

[134] I use here the word "metaphysical" in the same sense in which it is used by the adherents of the Copenhagen point of view, viz. in the sense of "neither mathematical nor empirical". That the Copenhagen Interpretation contains elements which are metaphysical in this sense has been asserted, in slightly different words, by Heisenberg, who declared in 1930 (*Die Physikalischen Grundlagen der Quantentheorie*, p. 15) that its adoption was "a question of taste". This he repeated 1958 in the now more fashionable linguistic terminology (cf. *Physics and Philosophy*, New York, 1958, pp. 29ff).

empirical premises are (apart from the conservation laws) (1) the dual character of light and matter; and (2) the existence of the quantum of action as expressed in the laws

$$p = h/\lambda \quad E = h\nu \qquad (15)^{135}$$

[223]I do not intend in this paper to discuss all the difficult considerations which finally led to the announcement of the dual character of light and matter. Although these considerations have sometimes been criticized as being inconclusive, they yet seem to me to be essentially sound. It is also beyond the scope of the present paper to explain how duality can be used for providing a coherent account of the numerous experimental results which form the confirmation basis of the contemporary quantum theory.[136] I shall merely state the principle of duality and make a few comments upon it. Duality means (cf. the diagram) that all the experimental results about light and matter divide into two classes. The facts of the first class, while contradicting any wave theory, can be completely and exhaustively explained in terms of the assumption that light (or matter) consists of particles. The facts of the second class, while contradicting any particle theory, can be completely and exhaustively explained in terms of the assumption that light (or matter) consists of waves. There exists, at least at the present moment, no system of physical concepts which can provide us with an explanation that covers and is compatible with *all* the facts about light and matter.

---

[135] The assumption of the existence of the quantum of action is very often given an interpretation that goes beyond these two equations; however, I agree with Landé and Kaila (*Zur Metatheorie der Quantemechanik*, Helsinki, 1950, p. 48) who have both pointed out, though with somewhat different reasons, that a more "substantial" interpretation of the quantum of action than is contained in these two equations is neither justified, nor tenable. The original view according to which the quantum of action is an indivisible "link" between interacting systems which is responsible for their mutual changes has been refuted by Einstein, Podolsky, and Rosen. For this cf. footnote 116.

[136] For an account of these results cf. the literature in footnote 60. It is worth pointing out, by the way, that duality is only one of various ordering principles that are needed to give a rational account of the facts upon the atomic level, [275]and especially of the properties of atomic spectra. It took some time to separate the facts relevant for the enunciation of the principle of duality from numerous other facts which had to be explained in a different manner, viz., on the basis of Pauli's exclusion principle and the assumption of an electronic spin.

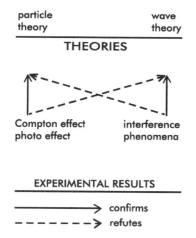

The following comments should be made. First: by a particle theory we understand, in the present context, any theory that works with entities of the following kind: they exert influence upon and are influenced by small regions of space only;[137] and they obey the principle of the conservation of momentum. No further assumptions are made about the nature of these particles and about the laws they obey. By a wave theory we understand, on the other hand, a theory that works with entities of the following kind: they are extended, their states at different places are correlated by a phase, and this phase obeys a (linear) superposition principle. It is the superposition principle that forms the core of all wave theories. What is refuted by either the Compton effect or by the photo-electric effect is not a *particular* wave theory (which may be characterized by a particular equation of motion for the waves) but the much more general assumption that light consists of extended and superimposable entities. Secondly it should be pointed out that the cross-relation between experimental evidence and theories, as indicated in the diagram, is essential for duality as well as for the idea of complementarity that is based upon it. I doubt whether anything like this exists in those domains in which complementarity has now become a kind of savior from trouble, such as in biology, psychology, sociology, and theology. Thirdly, it must be emphasized – and this remark will prove to be of great importance later on – that the wave theories (in the general sense explained above) [224]and the particle theories do not only serve as devices which allow us to *summarize*, and to unify, a host of experimental results in an

---

[137] This explanation is given by Heisenberg, *op. cit.*, p. 7.

economical way. Without the key terms of either theory these results *could neither be obtained, nor could they be stated.* To take an example: interference experiments work with coherent or partly coherent light only. Hence, in preparing them proper attention must be paid to the relative phases of the incoming wave train which means that we have to apply the wave theory already in the preparation of the experiment. On the other hand such facts as the localizability of interaction between light and matter and the conservation of momentum in these interactions cannot be properly described without the use of concepts which belong to some particle theory. Using the term "classical" for concepts of either a wave theory or a particle theory, we may therefore say that "only with the help of classical ideas is it possible to ascribe an unambiguous meaning to the results of observation".[138]

Duality is regarded by Bohr and by his followers as an experimental fact which must not be tampered with and upon which all future reasoning about microphysical events is to be based. As a physical theory is acceptable only if it is compatible with the relevant facts, and as "to object to a lesson of experience by appealing to metaphysical preconceptions is unscientific",[139] it follows that a microphysical theory will be adequate and acceptable only if it is compatible with the fact of duality, and that it must be discarded if it is not so compatible. This demand leads to a set of very general conditions to be satisfied by any microscopic theory. We are now going to state these conditions.

First of all the wave concepts and the particle concepts are the only concepts available for the description of the character of light and matter. Duality shows that these concepts cannot any more be applied generally, but can serve only for the description of what happens under certain experimental conditions. Using familiar terms of epistemology this means that the description of the *nature* of light and matter has now to be replaced by a description of the way in which light and matter appear under certain experimental conditions. Secondly, a change from conditions allowing for the application of, say, the wave picture to conditions allowing for the application of the particle picture will, in the absence of more general and more abstract concepts which apply under all conditions, have to be regarded as an *unpredictable jump*. The statistical laws connecting events in the first picture with events in the second picture will therefore not allow for a deterministic substratum, they will be

---

[138] Niels Bohr, *op. cit.*, p. 16.   [139] L. Rosenfeld in *Observation and interpretation*, p. 42.

irreducible. Thirdly, the combination of duality with the second set of [225]empirical premises introduced above (the Einstein–de Broglie relations) shows that the duality between the wave properties and the particle properties of matter may also be interpreted as a duality between two sets of variables (i.e., position and momentum), which in the classical theory are both necessary for the complete description of the state of a physical system. We are forced to say that a system can never be in a state in which all its classical variables possess sharp values. If we have determined with precision the position of a particle, then its momentum is not only undetermined, it is even meaningless to say that the particle possesses a well-defined momentum. Clearly, the *uncertainty relations* now indicate the domain of meaningful applicability of classical functors (such as the functor "position"), rather than the mean deviations of their otherwise well-defined values in large ensembles. This is nothing but Bohr's hypothesis of the indefiniteness of state descriptions. The relational character of state descriptions results from the need to restrict the application of any set of concepts to a certain experimental domain. This is how the more general point of view explained here is related to the two specific hypotheses which we discussed in the preceding sections.

Now it is important to realize that the above argument is quite generally valid. It follows, and this is Bohr's contention, that it will hold for any theory into which Planck's *h* enters in an essential way. Hence, any future microscopic theory will have to be descriptive of appearances only, it will contain irreducible probabilities, and it will have to work with commutation relations between variables which are only partly well-defined and meaningful. The development of microphysics can only lead to greater indeterminacy. It will never again return to a state of affairs where we are able to give a complete, objective, and deterministic description of the nature of physical systems and physical events. In the interest of economy of thought and effort theories of this kind should therefore be forever excluded from consideration.

I must repeat that in the above two paragraphs only a very sketchy outline has been given of the argument of Bohr and his followers and that it has not at all been shown what great variety of experimental facts is covered, and explained, by the two hypotheses which follow from this argument. This bare outline is not at all sufficient for making understandable the influence Bohr's ideas have had upon physicists and philosophers. But I think that it contains all the essential elements of the Copenhagen point of view, and that it will serve well as a starting point of criticism.

[226]The argument proceeds from what seems to be a mere truism; it proceeds from the assertion that, duality being an experimental fact, it must not be tampered with, but must be regarded as an unalterable basis for any further theoretical consideration. After all, facts are the building stones out of which a theory may be constructed and therefore they themselves neither can nor should be modified. To proceed in this way seems to be the truly scientific attitude, whereas any interference with the facts shows what can only be called the first step towards wild and unwarranted speculation. It is not surprising that this starting point of the argument is frequently taken for granted, as it seems to be the natural procedure to adopt for a scientist. Did not Galileo start modern science by eliminating speculation and by directly putting questions to nature? And do we not owe the existence of modern science to the fact that problems were finally dealt with in an empirical manner rather than on the basis of groundless speculation?

It is here, at the very beginning, that the position of the orthodox must be criticized. For what is regarded by them as a truism is neither correct nor reasonable; and their account of history, too, is at variance with the actual development. Things were just the other way round. It was the Aristotelian theory of motion which was defended by reference to experimental results and it was Galileo who was not prepared to take these results at their face value but who insisted that they be analyzed and be shown to be due to the interplay of various and as yet unknown factors. "[I]f we are seeking to understand [the] birth of modern science we must not imagine that everything is explained by the resort to an experimental mode of procedure, or even that experiments were any great novelty. It was commonly argued, even by the enemies of the Aristotelian system, that that system itself could never have been founded except on the footing of observation and experiment. ... We may [also] be surprised to note that in one of the dialogues of Galileo it is Simplicius, the spokesman of the Aristotelians – the butt of the whole piece – who defends the experimental method of Aristotle against what is described as the mathematical method of Galileo".[140] Indeed the whole tradition of science from Galileo (or even from Thales) up to Einstein and Bohm[141] is

---

[140] H. Butterfield, *The Origins of Modern Science*, London, 1957, p. 80. Butterfield's book contains a very valuable account of the role of the experimental method in the seventeenth century.

[141] Cf. the latter's *Causality and Chance in Modern Physics*, London, 1957.[141a] For a more detailed account cf. K. R. Popper, "The Aim of Science", in *Ratio*, Vol. I (1957), pp. 24ff as well as my paper referred to in fn. 84.

incompatible with the principle that "facts" should be regarded as the unalterable basis of any theorizing. In this tradition the results of experiment are not regarded as the unalterable and unanalyzable building stones of knowledge. They are regarded as capable of analysis, of improvement (after all, no observer, and no theoretician collecting observations is ever perfect), and it is assumed that such analysis and improvement is [227]absolutely necessary. What would be a more obvious observational fact than the difference between celestial motions (regularity) and terrestrial motions (irregularity)? Yet from the earliest times the attempt was made to explain both on the basis of the same laws. Again, what would be a more obvious observational fact than the great variety of substances and phenomena met on the surface of the earth? Yet from the very beginning of rational thinking it was attempted to explain this variety on the basis of the assumption that it was due to the working of a few simple laws and a few simple substances, or perhaps even a single substance. Also the new theory of motion which was developed by Galileo and Newton cannot possibly be understood as a device for establishing relations between our experiences or between laws which are directly founded upon our experiences, and this for the simple reason that the laws expressing these observable motions (such as the law of free fall, or Kepler's laws) were asserted to be incorrect by this theory.[141a] And this is quite in order. Our senses are not less fallible than our thoughts and not less capable of giving rise to deception. The Galilean tradition, as we may call it, therefore proceeds from the very reasonable point of view that our ideas *as well as* our experiences (complicated experimental results included) may be erroneous, and that the latter give us at most an *approximate* account of what is going on in reality. Hence, within this tradition the condition to be satisfied by a future theory of the microcosm is not that it be simply compatible with duality and the other laws used in the above argument, but that it be compatible with duality *to a certain degree of approximation* which will have to depend on the precision of the experiments used for establishing the "fact" of duality.[142]

A completely analogous remark holds for the assertion that Planck's constant will have to enter every microscopic theory in an essential way. After all, it is quite possible[143] that this constant has meaning only under certain well-defined conditions (just as the density of a fluid, or its

---

[142] For a more detailed account cf. again my paper referred to in fn. 84.
[143] Cf. Bohm, *op. cit.*, Ch. IV.

viscosity, or its diffusion constant can be meaningfully defined only for not too small a volume) and that all the experiments we have made so far explore only part of these conditions. Quite obviously the invariance of *h* in all *these* experiments cannot be used as an argument against such a possibility. But if neither the constancy of *h* nor duality can be guaranteed to hold in new domains of research, then the whole argument is bound to break down: it does not guarantee the persistence of the familiar features of complementarity, of probabilistic laws, of quantum jumps, of the commutation relations in future investigations.

[228]It ought to be pointed out, by the way, that the above two paragraphs cannot be regarded as a *refutation* of the principle that our theories must never contradict what at a certain time counts as an experimental fact. After all, it may well be possible (and it has been possible) to construct theories which satisfy this demand of maximal empirical adequacy with respect to a set of observations which are then removed from all analysis and criticism. Part of Aristotle's theory of motion was of this kind. However, it is very doubtful whether this restriction of research would ever allow for theories of the universality, the precision, and the formal accomplishment of Newton's celestial mechanics, or of Einstein's general theory of relativity, both of which lead to a correction of previously existing experimental laws.

Let us now assume, for the sake of argument, that a radically empiricistic point of view has been adopted, i.e., let us regard duality and the constancy of *h* as holding with absolute precision. Would then perhaps the argument be valid? This at once introduces the second "metaphysical" assumption that is used by Bohr and his followers. According to this second assumption the classical concepts are the only concepts which we possess. As we cannot construct a theory or a description of fact out of concepts which we do not possess; and as the classical concepts cannot any more be applied in an unrestricted way; for these two reasons it follows that we are stuck with the complementary mode of description. Against this argument which has been elaborated in some detail by Heisenberg[144] and by von

---

[144] *Physics and Philosophy*, New York, 1958, esp. p. 56: "It has sometimes been suggested that one should depart from the classical concepts altogether and that a radical change of the concepts used for describing the experiments might possibly lead back to a ... completely objective description of nature. This suggestion, however, rests upon a misunderstanding.... Our actual situation in science is such that we do use the classical concepts for the description of the experiments. There is no use discussing what could be done if we were other beings than we are".

Weizsäcker[145] it is sufficient to point out that introducing a set of concepts is not something that occurs independently of and prior to the construction of theories. Concepts are introduced as part of a theoretical framework, they are not introduced by themselves. However with respect to theories it must be asserted that man is not only capable of using them and the concepts which they embody for the construction of descriptions, experimental and otherwise, but that he is also capable of inventing them. How else could it have been possible, to mention only one example, to replace the Aristotelian physics and the Aristotelian cosmology by the new physics of Galileo and Newton? The only conceptual apparatus then available was the Aristotelian theory of change with its opposition of actual and potential properties, form and matter, the four causes, and the like. This conceptual apparatus was much more general and universal than the physical theories of today as it contained a general theory of change, spatio-temporal and otherwise. It also seems to be closer to everyday thinking and was therefore more firmly entrenched than any succeeding physical theory, classical physics [229]included. Within this tremendously involved conceptual scheme Galileo's (or rather Descartes') law of inertia does not make sense. Should, then, Galileo have tried to get on with the Aristotelian concepts as well as possible because these concepts were the only ones in actual use and as "there is no use discussing what could be done if we were other [i.e., more ingenious] beings than we are?". [146] By no means! What was needed was not improvement, or delimitation

This is an astounding argument indeed! It asserts, in fact, that a language that is used for describing observational results and that is fairly general cannot possibly be replaced by a different language. How, then, did it happen that the Aristotelian physics (which was much closer to the everyday idiom and to observation than the physics of Galileo and Newton!) was replaced by the point of view of the classical science? And how could the theory of witchcraft be replaced by reasonable psychology, based as it was upon innumerable direct observations of daemons and daemonic influence? (Think of the phenomenon of split personality which lends very direct support to the idea of daemonic influence!) On the other hand why should we not try to improve our situation and thereby indeed become "other beings than we are"? Is it assumed that the physicist has to remain content with the state of human thought and perception as it is given at a certain time and that he cannot (or should not) attempt to change, and to improve upon that state? Only the inductivistic prejudice that all a physicist can do is to assemble facts and present them in a formally satisfactory way, only such a prejudice can explain the defeatist attitude which is expressed in the above quotation.

[276][145] *Zum Weltbild*, etc., p. 110: "Every actual experiment we know is described with the help of classical terms and we do not know how to do it differently". The obvious reply is, of course: "Too bad; try again!"

[146] Cf. footnote 144.

of the Aristotelian concepts in order to "make room for new physical laws"; [147] what was needed was an *entirely new theory*. Now at the time of Galileo human beings were apparently able to do this extraordinary thing and become beings different from what they were before (and one should again realize that the conceptual change that was implied was a much more radical one than the conceptual change necessitated by the appearance of the quantum of action). Are there (apart from pessimism with respect to the abilities of contemporary physicists) any reasons to assume that what was possible in the 16th and 17th centuries will be impossible in the 20th century? As far as I can understand it is Bohr's contention that such reasons do indeed exist, that they are of a logical rather than of a sociological character, [148] and that they are connected with the peculiar nature of classical physics.

Bohr's first argument in favor of this contention proceeds from the situation, outlined above, that we need the classical concepts not only if we want to give a *summary* of facts, but that without these concepts the facts to be summarized could not be *stated* either. As already Kant before

---

[147] Cf. footnote 62.

[148] As will be evident from the quotation in footnote 144 Heisenberg and von Weizsäcker seem to base their argument upon the sociological fact that the majority of the contemporary physicists uses the language of classical physics as their observation language. Bohr seems to go further. He seems to assume that the attempt to use a different observation language can never succeed. His arguments in favor of this contention are very similar to the arguments by transcendental deduction used by Kant. The fact that Heisenberg and von Weizsäcker seem to represent a less dogmatic and more practical position has prompted Hanson to distinguish between two different wings, as it were, in the Copenhagen school; the extreme Right, represented by Bohr, which regards the attempt to introduce a new observation language as logically impossible; and the Center, represented by von Weizaecker and Heisenberg, where such an attempt is only regarded as being *practically* impossible. I deny that this distinction exists. First of all the difference between logical impossibility and *sociological* impossibility (or *practical* impossibility), although regarded with awe by a good many philosophers is too subtle to impress any physicist. Neither will the assertion of logical impossibility deter him from trying to achieve the impossible (for example, to achieve a relative theory of space and time); nor will he feel relieved when he is being offered practical impossibility instead of logical impossibility. But we find that the distinction which Hanson wants to draw between Heisenberg and Bohr is not really one which Heisenberg himself would recognize, or at least so it appears from his writings. For on p. 132 of *Physics and Philosophy* the possibility of an alternative to the Copenhagen point of view is equated with the possibility that 2 times 2 may equal five, that is, the issue is now made a matter of logic. For this point cf. also the discussion between Professor Hanson and myself in *Current Issues in the Philosophy of Science*, pp. 390–400 {reprinted at the end of Chapter 6 of the present collection}. There do, of course, exist some very decisive differences between Bohr's approach and Heisenberg's approach. But these differences lie in an entirely different field. Cf. for this footnotes 129 and 119.

him he observes that even our experimental statements are always formulated with the help of theoretical terms and that the elimination of these terms must lead, not to the "foundations of knowledge" as the positivists would have it, but to complete chaos. "Any experience", he asserts,[149] "makes its appearance within the frame of our customary points of view and forms of perception" and at the present moment the forms of perception are those of classical physics.

But does it follow, as is asserted by Bohr, that we can never go beyond the classical framework and that all our future microscopic theories must have duality built into them?

It is easily seen that the use of classical concepts for the description of experiments in contemporary physics can never justify such an assumption. For a theory may be found whose conceptual apparatus, when applied to the domain of validity of classical physics, would be just as comprehensive and useful as the classical apparatus without yet coinciding with it. Such a situation is by no means uncommon. The behavior of the [230]planets, of the sun, and of the satellites can be described both by the Newtonian concepts and by the concepts of general relativity. The order introduced into our experiences by Newton's theory is retained *and improved upon* by relativity. This means that the concepts of relativity are sufficiently rich for the formulation even of all the *facts* which were stated before with the help of Newtonian physics. Yet the two sets of concepts are completely different and bear no logical relation to each other.

An even more striking example is provided by the phenomena known as the "appearances of the devil". These phenomena are accounted for both by the assumption that the devil exists, and by some more recent psychological (and psychosociological) theories. [150] The concepts used by these two schemes of explanation are in no way related to each other. Nevertheless the abandonment of the idea that the devil exists does not lead to experiential chaos, as the psychological scheme is rich enough to account for the order already introduced.

To sum up: although in reporting our experiences we make use, and must make use, of certain theoretical terms, it does not follow that

---

[149] *Op. cit.*, p. 1.

[150] Cf. Huxley's highly interesting discussion of the merits of the Cartesian psychology as a means for the explanation of demonic appearances as well his account of what is and what is not unthinkable at a certain time and within a certain point of view in Ch. VII of his *Devils of Loudun*, New York, [277]1952. Cf. also my discussion of the self-petrifying influence of single theories in section 7 of my "Explanation, Reduction, and Empiricism" {reprinted in *PP1*, ch. 4}.

different terms will not do the job equally well, or perhaps even better, because more coherently. And as our argument was quite general, it seems to apply to the classical concepts as well.

This is where Bohr's second argument comes in. According to this second argument, which is quite ingenious, we shall have to stay with the classical concepts, as the human mind will never be able to invent a new and different conceptual scheme. As far as I can make out, the argument for this peculiar inability of the human mind rests upon the following *premises*: (a) we invent (or should use) only such ideas, concepts, theories, as are suggested by observation; "only by observation itself", writes Bohr, [151] "do we come to recognize those laws which grant us a comprehensive view of the diversity of phenomena". (b) because of the formation of appropriate habits any conceptual scheme employed for the explanation and prediction of facts will imprint itself upon our language, our experimental procedures, our expectations, as well as our experiences. (c) classical physics is a universal conceptual scheme, i.e., it is so general that no conceivable fact falls outside the domain of its application. (d) classical physics has been used long enough for the formation of habits, referred to under (b), to become operative. The *argument* itself runs as follows: if classical physics is a universal theory (premise c) and has been used long enough (premise d), then all our experiences will be classical (premise b) and we shall therefore be unable to conceive any concepts which fall outside the classical scheme (premise a). Hence the invention [231]of a new conceptual scheme which might enable us to circumvent duality is impossible.

That there must be something amiss with the argument is seen from the fact that all the premises except perhaps the first one apply also to the Aristotelian theory of motion. As a matter of fact the very generality of this theory would seem to make it a much stronger candidate for the argument than the classical physics could ever be. However the Aristotelian theory *has been* superseded by a very different conceptual apparatus. Clearly, this new conceptual apparatus was then not suggested by experience *as interpreted in the Aristotelian manner* and it was therefore a "free mental". [152] This refutes (a). That (b) needs modifying becomes clear when we consider that a scientist should always keep an open mind and that he should therefore always consider possible alternatives along with

---

[151] *Loc. cit.*
[152] Albert Einstein in *Ideas and Opinions*, London, 1954, p. 291 (reprint of an article that was first published in 1936); cf. also H. Butterfield, *op. cit.*

the theory he is favoring at a certain moment.[153] If this demand is satisfied, then the habits cannot form, or at least they will not any longer completely determine the actions of the scientist. Furthermore, it cannot be admitted that the classical scheme is universally valid. It is not applicable to such phenomena as the behavior of living organisms (which the Aristotelian scheme did cover), to personal consciousness, to the formation and the behavior of social groups, and to many other phenomena. We have to conclude, then, that Bohr's arguments against the possibility of alternatives to the point of view of complementarity are all inconclusive.

And this result is exactly as it should be. Any restrictive demand with respect to the form and the properties of future theories, any such demand can be justified only if an assertion is made to the effect that certain parts of the knowledge we possess are absolute and irrevocable. Dogmatism, however, should be alien to the spirit of scientific research, and this quite irrespective of whether it is now grounded upon "experience" or upon a different and more "aprioristic" kind of argument.

What has been refuted so far is the contention that complementarity is the *only possible point of view* in matters microphysical and that the only

---

[153] As J. Agassi has pointed out to me, this principle was consciously used by Faraday in his research work. Against the use of such a procedure it has been argued, by T. S. Kuhn (private communication) that the close fitting between the facts and the theory that is a necessary presupposition of the proper organization of the observational material can be achieved only by people who devote themselves to the investigation of one single theory to the exclusion of all alternatives. For this *psychological* reason he is prepared to defend the (dogmatic) rejection of novel ideas at a period when the theory which stands in the center of discussion is being built up. I cannot accept this argument. My first reason is that many great scientists seemed to be able to do better than just devote themselves to the development of one single theory. Einstein is the outstanding recent example. Faraday and Newton are notable examples in history. Kuhn seems to be thinking mainly of the average scientist who may well have difficulties when asked not only to work out the details of some fashionable theory but also to consider alternatives. However, even in this case I am not sure whether this inability is "innate", as it were, and incurable, or simply due to the fact that the *education* of the "average scientist" is in the hands of people who subscribe, implicitly, to Kuhn's doctrine of the necessity of concentration. My second reason for not being able to accept this argument is as follows: assume that it is indeed correct that human beings are not able at the same time to work out the details of one theory and to consider alternatives. Who says, then, that details are more important than alternatives which, after all, keep us from dogmatism and are a very concrete and lively warning of the limitations of all our knowledge? If I had to choose between a very detailed account of the fabric of the universe at the expense of not being able to see its limitations and between a less detailed account whose limitations however were very obvious, then I would at once choose the latter. The details I could gladly leave to those who are interested in practical application.

successful theories will be those which work with inbuilt uncertainties that are interpreted in accordance with Bohr's two hypotheses. Still, it has not been shown that complementarity is not *a possible point of view*. Quite on the contrary, we have tried to exhibit the advantages of Bohr's point of view and we have also defended this point of view against irrelevant attacks. It is now time to turn to the *difficulties* which beset the idea of complementarity even if it is not interpreted in the dogmatic manner which we have just criticized. These difficulties will be discussed in [232]Sections 10ff of the present paper. However, at first a few comments should be made on the results so far obtained.

8. THE ROLE OF SPECULATION IN PHYSICS

There are many physicists who, when presented with our above results, will point out, rather impatiently, that a general discussion of possibilities is of no use whatever as long as a well-developed and successful alternative to the present quantum theory is missing. These physicists will refer to the fact that, after all, *there exists* a very successful body of theory which is in agreement with the idea of complementarity whereas the other side, despite all the talk about possibilities, has not yet produced anything that would only vaguely resemble this theory in formal accomplishment and empirical accuracy. "What is the message", exclaims Professor Hanson, expressing the view of many of his physicist colleagues,[154] "Bohr ... wish[es] to carry to Copenhagen?". And he implies that there is no message as there is "no algebraically detailed, experimentally acceptable" *theory* to the present quantum theory. [155] In an earlier paper[156] Professor

---

[154] The following quotations are from Hanson's article "Five Cautions for the Copenhagen Critics", *Philosophy of Science*, Vol. 26 (1959), pp. 325–337. The present quotation is p. 337.

[155] *Loc. cit.*, p. 334.

[156] Reprinted in Danto-Morgenbesser (eds.), *Philosophy of Science*, New York, 1960, pp. 450–470. The present quotation is to be found on page 455. A few [278]lines above this quotation, Hanson asserts that "There is as yet no working alternative to the Copenhagen Interpretation". I do not quite understand this assertion. For quite obviously Bohm, Vigier and their collaborators have provided just this: an interpretation of the elementary theory, and of the field theories which does not any more work with irreducible probabilities and which is still compatible with the existing formalisms. Maybe Hanson still believes, as he did in 1958 (*Patterns of Discovery*, pp. 172ff) that von Neumann's proof can be used for eliminating any such interpretation. This belief can easily be shown to be incorrect. For this cf. my arguments in Section 2 and Section 9. Maybe Hanson is willing to accept an interpretation only if it is connected with a detailed and empirically satisfactory formalism. This is the pragmatic argument we are

Hanson allows for general speculations which are different from those contained in the Copenhagen picture. But he demands that a distinction be made between "those speculations which have proven themselves to *work* in theory and practice from those which have not yet been put to any test", [157] i.e., which have not yet led to the construction of detailed physical theories. And he again implies that for this reason the ideas of Bohm and Vigier should be regarded with scepticism. It is this pragmatic criticism of unpopular and fairly general speculations which I want to examine in the present section.

It is clear that such criticism will be well liked by the great majority of physicists, as it enables them to enjoy the riches (or apparent riches) they possess without forcing them to think of means either to enlarge their capital, or to improve upon the quality of their currency. However, it becomes a dangerous tool when it is elevated from an instrument giving security to those who think that their everyday life as practicing scientists is sufficiently troublesome to excuse them from additional metaphysical

about to refute. Besides, Hanson himself regards the elementary theory as "an arbitrarily delimited sub-theory of the more general quantum theory of fields" (*Philosophy of Science*, Vol. 26, p. 329) which latter "*simply does not exist*" as a mathematically sound theory (*loc. cit.*; his italics). From which it would follow that also the point of view of complementarity has not yet been connected with an "algebraically detailed, experimentally" satisfactory theory (an opinion, by the way, which I myself cannot wholly agree with; but this is not the point). A third possibility is that Hanson means by the "Copenhagen Interpretation" not just Bohr's general point of view but this point of view *taken together with the elementary theory and the field theories* (cf. for this *loc. cit.*, p. 336). This procedure has also been adopted by Heisenberg (cf. his contribution to *Niels Bohr and the Development of Physics*, pp. 18, 19); it is extremely misleading. For first of all, the full elementary theory was constructed not in Copenhagen, but by Schrödinger who held a philosophy that was completely different from the philosophy of Heisenberg, Weizsäcker, and others. And secondly the most satisfactory forms of the field theory do not any more agree with the principle of correspondence, but are developed on an independent basis. In no case it is possible to agree with Hanson's assertion, even if it should be supported by the "next synchrotron operator" (Danto-Morgenbesser, *op. cit.*, p. 455). However, after all this has been said, it should be added that the point of view of Bohm and Vigier has already been developed in much greater detail than commonly supposed by most of the opponents. Thus Vigier has been able to find a classical model the quantization of which gives at one stroke the four and only four possible interactions (gravitation, electromagnetic field, weak and strong interaction) and represents such abstract notions as isospin as the result of quantization in ordinary space-time. The Gell-Mann scheme also emerges and the calculation of the correct masses now only depends on the proper calculation of the interactions (a problem that also exists in the [279]orthodox theories). The fact that only four interactions emerge from the model makes a very strong one and much less *ad hoc* than the usual theories which can accommodate any kind of field in a simple additive manner.

[157] Danto-Morgenbesser, *op. cit.*, p. 455.

worries, when it is elevated from such a psychological crutch into a philosophical principle. First of all it is somewhat doubtful whether even the quantum theory of today is adequately represented by the idea of complementarity. It is quite true that this idea gives a correct account of some general characteristics of the elementary theory; however, difficulties arise as soon as we either consider details, or leave the elementary theory and proceed to an analysis of the more recent field theories. This point [233]will be elaborated later on. Secondly, the argument which praises the idea of complementarity for having given rise to a very valuable physical theory overlooks the fact that the full Copenhagen Interpretation was only completed *after* the introduction of wave mechanics. And wave mechanics, or the elementary theory, was completed by Schrödinger whose general philosophy was very different from the ideas which originated in Copenhagen.[158] However, quite apart from more detailed criticisms of this kind, it must be asserted that the discussion of possibilities and of alternatives to a current theory plays a most important role in the development of our physical knowledge. After all, a physicist who has been convinced by Bohr's arguments (which, as we have shown, are invalid) will exclude from consideration any theory that does not work with inbuilt uncertainties. He will thereby severely restrict his domain of research and he will do so because he thinks that what lies outside this domain is of no empirical value whatever. He will support his belief by arguments of the kind we have outlined above. Now it is very important to realize that it is always possible first to restrict oneself to theories which satisfy certain requirements and then to "save the phenomena" in one way or another. The impetus theory was such an attempt to save the Aristotelian theory of motion from refutation by new theories concerning the motion of projectiles.[159] It often turns out that such a procedure is possible only at the expense of simplicity, comprehensiveness, and intuitive appeal. As long as the belief in the uniqueness of the point of view adopted continues to influence the scientists, these complications will be felt to be unavoidable features of nature rather than avoidable features of their theories and they will be suffered with a patient shrug. At such a stage of complication and confusion a hint to the effect that this uniqueness is neither justifiable nor desired may be of paramount importance. It may give rise to the hope that there is a more direct way of

---

[158] Cf. also footnote 156.
[159] For an excellent account and sources (in the original Latin with English translation) cf. M. Clagett, *The Science of Mechanics in the Middle Ages*, Madison, 1954.

attacking the difficulties created by the increasing complications of the theory and it may also encourage new ways of thinking. Think what would have happened if the idea that the sun alone possesses the power to influence the planets had been proposed, and then dogmatically retained. The mutual disturbances of the planets would soon have been discovered – but one would have tried to account for them by further complicating the arrangement of epicycles that Copernicus still used. The idea that any object may attract any other object opened up a completely new way of accounting of these irregularities. There are of course many physicists who will point out that their theories have been derived from experiment and do therefore not admit of alternatives. They overlook that experi[234]mental results only possess approximate validity and therefore admit of different, and even mutually inconsistent interpretations. *There is no way of singling out one and only one theory on the basis of observation.*[160]

However, if general principles such as those underlying the Copenhagen point of view are liable unduly and unjustifiably to restrict future research, is it then not better to omit them altogether and to be content with one's physics alone? The reply is that a complicated physical theory cannot be invented in its full formal splendor without some preparation. Consider for example the astronomical system of Ptolemy with its elaborate and delicate machinery of deferents, epicycles, excenters, and the like. Is it likely that the transition from "experience" to this theory can be made without intermediate steps? After all, what "experience" tells us is that the behavior of the planets is very complicated and quite different from the behavior of the fixed stars. The idea that both may be subjected to the same laws of circular motion could therefore not possibly have been suggested by what we see with our eyes; quite the contrary, this idea is to a certain extent even *contradicted* by the crude experience which was available to the ancient astronomers. Still it had to be used if a coherent treatment of both planets and fixed stars was to be possible. And as it was in disagreement with prior *prima facie* observations it had to be introduced as a metaphysical hypothesis, i.e., as a hypothesis about features of the world that are not accessible to direct observation. This metaphysical hypothesis was then the guiding principle of the planetary astronomy from Anaximander to Copernicus, and even Galileo. Without this idea it would perhaps have been possible to accumulate numerous useful

---

[160] Cf. again my "Explanation, Reduction, and Empiricism" {reprinted in *PP1*, ch. 4}.

empirical regularities about the planets, but it would have been quite impossible to devise a theory of the formal accomplishment and the empirical accuracy of the Ptolemaean astronomy. This shows very clearly that the pragmatic criticism which we presented at the beginning of the present section altogether puts the cart before the horse when demanding that the consideration of fully fledged scientific theories should come first and that the discussion of possibilities was only of secondary importance and should be carried out later, or perhaps not at all. As a second example of a metaphysical idea, take the idea that celestial and terrestrial events are guided by the same laws. Again, at the time when it was conceived this idea could not be called empirical as it was so obviously contradicted by the observable difference between the regularity and apparent purity of the celestial matter (astronomy), and the irregularity and sluggishness of the terrestrial matter (meteorology). Yet, how could the new mechanics of the heavens have been developed [235]without a firm belief, not in one's senses, but in this extraordinary hypothesis? The best example, however, seems to me to be the atomic theory itself. For here it is shown more clearly than in other cases that a theory largely metaphysical need not on that account be irrational and arbitrary. As we know, the atomic theory was developed with the purpose of solving the following difficulty: according to the ideas about matter of Thales, Anaximander, and the other early Ionian monists the things in the universe consisted basically of a single kind of substance. From this premise Parmenides derived that change, being a transition from one thing to a thing of a different kind, could not exist in such a monistic universe. This derivation taken together with the fact that there *is* change was regarded by the atomists as a refutation of monism. They therefore replaced monism by their pluralistic atomic theories.[161] Now the arguments leading up to this replacement, although dealing with matters which are not all directly observable, are clear and easy to comprehend. This example and our above examination of the point of view of complementarity should be sufficient to dispel the notion that metaphysical considerations may perhaps play an important role in the development of scientific theories, but that they must be classified together with such other important, but irrational, factors of theory-construction as intelligence and absence of fatigue. The difference between a metaphysical point of view and a scientific theory does not

---

[161] The atomic theory was not the only reply that was given to the Parmenidean argument. Aristotle's physics is another such reply and it was this theory that was adopted throughout the Middle Ages.

consist in the fact that the former is utterly irrational and arbitrary whereas only the latter can be reasonably discussed. It rather consists in the fact that experience plays a smaller role in the discussion of the former, and that apparently adverse experiences are sometimes disregarded pending a more detailed development of the point of view: it took about 2,000 years until the atomic theory was sufficiently developed to lead to predictions that could be tested and compared with predictions made by alternative theories (Brownian motion). During this time the theory was frequently attacked the basis of "experience". It was so attacked by the Aristotelians (whose theory of motion was more developed and more sophisticated than the theory of motion connected with the atomic theory), and it was so attacked, more recently, when it was shown to be inconsistent with a highly confirmed and formally highly developed physical theory, viz. thermodynamics (reversibility objection; recurrence objection). One can easily imagine what would have happened had the pragmatists and the radical empiricists had their way. The former could have pointed out, already in the time of Aristotle, that the atomic theory had not yet led to any "... detailed, experimentally acceptable" dynamics[162] that the [236]speculations of the atomists had not "proven themselves to *work* in theory and practice"[163] and that they should therefore not be taken too seriously. And the latter, pointing to the strong empirical backing of thermodynamics, could have altogether dismissed the atomic theory as being in disagreement with experiment.[164] What would our present situation be under these circumstances? We would now be working with a host of empirical generalizations such as Balmer's formulae, the rules of Ritz, rules concerning the fine structure of spectra, and we would not possess a coherent account of spectra, the motion of small particles, electric conductivity and the like. I conclude, then, that the development of comprehensive scientific theories essentially depends upon the development, through argument and discussion, of metaphysical points of view together with the attempt to make these points of view more and more specific until their truth can finally be decided by experiment. Also results should not be expected too quickly. It took about 2,000 years before the metaphysical idea of atomism had been transformed into an independently testable scientific theory. During these 2,000 years the atomists were frequently attacked by opponents who contrasted their

---

[162] Cf. footnote 154.   [163] Cf. footnote 156.

[164] I am here thinking mainly of the reversibility objection that for some physicists constituted a serious objection against the kinetic theory.

own detailed physics (i.e., thermodynamics) with the "idle speculations" of the atomists and who thought that such a remark was a good argument against the further pursuit of atomism. The final success of the atomic philosophy shows how little such pragmatic arguments count and how important it is to pursue a reasonable idea, even if practical results in the form of mathematical formalisms or empirical predictions are not immediately forthcoming.

9. VON NEUMANN'S INVESTIGATIONS

We have not yet dealt with all the arguments against the possibility of alternatives to the point of view of complementarity. There are many physicists who would readily admit that Bohr's reasoning is not very convincing and that it may even be invalid. But they will point out that there exists a much better way of arriving at its *result*, viz. von Neumann's proof to the effect that the elementary quantum theory is incompatible with hidden variables. This proof has not only been utilized by those who found the metaphysical elements in Bohr's philosophy not to their taste, it has also been used by members of the Copenhagen school in order to show that what Bohr had derived on the basis of qualitative arguments could be proved in a rigorous way. However, one ought to keep in mind that the relation between the point of view of Bohr and the point of view of von Neumann is by no means very close. For example [237]Bohr has repeatedly emphasized that the measuring device must be described in *classical* terms[165] whereas it is essential for von Neumann's theory of measurement that both the system investigated and the measuring device be described with the help of a ψ-function. The latter procedure leads to difficulties which do not arise in Bohr's treatment. Hence, when dealing with von Neumann's investigations, we are not dealing with a refinement, as it were, of the arguments of Bohr – we are dealing with a completely different approach.

In Section 3 we described the proof itself and we had then occasion to point out that it involves an illegitimate transition from the properties of ensembles to the properties of the elements of these ensembles. In the present section we shall assume the proof to be correct and we shall point out that even then it cannot be used as an argument to the effect that the atomic theory will forever have to work with inbuilt uncertainties.

---

[165] Cf. footnote 119. Cf. also Bohr's definition of a "phenomenon" in *Dialectica*, 7/8 (1948), p. 317 as well as in *Albert Einstein, Philosopher–Scientist*, pp. 237f.

The proof consists in the derivation of a certain result from the quantum theory (the elementary theory) in its present form and interpretation. It follows at once that even if the result were the one claimed by von Neumann it could not be used for excluding a theory according to which the present theory is only approximately correct, i.e., agrees with experiments in some respects but not in others. However simple this argument – the fact that the present theory is confirmed at all has created such a bias in its favor that a little more explanation seems to be required.[166] Assume for that purpose that somebody tries to utilize von Neumann's proof in order to show that any future theory of the microcosm will have to work with irreducible probabilities. If he wants to do this then he must quite obviously assume that the principles upon which von Neumann bases his result are valid under *all* circumstances future research might uncover. Now the assertion of the absolute validity of a physical principle implies the denial of any theory that contains its negation. For example, the assertion of the absolute validity of von Neumann's premises implies the denial of any theory that ascribes to these premises a limited validity in a restricted domain only. But how could such a denial be justified by *experience* if the denied theory is constructed in such a way that it gives the same predictions as the defended one wherever the latter has been found to be in agreement with experiment? And that theories of the kind described can indeed be constructed has been shown, most clearly, by Professor D. Bohm.[167]

Apart from his error with respect to the result of his proof von Neumann himself was completely aware of the limitations of this alleged result. "It [238]would be an exaggeration," writes he, [168] "to maintain that causality has thereby [i.e., by the proof of the two theorems referred to in Section 3] been done away with: quantum mechanics has, in its present form, several serious lacunae and it may even be that it is false". Not all physicists have shared this detached attitude. Thus having outlined the proof, Max Born[169] makes the following comment: "Hence, if any future theory should be deterministic, it cannot be a modification of the present one, but must be essentially different. How this should be possible without sacrificing a whole treasure of well-established results I leave the determinist to worry about". Does he not realize that precisely that same

---

[166] For details cf. again my "Explanation, Reduction, and Empiricism", 167 {reprinted in *PP1*, ch. 4}.

[167] Cf. the references in fn. 15 and in fn. 141.   [168] *Op. cit.*, p. 327.

[169] *Natural Philosophy of Cause and Chance*, Oxford, 1948, p. 109.

argument could be used for the retention of absolute space in mechanics, or against the introduction of the statistical version of the second law? And has the fact that very different theories, such as the Newtonian mechanics on the one side and general relativity on the other, can be used for describing the same facts (for example, the path of Jupiter) not already made it clear that theories can be "essentially different" without a "sacrifice of a whole treasure of well-established results" being involved? This being the case there is no reason whatever why a future atomic theory should not return to a more classical outlook without contradicting actual experiment, or without leaving out facts already known and accounted for by wave mechanics. It follows that von Neumann's imaginary results cannot in any way be used as an argument against the application at the microlevel of theories of a certain type (for example, deterministic theories).

Professor Hanson's attitude is still less comprehensible. He, too, tries to defend indeterminism and the absence of hidden parameters by a combined reference to von Neumann's proof and "nature". [170] But he also realizes, as did von Neumann, that the elementary theory on which the proof is based is "but a programmatic sketch of something more comprehensive"[171] and that it is empirically unsatisfactory. He even admits that a more comprehensive and really satisfactory theory "simply does not exist". [172] Now if all that is granted – how then can he still try to make use of von Neumann's argument whose result will be correct and satisfactory only if the premises are correct, satisfactory, and complete, i.e., only if the elementary theory is correct, satisfactory and complete? After all, who can now say that the observational and other difficulties of the elementary theory are *not* due to the fact that hidden parameters do exist and have been omitted from consideration?

What we have shown so far is that all the arguments which have been used in the literature against alternatives to complementarity are invalid. [239]There does not exist the slightest reason why we should assume that the proper road to future progress will consist in devising theories which are even more indeterministic than wave mechanics, and that the appropriate formalism will forever have to be one with inbuilt commutation relations. All the way through the question has been left undecided as to whether the more general ideas of the point of view of complementarity give an adequate account of the *existing theories*, i.e., whether these ideas

---

[170] "Five Cautions", etc. *loc. cit.*, p. 332.   [171] *Loc. cit.*, p. 329.   [172] *Loc. cit.*

give an adequate account of the elementary theory and of the field theories. The answer to this question, which will exhibit various difficulties, will be given in the remaining sections of the paper, where we shall also have an opportunity to consider some of the more formal alternatives to the ideas of Bohr.

10. OBSERVATIONAL COMPLETENESS

It was the intention of Bohr and Heisenberg, but notably of the latter, to develop a theory which was thoroughly observational in the sense that a sentence expressing an unobservable state of affairs could not be formulated in it. According to the point of view of complementarity, the mathematics of the theory is to be regarded only as a means for transforming statements about observable events into statements about other observable events, and it has no meaning over and above that function. This is not the case with either our everyday language, or with classical physics. Both allow for the existence of physical situations which cannot be discovered by any observation whatever. As an example,[173] we consider the case of two banknotes, both printed with the help of the same printing press, the one under legal circumstances, the other by a gang of counterfeiters who used the same press at night, and illegally. If we assume that the banknotes have been printed within a very short interval of time and that they show the same numbers, and if we further assume that they somehow got mixed up, then we shall have to say that by virtue of their different history they possess certain properties, different for both, which we shall never be able to distinguish. Another example frequently referred to is the intensity of an electromagnetic field at a certain point.[174] The usual methods of measuring this field use bodies of finite extension and finite charge and they can therefore inform us only about average values, but not about the exact values of the field components. As there exist laws of nature according to which there is a lower limit to the size of test bodies, it is even physically impossible to perform a measurement which would result in such information. A third example which should be even more instructive is the disappearance of historical evidence in the course of time. That Caesar sneezed twice on the morning of April 5, 67 B.C. should [240]either be true or false. However as it is very unlikely that this event was recorded by any contemporary writer, and as the physical

---

[173] The example is due to K. R. Popper.   [174] Cf. E. Kaila, *op. cit.*, p. 34.

traces it left in the surroundings as well as the memory traces it left in the brains of the bystanders have long since disappeared (in accordance, among other things, with the second law of thermodynamics) we now possess no evidence whatsoever. Again we are presented with a physical situation which exists (existed), and yet cannot be discovered by any observational means.

A physicist or a philosopher who is biased in favor of a radical empiricism will quite naturally regard such a situation as unsatisfactory. He will be inclined to reject statements such as those contained in our examples by pointing out that they are observationally insignificant, and in doing so he will be guided by the demand that one should not allow talk about situations which can be shown to be inaccessible to observation. Classical physics does not satisfy this demand automatically. It allows for the consistent formulation of sentences with no observational consequences, *together with* the assertion that such consequences do not exist. An attempt to enforce the radical empiricist's demand will therefore have to consist in an *interpretation* of classical physics according to which some of its statements are cognitively meaningful, whereas others are not. This means that the exclusion of the undesirable sentences will have to be achieved by a philosophical manoeuvre which is superimposed upon physics. Classical physics itself does not provide means for excluding them. [175]

There exist, however, *philosophical* theories which possess exactly this character. An example is Berkeley's theory of matter (if we omit the *ad hoc* hypothesis that objects unperceived by human beings are still being perceived by God). According to this theory material objects are bundles of sensations and their existence consists in their being perceived or observed. If this theory is developed in a formally satisfactory manner then it does not allow for the consistent formulation of any statement about material objects in which it is asserted that there is a situation which is not accessible to perception. One may call such a theory *observationally complete*. When formulating matrix mechanics, Heisenberg had the intention of constructing a *physical* theory that was observationally complete in exactly this sense, observation with the help of classically well-defined apparatus replacing the more direct form of observation with the help of one's senses. It is assumed in the more general ideas held by the members of the Copenhagen school, and notably by Bohr, that the elementary quantum theory in its present form and interpretation corresponds with

---

[175] It is worthwhile pointing out that the Aristotelian physics was much closer to the crude experiences of everyday life than is the classical physics.

this intention. It will turn out that this assumption is not justified or, at [241]least, nobody has as yet shown it to be correct. However, let us first examine an apparently very strong argument in its favor.

Consider a state $|\Phi>$ which is such that it cannot be characterized by the values of any complete set of commuting observables. Such a state would be truly unobservable. For first of all there is no measurement (in the sophisticated sense of the quantum theory) which can bring it about; and secondly there is no measurement which on immediate repetition would lead to the result characteristic for this state as we have assumed that there is no such result. If we still want to assert the existence of such a state then we must regard it as an element of Hubert space (we adopt von Neumann's formalism) and it must be possible to represent it in the form

$$|\Phi> = \sum_i c_i |a_i>$$

where the $|a_i>$ form a complete orthonormal set which is connected with the complete commuting set of observables α. Now incorporate $|\Phi>$ into an orthonormal set $\{|\Phi_i>\}$ in such a manner that the set $\{|\Phi>\} + \sum_i \{|\Phi_i>\}$ is complete. Then for any $|a_k>$,

$$|a_k> = \sum_i |\Phi|\Phi_i><\Phi_i|a_k> + |\Phi><\Phi|a_k>$$

which on measurement of the observable corresponding to the set $\{|\Phi>\} + \sum_i \{|\Phi_i>\}$ would yield $|\Phi>$ unless $<\Phi|a_k> = <a_k|\Phi>^* = c_k^* = 0$. Now as $<a_k|$ may be any eigenstates of α it follows that $|\Phi> = 0$: *states which are not accessible to observation do not exist.*

Now if this argument is supposed to prove observational completeness with respect to *classical states of affairs* then the formal scheme of it must be filled with empirical content. More especially, we must make the following assumptions. First, it must be assumed that there exist changes of states which can transform any state into a mixture of the eigenstates of any observable α, or into an α-mixture as we shall call it. This demand is a purely theoretical demand which must be satisfied by the *formalism* of the theory and which is independent of the interpretation of this formalism. Secondly, it must be assumed that states or observables can be characterized in a purely classical manner. Thirdly, we must demand that for any observable thus interpreted there exists a classical device capable of transforming any state into an α-mixture (again classically interpreted). Finally it must also be the case that the methods of measurement in actual use today produce α-mixtures with respect to the observables they are supposed to measure, or else the numbers obtained are of no relevance whatever.

Now, as regards the first assumption, it must be pointed out that it can be discussed only if a definite meaning has been given to the phrase [242]"any state", i.e., if the class of all states has been well defined. As is well known, there is no unanimity on this point. The usual attitude is altogether to neglect the question and to decide it differently in different concrete cases. The trouble with this procedure is, of course, that it must lead to a breakdown of the universal applicability of the Born-interpretation in the sense that no theoretical justification will be available for the comparison of probabilities that have been obtained in different cases, or even in different treatments of one and the same case.[176] On the other hand, the only presentation of the theory which gives a definite account of the manifold of states to be used, von Neumann's presentation, has sometimes been regarded as too narrow as it excludes as illegitimate procedures for which it provides no equivalent whatever, and which yet seem to be necessary for the calculation of some of the most important experimental applications of the theory (problem of scattering).[177] We see that one of the basic presuppositions of assumption one is still far from clear. However, suppose that a satisfactory way has been found of delimiting the totality of allowed states. Is it then possible to justify in a theoretical way the assumption that there exist changes from pure states into mixtures?

There exist two attempts at such a justification. The first attempt is based upon the Born interpretation. This interpretation associates a certain probability to each transition from originally given states into one of the eigenstates of the observable measured (let us assume that α is this observable). These probabilities will have to be obtained on the basis of counting all those systems which after measurement possess identical eigenvalues α', α'', α''', etc. From these two assumptions it follows at once that immediately after a measurement of a state that is not an eigenstate of α, the state of the system has turned into an α-mixture (this consequence has sometimes been called the *projection postulate*). Now it must be realized that this "transcendental deduction" of what was above called

---

[176] Cf. E. L. Hill, *Phys. Rev.*, Vol. 104 (1956), pp. 1173ff.

[177] J. M. Cook (*Journal Math. Phys.*, Vol. 36 (1957), pp. 82ff) shows that within Hilbert space the scattering problem can be solved only if $\iiint |V(xyz)|^2 dx dy dz < \infty$ which excludes the Coulomb case. I owe this reference to Professor E. L. Hill. In a discussion of the above paper, Bolsterli (Univ. of Minnesota) has pointed out that this does not invalidate the applicability of von Neumann's approach to the problem of scattering. The solution lies in not working with the complete Coulomb field, but in using a suitable cutoff. This is, of course, a possible procedure; but it is very ad hoc and not at all satisfactory as long as a general procedure for determining the size of the cutoff is not available.

the second assumption works only on condition that Born's statistical interpretation is universally applicable, and this is exactly what we want to find out. For it is the universal applicability of Born's rules which guarantees the observational completeness of the theory. What we have shown is that this universal applicability can be guaranteed only if we add the projection postulate to the theory: the projection postulate is a necessary condition of the observational completeness of the theory. However, is it possible to justify this postulate in an independent way?

An attempt at an empirical justification of the postulate which also brings in the empirical considerations demanded by assumption three [243]is due to von Neumann. [178] Von Neumann interprets the Compton effect as a quantum mechanical measurement. The quantity to be measured is any coordinate of the place of collision P. One way of measuring P (method 1) consists in determining the path of the light quantum after the collision. Another way of measuring P (method 2) consists in determining the path of the electron after the collision. Now assume that M_I has been performed. Before we could only make statistical assumptions about its outcome (i.e., about ϑ or about P). But as tg $\alpha = \frac{\lambda}{\lambda + \lambda_c}$ tg $\frac{\vartheta}{2}$ (λ the wave length of the incident photon which is assumed to be known, λ_c the Compton wave length, α and ϑ as indicated) the outcome of M_2 is certain once M_I has been performed and the result used: M_2 leads to exactly the same result as M_I. It follows that a state in which the value of P was not well defined is transformed by M_I, i.e., by a measurement of this very quantity, into a state in which its value is well defined. Or, by generalization: a general state is transformed into an eigenstate of the quantity measured.

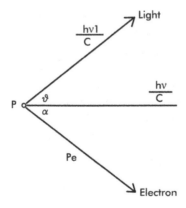

[280][178] *Op. cit.*, p. 212.

The simplicity of this argument and the force derived from it is only apparent. It is due to the fact that a rather simple way has been used of describing what happens before and after the interaction of the electron and the photon. The application of wave mechanics[179] to the problem shows that the interpretation of the result in terms of quantum jumps is only a first approximation unless one has already introduced this hypo[244]thesis from the very beginning and used it during the calculation. Also the detailed account is much too complicated to allow for a simple argument such as the one just presented. We know too little about processes of interaction to be able to make any experimental result the basis for an argument with regard to certain features of the theory. Another, completely different argument against the validity of von Neumann's empirical derivation of quantum jumps is the following: the state of the system [electron + photon] after the interaction is a state of type (2) (Section 4) with α and ϑ being correlated in a manner similar to the manner in which α and β are correlated in (2). Now an observation of a specific value of a can lead to the prediction of the correlated value of ϑ (or of β in the example of Section 4) only if immediately after the observation the state of the system [electron + photon] has been reduced to a state in which both α and ϑ have sharp values, i.e., only if the projection postulate has already been applied. Which shows that even if von Neumann's rather simple assumptions about the interaction process were acceptable, that even then the argument would have to be rejected as being circular.

However we have not yet discussed all the difficulties. Even if the argument were admissible, and even if it represented the situation in a sufficiently detailed way, and without circularity, even then it would only show that the projection postulate is *compatible with experience*; it would not remove the *theoretical difficulties* which are connected with it. These theoretical difficulties which I shall call the *fundamental problem of the quantum theory of measurement* consist in the following facts: (1) the Schrödinger equation transforms pure states into pure states; (2) in general the situation described by a mixture cannot be described by a single wave function; hence, (3) the projection postulate cannot be explained on the basis of Schrödinger's equation alone, and it is even inconsistent with it if we assume that this equation is a process equation which governs all physical processes at the microlevel. The discussion of this fundamental problem leads straight into the quantum theory of measurement.

---

[179] Cf. i.e., Sommerfeld, *Atombau und Spektrallinien*, Braunschweig, 1939, Ch. VIII; W. Heitler, *Quantum Theory of Radiation*, 3rd edition, sec. 22ff.

11. MEASUREMENT

A measurement is a physical process which has been arranged either for the purpose of testing a theory or for the purpose of determining some as yet unknown constants of a theory. A complete account of measurement will give rise to at least three sets of problems. First of all, there is the problem of whether the statements obtained with the help of the [245]experiment are relevant, i.e., whether they indeed concern the theory or the constants in question, and under what conditions. Problems of this kind may be called *problems of confirmation*. In the present paper they will not be dealt with in a systematic manner. Another set of problems concerns the question whether the observable elements of the process (or of the equilibrium state in which it usually terminates) stand in a one-to-one relationship to the (not necessarily directly observable) elements whose properties we are investigating when performing the measurement. This set of problems may be split into two parts viz. (1) the question under what circumstances a situation may be called observable; and (2) the question whether it is possible *by physical means* to bring about a situation in which observable states are correlated with nonobservable states in such a manner that an inference is possible from the structure of the first to the structure of the second. Question (1) is a question of psychology. It is the question whether and how human beings, assisted or not assisted by instruments, react towards situations of a certain kind. Problems connected with question (1) will be called *observer problems*. Question (2) is a question of physics. It is the question whether the physical conditions which must be satisfied by a well-designed measurement are compatible with the laws of physics. Problems connected with question (2) will be called *physical problems of measurement*. Quite obviously the physical problems admit of a solution only if first a physical characterization, in terms of some theory, is available of those states of affairs which are observable.

In the present section we shall be mainly concerned with the physical problems of measurement in the quantum theory, although we may occasionally also deal with observer problems and problems of confirmation. More especially, we shall be dealing with the question whether and how the physical conditions which must be satisfied by any well-designed measurement of a quantum mechanical observable can be made compatible with the dynamical laws, and especially with Schrödinger's equation. This essentially is our problem: the theory is observationally complete only if the projection postulate is added to its basic postulates. The projection postulate is incompatible with the unrestricted validity of Schrödinger's equation. How can this apparent inconsistency be resolved?

According to Popper this problem is only an apparent one and it "arises in all probability contexts".[180] For example "assume that we have tossed [a] penny and that we are shortsighted and have to bend down before we can observe which side is upmost. The probability formalism tells us then that each of the possible states has a probability of ½. So we can say [246]that the penny is half in one state and half in the other. And when we bend down to observe it the Copenhagen spirit will inspire the penny to make a quantum jump into one of its two eigenstates. For nowadays a quantum jump is said ... to be the same as a reduction of the wave packet. And by 'observing' the penny we induce exactly what in Copenhagen is called a 'reduction of the wave packet'".[181]

This is a very seductive proposal indeed, for there is no uneasiness combined with the classical case. But this is due to the fact that *in the classical case we are not dealing with wave packets*. What we are dealing with are mutually exclusive alternatives only one of which, *and this we know*, will be realized in the end. The classical description allows for this possibility for it is constructed in such a manner that the statement "the probability that the outcome is a head is ½" is compatible with "the outcome is actually a head". We may therefore interpret the transition from the first statement to the second statement as the transition from a less definite description to a more definite one, as a transition that is due to a change in our knowledge but that has no implications whatever as regards the actual physical state of the system which we are describing. The second statement does not assert the occurrence of a process that is denied by the first statement. Hence, the "jump" that occurs is a purely subjective phenomenon and a rather harmless phenomenon at that.

This is not so in the quantum mechanical case. Consider for this purpose a measurement whose possible outcomes are represented by the states Φ' and Φ'' and which occurs when the system is in a state $\Phi = \Phi' + \Phi''$. In this case Φ cannot be regarded as an assertion to the effect that one or two mutually exclusive alternatives, Φ' or Φ'', occurs, for when Φ

---

[180] *Observation and Interpretation*, p. 88.   [181] *Op. cit.*, p. 69.

is realized physical processes may occur which do not occur either when Φ' is realized or when Φ'' is realized. This, after all, is what interference amounts to. The transition, on measurement, from Φ to, say, Φ'', is therefore accompanied by a change in physical conditions which does not take place in the classical case. Even worse: what the Schrödinger equation yields when applied to the system Φ + a suitable measuring apparatus is, strictly speaking, another pure state $\Psi = A\Phi' + B\Phi''$ so that now *a mere look* at the apparatus (after which we assert that we have found Φ'') seems to lead to a physical change, namely to the destruction of interference between $A\Phi'$ and $B\Phi''$. It is this characteristic of the quantum mechanical case that is completely overlooked in Popper's analysis.[182]

Another fairly simple suggestion is due to Landé. Landé tries to solve what we have above called the fundamental problem of measurement by [247]denying that Schrödinger's equation is correctly interpreted as a process equation. In order to show this, [183] he proceeds from the usual interpretation of $<\phi|\alpha_i>$ and $<\phi|\beta_k>$ (α and β two complete sets of commuting observables) as different expectation functions belonging to *one and the same state*, and he suggests that this interpretation be also used in the case of the temporal development of states, i.e., he suggests that also $\Phi(t')$ and $\Phi(t'')$ be interpreted as two different expectation functions belonging to the same state rather than as two different states. If this interpretation is adopted,[184] then Schrödinger's equation no longer plays the role of a process equation which transforms states into other ("later") states, but it then plays the role of an equation which transforms expectation functions of one state into other expectation functions *of the very same state*. In this interpretation, states never change unless we perform a measurement in which latter case there occurs a sudden transition from a state into one of the eigenstates of the observable measured. This procedure only apparently removes the temporal changes. Strictly speaking it is nothing but a verbal manoeuvre. By pushing the temporal changes into the representatives it now makes the dynamical variables time-dependent, whereas in the usual presentation, which is criticized by Landé, the variables do not change in time. However a representation of the quantum theory in terms of stable states

---

[182] I discussed this difficulty with Professor Popper on a beautiful summer morning driving from London to Glyndebourne and he seemed then to agree with my arguments (1958).

[183] *Am. Journal for Physics*, Vol. 27 (1959), Nr 6 as well as *Zeitschrift für Physik*, Vol. 153 (1959), pp. 389–393.

[184] Cf. my critical note in *Am. Journal of Physics*, Vol. 28 (1960), Nr 5 {reprinted as Chapter 19 of the present collection}.

and moving variables is well known: it is the Heisenberg representation.[185] And as this latter representation can be shown to be equivalent to the one Landé wants to abandon (the Schrödinger representation) it follows that his criticism, and his alternative suggestion completely lose their point.

We now turn to a very brief examination of von Neumann's theory of measurement. [186] In von Neumann's investigations the projection postulate and Schrödinger's equation are given equal importance. The process of measurement itself is regarded as an *interaction* between a macrosystem (represented by some wave function) and a microsystem (represented by another wave function). The main result is that the projection postulate is compatible with the formalism of wave mechanics and Born's interpretation. This theory has been attacked for epistemological, physical, and mathematical reasons, and it seems now definitely established that it cannot be regarded as a satisfactory account of the process of measurement in the quantum theory. The *epistemological* difficulty of von Neumann's theory[187] consists in this: the theory allows for the application of the projection postulate even on the macrolevel, and it then leads to the paradoxical result that by simply taking notice of a macroscopic trace the observer may destroy interferences and thus [248]influence the physical course of events. [188]

---

[185] Landé's very valuable investigations (cf. especially *Foundations of Quantum Theory, A Study in Continuity and Symmetry*, New Haven, 1955) are much more adapted to the Heisenberg representation, and therefore much closer to the Copenhagen point of view than are the interpretations which proceed from the Schrödinger representation. They may be regarded as an attempt to derive the Born interpretation and the completeness assumption (which Landé uses in the above-mentioned book, pp. 24f, but which he unfortunately drops later on) as well as some other characteristic features of the quantum level (quantum jumps; superposition) from purely thermodynamical considerations and some very plausible philosophical assumptions. At this stage the result of Landé's investigations is much closer to the Copenhagen point of view than both Landé and the orthodox are prepared to admit (cf. a similar judgment by H. Mehlberg, *Current Issues in the Philosophy of Science*, p. 368.) This is only one of the many instances in the history of the quantum theory where people passionately attack each other when they are in fact doing the same thing.

[186] *Op. cit.*, Ch. VI.

[187] For a more detailed discussion cf. my papers "On the Quantum Theory of Measurement", in *Observation and Interpretation*, pp. 121–130 (reprinted in *PP1*, ch. 13), and the corresponding paper in *Zs. Physik*, Vol. 148 (1957), pp. 551ff.

[188] It was E. Schrödinger who first drew attention to this paradoxical consequence of the theory. Cf. his article "Die gegenwärtige Lage in der Quantenmechanik", *Naturwissenschaften*, Vol. 23 (1935), p. 812. Cf. also the discussion by P. Jordan in *Philosophy of Science*, Vol. 16 (1949), pp. 269ff.

The source of this difficulty is easily seen: it lies in the fact that a *microaccount* is given both of the measuring apparatus and of the system investigated, although only the latter can be said to have microscopic dimensions and although only the latter is investigated in such a detailed manner that its microscopic features and their dual nature become apparent. Clearly such a procedure will not reflect properly the behavior of macroobjects *as seen by a macroobserver* as it does not contain the approximations which are necessary for a return to the classical level. One of the most obvious physical consequences is that the entropy of the total system [micro + macrosystem] remains unchanged as long as the projection postulate is not applied. This is very different from the corresponding result in the classical case, a difference which cannot be ascribed to the appearance on the microlevel of the quantum of action.[189] It seems therefore advisable to employ a greater latency of description when discussing the process of measurement. A very similar suggestion emerges from some considerations which are due to Elsasser:[190] The observation of the trace, or of the macroscopic movement which terminates the process of measurement, will usually take some time interval Δt of macroscopic dimensions. During this time the measuring apparatus is supposed to retain its main *classical* properties, or, to express it differently, it is supposed to remain an element of a statistical ensemble which is defined in a way which depends on the imprecision of macroscopic operations. Now if we assume that the measuring apparatus is in a pure state in which variables complementary to variables on which the main apparatus variables depend in a decisive manner possess sharp values, then such constancy of classical properties over a classically reasonable time interval cannot any more be guaranteed. Suppose, for example, the pure state in which the apparatus allegedly dwells is one in which all the elementary constituents of the apparatus possess a well-defined position. Then formula (1) will predict that the corresponding momenta will range over all possible values, i.e., the system will disintegrate in a very short time.[191] Hence, "if systems of many degrees of freedom are

---

[189] As a matter of fact, we at once obtain an H-theorem when we introduce the usual subdivision of phase space expressing the limitations of measurement on the macrolevel. For a first application of a procedure of this kind cf. J. von Neumann, *Zs. Physik*, Vol. 57 (1929), pp. 8off.

[190] *Phys. Rev.*, Vol. 52 (1937), pp. 987ff.

[281][191] Cf. also Ludwig's considerations concerning the possibilities of measurement and of actually creating such a state in *Grundlagen der Quantenmechanik*, pp. 171f.

involved the possibility of giving a unique quantum mechanical representation of a system by a pure state, and the possibility of leaving it in approximately the conditions under which it appears as sample of a given collective, will in general exclude each other". [192] Finally I would like to draw attention to the *mathematical* fact, first pointed out by Wigner,[193] that only an approximate measurement is possible of operators which do not commute with a conserved quantity. [194] All these results taken together make it very clear that the [249]problem of measurement demands application of the methods of statistical mechanics *in addition* to the laws of the elementary theory. A similar suggestion seems to emerge from the analysis of P. Jordan[195] and of H. Margenau.[196] In a very suggestive paper, Jordan has pointed out that the application of statistical considerations may lead to the elimination, on the macroscopic level, of the very troublesome interference terms which in von Neumann's account were removed with the help of the projection postulate. Margenau, on the other hand, has drawn attention to the fact that no stage of a real measurement is correctly described by the projection postulate. The postulate does not correctly describe the state of the system investigated after the result of the measurement has been recorded (in the form of, i.e., a macroscopic trace on a photographic plate). The reason is that the process of recording very often destroys the system. [197] Nor does the postulate describe the state of the system before the recording. The reason is that at this moment the beams corresponding to the various eigen-functions of the observable measured are still capable of interfering so that the system cannot be said to dwell in any one of them to the complete

---

[192] *Loc. cit.*, p. 989.   [193] *Zs. Physik*, Vol. 131 (1952), pp. 101ff.

[194] For a more detailed account cf. the paper "On the measurement of Quantum Mechanical Operators" by H. Araki and M. M. Yanase which the latter author was kind enough to let me have prior to its publication.

[195] *Loc. cit.*

[196] *Physics Today*, Vol. 7 (1957); *Philosophy of Science*, Vol. 25 (1958), pp. 23–33. Cf. also John McKnight "The Quantum Theoretical Concept of Measurement", *Phil. of Science*, Vol. 24 (1957), pp. 321–330 and Loyal Durand III, Princeton 1958, *On the Theory and interpretation of Measurement in Quantum Mechanical Systems* (available in stencilled form).

[197] I have not been convinced that this difficulty which plays a central role in Margenau's considerations is a difficulty of principle rather than a technical difficulty that can be superseded by the construction of a more efficient measuring apparatus. Margenau himself has indicated that in the case of the position measurement of either a photon or an electron the Compton effect may be used in such a manner that the system investigated is not destroyed.

exclusion of dwelling in a different one.[198] Taking all this into account, Margenau drops the projection postulate and assumes that state functions are objectively real probabilities which are *tested* by a measurement without being *transformed* by it into a different state function. What is important in such a test is not the fact that the state of a system has become an eigenstate of the observable measured; what is important is the emergence of a set of numbers which is the one and only result of the measurement and its only point of interest.

It seems to me that none of the objections I have reported in the last paragraph can be raised against a theory that has been developed by G. Ludwig of the University of Berlin.[199] Ludwig's account is based upon

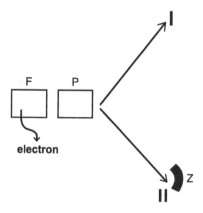

We may also use, at least theoretically, the mutual scattering of photons or the correlations between polarization which exist in the radiation created by the annihilation of a positron–electron pair (cf. the literature in footnote 94). Concerning the use of a polariser the following procedure suggests itself: take the example of a photon and examine whether it left the polariser P (which we shall assume to [282] be capable of transmitting photons of mutually perpendicular polarization in two different directions, I and II) in direction I. Assume furthermore that only such photons are counted whose passage through a filter F (and through P) can be ascertained by catching Compton recoil electrons that have been emitted by F. Of those photons we know with certainty that they have passed P and will therefore travel along I and II. Now if immediately after the capture of a recoil electron no reaction is observed in a photomultiplier Z that is located in II then we may be certain that the photon left in direction I with the opposite polarization – and this without any interference with the photon itself *after* its passage through P. (This example is an elaboration of a suggestion made by E. Schrödinger in *Naturwissenschaften*, Vol. 22 (1934), p. 341.)

[198] I would like to point out that this is an attempt, on my part, to reconstruct Margenau's argument which is not too clear on this second point.

[199] *Die Grundlagen der Quantenmechanik*, Ch. V.

the Schrödinger equation and certain assumptions concerning the formal features (in terms of the elementary theory) of the macroscopic level. The following result is obtained: measurements are complicated thermo-dynamic processes which terminate in a state in which the macroobser-vables of the measuring instrument M which are correlated with the microproperties of the system S under investigation possess fixed values, i.e., values that are independent of the nature of the macroprocedure which has led to their determination. The projection postulate is not used anywhere in the calculation of the expected changes of either S or of M. Yet the theory results in something very close to this postulate; for in the equilibrium state that terminates the measurement the expectation [250]values of obtaining certain macroscopic results are identical with the expected values for the correlated microscopic properties. It is for this reason that Ludwig regards the projection postulate as an "abbreviated account of a very complicated process".[200] It would seem that a very careful interpretation is needed of this result. It cannot mean that the measurement transforms the state of the system S into a state that is very close to, though not identical with, the state predicted by the projection postulate. This interpretation is excluded by virtue of the fact that the measurement leads in many cases to a destruction of the system investi-gated. All we can say in *these* cases is that the final macroscopic situation adequately mirrors the number of the eigenstates of the observable meas-ured and their relative weights in the state of S *before* the measurement commenced. Such an interpretation which is also defended by Margenau would be very close to Bohr's where "the 'properties' of a microscopic object are nothing but possible changes in various macroscopic systems".[201] Indeed, I must confess that I cannot see a very great

---

[200] "Die Stellung des Subjekts in der Quantentheorie", in *Veritas, Iustitia, Libertas*, Fest-schrift zur 200 Jahresfeier der Columbia University; Berlin 1952, p. 266. This paper also contains a popular account of Ludwig's theory and a comparison of this theory with the customary point of view as expressed by von Weizsäcker.

[201] *Grundlagen*, p. 170. Other accounts which regard the projection postulate as an approximate description of a more complicated process have been given by Bohm, *Quantum Theory*, Princeton, 1951, Ch. 22; Green, *Nuovo Cimento*, Vol. IX (1958), pp. 88off; Daneri-Loinger, *Quantum Theory of Measurement*, Pubblicazioni dell'Istituto Nazionale di Fisica Nucleare, 1959, esp. sec. 9. All these alternative accounts however, use examples and by no means as general and detailed as Ludwig's account.

[201a] In their most recent communication (*Quantum Theory of Measurement and Ergodicity Conditions*) A. Daneri, A. Loinger, and G. M. Prosperi quote a remark by Professor L. Rosenfeld to the effect that "the conception of Jordan and Ludwig is in harmony with the ideas of Bohr". As regards Jordan's intuitive approach this may well be the case. But Ludwig's ideas on measurement most definitely deviate from those of

difference between Margenau's suggestions and the theory of Bohr where the microsystem and the macrosystem are supposed to form an indivisible block the only changes of which are those that can be described in classical terms. As the above quotations show Ludwig himself regards his own theory as an attempt at a more formal presentation of Bohr's point of view. And this it is but only to a certain extent.[201a] For whereas in Ludwig's account the properties of the macroscopic level agree with the properties required by the classical physics only to a certain degree of approximation, such a theoretical (and practically negligible) difference is not admitted by Bohr: according to Bohr the measuring instrument is *fully classical* and restrictions occur only if we try to understand the microsystems, per analogy, in classical terms. Also Bohr's account is not beset by the mathematical[202] and philosophical[203] difficulties that are still present in Ludwig's theory. Altogether his semiqualitative ideas still seem to be preferable to all those very sophisticated mathematical accounts (including Ludwig's) where measurement is regarded as an *interaction* between systems that can be described, either exactly, or to a certain degree of approximation, by the formalism of the elementary theory. Let us briefly recall the main features of Bohr's theory: we are concerned with macrosystems which are described in classical terms, and with the calculation of expectation values *in these systems only*.[204] The properties of a microobject are nothing but possible changes in these macroscopic systems. This interpretation of measurement removes most of the unsatisfactory features of theories of interaction. However for this [251]interpretation of the process of measurement – and with this remark we resume our discussion of the observational completeness of the quantum theory – the truth of what we called above the second and the third assumption

---

Bohr who would never dream of representing the state of a macrosystem by a statistical operator. Considering the difficulties of Ludwig's approach, such an identification would also seem to be unfair to Bohr.

[202] One difficulty which has been pointed out by Ludwig himself (*Z. Naturforschung*, Vol. 12a (1957), pp. 662ff) and which is also discussed in Daneri-Loinger, *op. cit.*, is that no satisfactory definition has been given of macroobservables.

[283][203] According to Hilary Putnam, Ludwig's theory leads to the following difficulty: the final stage, in Ludwig's theory, is interpreted as a purely classical mixture only one of whose elements may be assumed to really exist. The derivation, from the formalism of the quantum theory, does *not* give us such a mixture. It rather provides us with a mixture all of whose elements have an equal claim to existence. Thus the transition to the usual classical interpretation of the resulting mixture is completely unaccounted for.

[204] For a detailed account along these lines cf. H. J. Groenewold's essay on measurement in *Observation and Interpretation*.

now becomes of paramount importance: if all statements of the theory are to be about macroscopic situations then it is decisive indeed to show (a) that for every observable α there exists a classical device capable of transforming any state into an α-mixture; and (b) that the elements of the resulting mixture can again be observed as macroscopic modifications of measuring instruments. And it is equally important to show that all the magnitudes that are customarily used for the description of quantum mechanical systems are observables in the sense of the theory (or hyper-maximal operators, if von Neumann's approach is adopted). Needless to say, this problem is far from being solved. But the situation is even worse. The difficulties of the problem of observation in the quantum theory seem to be much greater than the difficulties of the analogous problem within say, classical point mechanics and this despite the fact that a great deal of the former theory was constructed with the explicit purpose of not admitting anything unobservable. The reason can be easily seen. Fundamentally any property of a classical system of point particles can be calculated from the positions and momenta of the elements. In the quantum theory, the existence of the commutation relations necessitates the use of a new instrument for any function of non-commuting variables. This greatly increases the number of measuring instruments required for giving meaning to the main terms of the theory. It can be shown that this number must be Aleph One. If we now realize that so far measuring instruments have been found only for the simplest magnitudes, and that there does not seem to exist any way of finding instruments for the measurement of more complicated magnitudes such as, for example, the angle between two mutually inclined surfaces of a crystal[205] then we must admit that the idea of the observational completeness of the quantum theory is not far from being a myth. Also the empirical content of a theory usually contains the preceding theories as approximations. Despite the many assertions to the contrary and despite the fact that the idea of a "rational generalization" is built in such a manner that a transition to the classical level seems to be an almost trivial affair[206] no proof is as yet available to the effect that the *existing* theories contain the classical point mechanics as a special case. This further reduces their empirical content.

Taking all this into account we seem to arrive at the following paradoxical result: more than any other theory in the history of physics (the

---

[205] Cf. E. Schrödinger in *Nature*, Vol. 173 (1954), pp. 442.   [206] Cf. footnote 62.

[252]Aristotelian physics perhaps, excluded) the quantum theory has been connected with a radically empiricistic outlook. It has been asserted that we have here finally arrived at a theory which directly deals with observations (observations of the classical kind, that is). It now turns out that this theory is much further removed from what it regards as its own empirical basis, viz. classical observational results, than was any of the theories which preceded it. "Quantum mechanics" writes Schrödinger, [207] "claims that it deals ultimately, and directly, with nothing but actual observations since they are the only real things, the only source of information, which is only about *them*. The theory of measurement is carefully phrased so as to make it epistemologically unassailable. But what is all this epistemological fuss about if we have not to do with actual, real findings 'in the flesh', but only with imagined findings?". A similar sentiment is expressed by Bridgman.[208] According to him, a first glance at the quantum theory seems to show that it is a "thoroughly operational theory" which impression "is achieved by labelling some of the mathematical symbols 'operators', 'observables', etc. But in spite of the existence of a mathematical symbolism of that sort, the exact corresponding physical manipulations are … obscure, at least in the sense that it is not obvious how one would construct an idealized laboratory apparatus for making any desired sort of measurement".

It is therefore not only incorrect, and dogmatic, to say that complementarity is the *only possible* point of view in matters microphysical, there also exist grave doubts as to whether it is even a *possible* point of view, i.e., there exists grave doubt as to whether it accurately represents the one fully developed quantum theory of today, viz. the elementary quantum theory of Schrödinger and Heisenberg. And we may also say that the empiricistic and positivistic objections which some of the followers of the Copenhagen point of view have raised against alternative interpretations apply with full force to the elementary theory which, they claim, is correctly represented by complementarity. This does not diminish the great merits of this interpretation as regards our understanding of the microscopic level. It only shows that like so many other things it has its faults and should therefore not be regarded as the last, the final, and the only possible word in matters microphysical.

---

[207] *Nuovo Cimento*, 1955, p. 3.
[208] *The Nature of Physical Theory*, Dover, 1936, pp. 188f.

12. RELATIVITY AND QUANTUM MECHANICS;
THE FIELD THEORIES

In the present section I want to deal, in a very sketchy manner, with the question as to whether the actual development of the field theories has [253]confirmed Bohr's predictions. It is difficult to see this question in its proper light. After all, the great influence of Bohr's ideas which has been transmitted in various forms to what one might call the "younger generation" has led to the tendency to stick to theories of a certain kind, come what may. However, I think there are decisive arguments to the effect that this will not do. Also the field theories have started moving away from the correspondence principle and to develop formalisms which start in a fairly independent manner. It is quite clear why this must be so: the methods of quantization which have led to such tremendous successes in the quantum theory of particles and fields will continue to succeed only as long as there exist classical theories which after proper reformulation can be subject to quantization. As soon as the reservoir of classical theories is exhausted, the development of the quantum theory will either have to stop, or it will have to free itself from the correspondence principle which has been the physical rationale behind the point of view of complementarity. Such a development would also seem to be desirable in order to guarantee the symmetry between space and time coordinates which is demanded by relativity. [209] A further reason for trying to develop the quantum theory independently of the principle of correspondence which has been explained by Hill[210] seems to be this: "The classification of atomic and nuclear energy states depends heavily on the ... vector coupling model. The physical meaning of this model is that it expresses the conservation of angular momentum. In a mathematical sense it represents the Lie algebra of the 3 dimensional continuous rotation group of ordinary Euclidean space. The usual classification scheme of atomic spectra therefore becomes direct evidence for the Euclidean character of space, at least on the level of atomic dimensions. If we adjoin the continuous rotation group and the crystalline groups, we can generate the full continuous group of translations and rotations as the characteristic symmetry group of 3-dimensional Euclidean space *in the large*". This means

---

[209] Cf. J. Schwinger, *Quantum Electrodynamics*, Dover, 1959, p. XIV.

[210] "Quantum Physics and the Relativity Theory", in *Current Issues*, etc., pp. 429–441. Cf. also my comments on this paper, *op. cit.* {reprinted as Chapter 21 of the present collection}, and Hill's reply.

that the synthesis of the quantum theory and of general relativity will necessitate a complete recasting of the structure of either general relativity, or of the quantum theory. The results we have obtained so far imply that such attempts will have to be judged by their fruits, i.e., by the predictions they may produce at some future time rather than by a point of view which, despite its obvious merits, is neither infallible, nor the only possible microphilosophy.

8

About Conservative Traits in the Sciences, and Especially in Quantum Theory, and Their Elimination (1963)

It is often assumed that the development of the sciences consists in long periods of success, which are interrupted by short, but severe, crises. Successful periods are dominated by either a single theory or a few theories that complement one another. These theories are used to explain known facts as well as to predict new facts. The attempt to explain a given fact may lead to difficulties, and stronger efforts may be needed in order to overcome them. Doubt about the correctness of the theory nevertheless remains small. The researcher is confident that one day it will be possible to discover the right solution: difficulties do not count as fundamental. This is why Prof. T. S. Kuhn, who has investigated this situation in a few individual cases in his recently published book *The Structure of Scientific Revolutions*, calls such difficulties "puzzles" and not "problems". The solution of a "puzzle" may be difficult and take some time; but no change in the fundamental principles of the accepted theory is necessary.

Crises have a completely different character. The optimism that marks a successful period disappears. Problems turn up that suggest a need for of fundamental change. Numerous suggestions are made, a large number of diverse theories appear on the horizon, examined, modified, shoved aside. While the conspicuous traits of a successful period of scientific development lie in the fact that a *single standpoint* (which can consist of various theories united with each other) prevails, whose principles are strictly followed, a crisis leads to many different theories in mutual contradiction. A successful period or a *normal* period, as it is also called, is *monistic*; crises are *pluralistic*.

[281]This *factual report* about the history of the sciences and especially about the history of physics is almost always accompanied by an *evaluation*. According to this evaluation, normal periods are desired and crises are not desired. Naturally, it is acknowledged that a crisis uncovers fundamental weaknesses in the accepted theories, and that afterward, as a consequence, we progress to new and better theories. This radical progress, however, counts only as a *stage* on the way to *normal* science. Normal science is still always the *aim*. Secondly, it is often argued that, had previous theories been formulated more cautiously, a crisis would not have been necessary. This means, however, that science is basically identified with normal science. Crises are embarrassments, they are periods of confusion, which should be left behind as quickly as possible, and whose discussion is a sign of bad breeding.

The general aversion to crises has a large influence on the "ethics" of the community of scientists; it also quite decisively affects the manner and the way in which historians regard science. Let's start with this! Historians of science normally extend a period of normal science so far that it covers the whole history of science from the Renaissance until around 1900. (That is true particularly for the history of physics.) It is legitimate to interpret the scientific revolution of the 16th and 17th centuries as a crisis – after all, it marked the beginning of the empirical sciences, of the separation of empirical knowledge from metaphysics and false speculation. It is still possible for historians to ignore the fundamental feature of recent changes in physics, namely, the theory of relativity and quantum theory. But the period *between* these two points in time is normally portrayed as a period of continuous success, as a period in which more and more *known* facts are dealt with theoretically, and more and more *new* facts are discovered.

The influence of a predilection for theoretical monism on the community of scientists is perhaps even more evident. Every method that *lessens* existing difficulties and brings strange phenomena closer to accepted theories is welcome. Every method that *broadens* these difficulties and suggests that a fundamental revision, an overturn of the accepted picture of the world, is necessary, is viewed with uneasiness. The treatment of thinkers who use such methods is often very radical. Violent removal is naturally no longer possible – the community of scholars is too civilized for this. But refusal *to publish* certain researches, or refusal *to take certain* published [282]results *seriously*, is, for a *thinker*, not very different from physical death. Most outstanding scientists had to suffer from the conservatism of their colleagues. Conservatism is not the whole story, of course.

There is scarcely a field of our knowledge that is so full of daring renovations and unusual ideas as science. Even the amazing speculations of metaphysics and the fantasy of poets cannot compete with the imaginativeness of scientific theories (even though scientists possess far less freedom since they have to work with the results of observation). Nevertheless, we cannot afford to overlook the conservative traits we find in science. They affect official doctrine and the way in which physicists analyze existing difficulties. This has direct application to some fundamental problems of microphysics.

2.

The history of quantum theory is one of the most interesting and least thoroughly researched field in the history of the sciences. We have here a really outstanding example of the way in which philosophical speculation, empirical inquiry, and mathematical invention can contribute *together* to the development of a physical theory. In order to make clear contemporary *interpretation* of this theory, a brief sketch of the history of science from about 1600 to about 1920 is necessary. Obviously this sketch will be very superficial. Nevertheless, I hope it makes it possible to understand some principles that today count as the basic elements of scientific method.

We have to distinguish three different stages of development. The *first stage* is dominated by Aristotelian philosophy. This philosophy contains a hair-splitting physics supported by a radical empirical theory of knowledge. The Aristotelians at the time of Galileo are still invariably depreciated by many historians of science as unintelligent dogmatists, who in defiance of clear facts of observation hold tightly to their metaphysical speculations. That is unfair *and false*. If anyone did speculate, it was Galileo. Besides, the Aristotelians could support their aversion to the Copernican system with physical arguments that appeared to come quite directly from experience. The *second stage* is the classical physics of Galileo, Newton, Faraday, and Maxwell. This was one of the strangest chapters in the history of our knowledge. Empiricism was still the official epistemological theory – and a very militant empiricism, which is still defended. Abstract speculation was criticized, [283]hypotheses looked on suspiciously. Experiments and the mathematical deduction of laws from the results of observations counted as the only admissible methods for the achievement of physical knowledge. During this time the field of known facts widened considerably. Theories based on this new material belong to the subtlest inventions of the human spirit. And they appear to fit like a glove to the empirical ideal – observation and mathematical derivation from facts of observation: almost all these theories follow from the facts by way of a strict mathematical derivation – or, at least, they *appear* to stand in this relationship. Today we know, on the basis of the work of Duhem, Einstein, and Popper, that none of these alleged deductions are correct, that the theories *go beyond* the observed facts, and that they also *contradict* these facts. It is quite an amazing play. Physicists of this period invented new and brave theories, they were revolutionary, but they believed that their theories were nothing other than a direct repetition of observable facts. They supported this belief with an alleged *deduction* from the facts. In this way they deceived themselves as well as their contemporaries, and they talked them into believing that they had followed a strict empiricist ideology. A kind of *schizophrenia* appears to have seized these thinkers, a schizophrenia characterized by a complete break between the epistemological theory they defended, which is conservative, and their revolutionary scientific practice. Naturally, there were thinkers who noticed the difference between philosophical ideal and scientific practice. Berkeley and Hume are examples. But the success of classical physics counted as a sufficient refutation of their "typical philosophical hair-splitting". This stage ended with the crisis in physics in the 20th century.

It is scarcely possible to overestimate the shock brought forth by this crisis. For 250 years physicists had believed that they had the correct method for finding the truth, that they had applied this method correctly, and that it had lead to valuable, certain, trustworthy knowledge. Kant had even proved the *necessity* of this knowledge! Naturally, changes occurred – but these changes merely concerned details. They could even be left out of discussions of the world picture. Now it turned out that during the entire time physicists had worked with false principles. How did they react to this discovery?

Einstein drew the right consequence: science is irreconcilable with the empirical method in which many classical physicists believed (without exactly following it). A scientist invents his theories [284]in an intuitive way. Theories always extend much farther than the known facts. As a consequence, they may be refuted by future research. The fact that a theory or a general standpoint breaks down is not a sign of a defective methodology: the possibility of such an event is essential for science. Einstein has broken quite explicitly with a highly influential tradition of the *portrayal* of scientific theories. According to this tradition a scientist must present his facts before showing how the theory can be derived from the facts. However, in his first publication on the theory of relativity,

"Zur Elektrodynamik bewegter Körper", Einstein does not begin with a list of *facts*, but with general *principles* (constancy of the speed of light in all inertial systems, special principle of relativity).

Developments within *quantum theory* were based on a completely different philosophy. First came a period of trials, in which nearly no basic law remained untouched. "Every student of physics was instructed in the art of refuting basic laws", writes the philosopher Hugo Dingler in his book *Der Zusammenbruch der Wissenschaft und der Primat der Philosophie* about this period, which was not very attractive for him. Research carried out during this period (especially under the leadership of Niels Bohr) led to a very curious result: there were classical laws, which remain *exactly valid* in the microphysical field. This result suggested that although classical physics was inadequate, nevertheless it could not be *completely false*. It seemed to retain a core of factual knowledge, which had to be freed from its speculative shell. This suspicion was further strengthened by the inclination of most physicists toward classical empiricism. The breakdown of classical physics was for them a proof that that empiricism had not been applied carefully enough. Or, to put it in other words, the breakdown of classical physics was proof that classical physics was *only in part* physics. It also contained metaphysical components. The combination of this *philosophical* conviction with the *factual* discovery of classical laws that retained exact validity led to the attempt (1) to give a comprehensive portrayal of the laws of classical physics that could still be used, (2) to add to this portrayal the characteristics of atomic events that had their source in quantum effects, and (3) to find an appropriate *formal* apparatus for combining (1) and (2) in a shortened mathematical way. The correspondence principle was used for the purpose of discovering the parts of this formalism; it was later replaced by more formal methods of quantification.

[285]It is now very important to know the difference between a theory such as that just mentioned, which was obtained through the *restriction* of a previous theory, and a general theory, which suggests a conceptual system *ab ovo* without any restriction. Newton's celestial mechanics and Einstein's theory of relativity are theories of the latter kind. The concepts of Einstein's theory, for example, can be applied to the world without qualification. Of course, they are relations; but these relations are objective and independent of the special apparatus employed to observe them. The theory of relativity allows, then, for a *realistic interpretation*. Such an interpretation indicates that it is a description of the objective (relational) features of the world in which we live, and this description is either true or false regardless of whether one conducts experiments or not.

The new quantum theory, which Bohr and his colleagues wanted to create, is in principle constructed in a different way. True, its main concepts are always the concepts of classical physics, but these concepts were "cleansed" of their metaphysical elements. The purging forbids their application independently of experiments. A concept, such as, for example, *position*, was previously deemed to be universally applicable: classical particles always had some well-determined position (which did not have to be known). But now conceptual cleansing, which is closely linked to the attempt to discover the empirical core of classical physics, eliminates the possibility of general application. We can apply the concept of position or the more general concept of a particle only when certain experimental conditions are fulfilled. Many descriptive concepts are limited in a similar way. Theory consequently cannot be interpreted as a description of a microworld, which exists independently of observation; it can only say what happens when certain experimental conditions are fulfilled. It says nothing about the processes *between* two experiments. It is, therefore, nothing but a machine to predict observable facts. Bohr's principle of complementarity gives us a clear picture of the way in which this machine works, and thereby makes it possible for us to conceive, at least *partially*, processes in the microworld.

The interpretation of quantum theory just sketched was used by physicists to predict not only the behavior of atomic objects, but also the future development of microphysics itself. The prediction was the following: the fundamental structure of quantum theory, the rise of formalisms with noncommutable variables in a Hilbert space or of a purposeful extension of one such space, and the associated intuitive picture of complementarity are the result of an analysis that has uncovered the core of the facts of classical physics. This structure, this picture, is firmly grounded on experience. Both [286]elements must, therefore, be viewed as final and irrefutable. We are not faced here "with a standpoint, which one can accept or let drop, depending on whether he agrees with a particular philosophical standard or not. Rather, it has unambiguously to do with the completely evident result of an accommodation of our ideas to a new experimental situation", writes Professor Rosenfeld from Copenhagen. Although it is acknowledged that quantum theory has not yet taken its final form and that future experiments will lead to new concepts (trying out new formalisms, such as the S-matrix theory or Heisenberg's uniform field theory, is entirely legitimate), the fundamental elements remain untouched and necessary for the interpretation of formalisms. (The situation is similar in certain respects to the development of classical celestial mechanics. In this discipline, too, we may differentiate between an intuitive picture and differing formalisms. The intuitive picture consists of particles that move under the influence of forces in a three-dimensional space and have certain inertial features. The formalisms, such as Newton's, Lagrange's, or Hamilton's, the minimum principles – in the end, they are all based on the same intuitive picture, and derive their physical meaning from it.) Future development can only increase the distance between the trusted pictures of classical physics and the new ideas necessary to order the observable facts in the atomic field. A return to the determinism and objectivism of classical physics is completely impossible. This attitude has a decisive influence on the way scientists interpret existing microphysical problems.

3.

These problems will appear as relatively insignificant "puzzles", which do not call for a revision of the fundamental structure of quantum theory. We stand, therefore, at the beginning of a new period of *normal* research. The circumstances that characterize such a period and that we have briefly indicated in the first section are all present: alternatives, which work with deterministic concepts, are rejected as hopeless follies. This new period of normal research has every prospect of lasting forever. The crises of the past, after all, occurred because scientists did not hold to the facts and allowed metaphysical ideas to enter physics. Such ideas are now eliminated. Future de[287]velopment will consist in the addition of new facts and the gradual construction of a tower of facts upon a solid basis. A reconstruction of the foundations, on which the new tower is grounded, will never again be necessary.

No great philosophical talent is needed to see that the reasons provided for this prophecy are not sound. One of the most important reasons is the circumstance – the *alleged* circumstance – that the indeterministic framework and the idea of complementarity directly express experimental facts. That gives them no privileges. Experiments transfer only an *approximate* validity to the laws necessary for their description. Whenever some experimental facts are given, it is always possible to formulate alternative laws, which agree with these facts within the limits of precision of the apparatus employed. The fundamental laws defended by the orthodoxy in quantum theory are, therefore, only one possibility among many others.

Another argument advanced to defend the orthodox viewpoint is that classical concepts are required to describe experimental facts and that the idea of complementarity follows from this circumstance alone. The answer to this argument is that a new theory must naturally lead to a new interpretation and to a new description of experimental facts. It is nevertheless completely unreasonable to assume that observations are more reliable than theories, and that we can replace theoretical concepts but not observational concepts. Observational concepts are normally the concepts of a *theory* that we have so often applied, and in such well-known circumstances, that now they are well *trusted*. But familiarity is no guarantee of usefulness. Quite the contrary, often observational concepts, like those of Aristotelian physics or of the demonology of the 15th and 16th centuries (which were observational concepts!), turned out to be inadequate and were replaced with better concepts. The transition from Aristotelian physics to the classical physics of Galileo and Newton would never have occurred without such a replacement. It *is* thus possible to get rid of observational concepts and replace them with new ones. This circumstance refutes, by the way, the argument of some contemporary physicists, above all Heisenberg's and von Weizsäcker's, that the *factual* use of classical concepts hinders physicists from picturing macroscopic experimental facts in a nonclassical way. A use of this principle at the time of Galileo would have delighted the Aristotelians.

Thus it *is* possible to give up certain observational concepts and replace them with other concepts. But is such a procedure *desirable*? [288]The answer to this question clearly depends on whether we believe existing terminology to be adequate or not. Only very *fundamental* difficulties can convince us of the necessity of a deep revision. We must investigate whether contemporary quantum theory really suffers from fundamental difficulties.

It is now clear that the difference between normal periods and periods of crisis cannot help us answer this question. This difference describes the situation *from outside*, and *after* scientists have resolved it. An individual scientist, whose individual decision contributes to the phenomenon Professor Kuhn has so excellently and so clearly described, certainly cannot base his decision upon the *result* of his decision. On this point Professor Kuhn's book seems to say – even if, perhaps, not in accord with its author's intentions – that the *historical situation* may provide a valuable guide. The book suggests that a crisis announces itself, so to say, in an accumulation of unsolved problems, together with a feeling of unrest and pessimism, and it needs no independent analysis of the theoretical situation on the part of the individual scientist. This idea may be said to, bridle the horse with its tail. A mere accumulation of "puzzles" is neither necessary nor sufficient to pose *real* problems. Newton's astronomy was initially confronted with various difficulties, but only very few physicists considered this to be a sign of uselessness. Confidence flowed from past triumphs of the theory as well as from the new view the theory provided. The theory of relativity, by contrast, was not the result of an *accumulation* of difficulties in matters concerning space and time. To be sure, Michelson's experiment was there. But, first, Einstein probably did not know of this experiment when he was developing his thoughts about relativity. Secondly, this experiment was generally viewed as an uninteresting failure. It assumed the character of a *problem* only on the basis of the new principles Einstein introduced, which required the revision of the fundamental concepts of classical physics. And this is really very often the case: it is very often the case that an accumulation of "puzzles", and under some conditions a single "puzzle", *assumes the character of a fundamental problem only when seen from a new viewpoint*. It cannot be otherwise. An accumulation of small difficulties does not make a big difficulty. But the organization of these small difficulties on the basis of a new idea, which contradicts the accepted theory, can yield a big difficulty, a real problem. Whether a problem counts as fundamental or not depends, therefore, on the existence of alternatives to the existing viewpoint: [289]*it depends on the fact that we think in a pluralistic, and not in a monistic, way* (see Section 1). Of course, it is always possible future research will refute the alternatives. But without testing alternatives, there is no chance of fundamental progress. It is also clear that new theories cannot be immediately developed in full detail. We always begin with general ideas, with "metaphysical" ideas, and only later develop these ideas in more concrete, formally irrefutable, form. It follows that the development of metaphysical ideas, which contradict accepted and "successful" scientific theories, is prerequisite for scientific progress and the discovery of fundamental errors in accepted theories. The exclusiveness of normal science, the strict prohibition against the use of new ideas, the refusal to scrutinize a successful theory from outside, tend to mask existing difficulties and, perhaps, to make them completely invisible. The standard ideology of normal science hinders scientific progress. And the standard argument of contemporary empiricists, namely, that we should not obscure with metaphysical fog a situation controlled by an empirically successful theory, results in the consequence that successful theories are immune from criticism and turned into a new myth. It is, therefore, naïve to assume, as Professor Dirac recently did in a lecture, that the so-called "philosophical" difficulties of a theory do not concern physicists, and that the progress of the sciences must automatically lead to their resolution. Philosophical difficulties arise from the comparison of a physical theory with a more general principle, such as the principle of determinism or of objectivity, or with a principle that has not yet been developed in detail and contradicts the fundamental doctrines of the theory. As I have just hinted, such a comparison is the motor that drives science forward, and solving philosophical difficulties is necessary if we want to achieve fundamental progress.

To sum up, we may say that theoretical monism, which marks normal science and is currently defended by numerous physicists on the basis of "experience", hinders progress. Theoretical pluralism is a necessary condition for progress. It is theoretical pluralism, not the accumulation of "puzzles", that drives science forward and prevents it from becoming an embalmed corpse of past knowledge. The consequence, of course, is that the battle cry of science must be *revolution in permanence!*

4.

[290]What has been said in the previous three sections leads to a very interesting problem. In the past, empiricism has often been linked to a progressive and liberal stance. Empiricists have sought to overcome fear and prejudices, they were among the most enthusiastic fighters for progress and enlightenment. That began with the physicians of the Kos school, continued with the brave and humane physicians who, at the beginning of the modern age, opposed witch-hunting by putting their own lives in danger, and had its high point in the research of Galileo, Newton, Laplace, Kant (his cosmogony), Darwin, Lyell, and others. Yet we have seen that a conservative tendency is inherent in empiricism and that it has used considerations about experience to support useless systems and reject better ones. *One* reason for this rift lays in the *vagueness* of the concept of experience, which allows for the defence of nearly any system by appealing to it. (The Aristotelians defended their system by referring to experience, and attacks on belief in witches and demons have often been rejected by appealing to experience.) But there is also a second element, which we now have to consider in more detail.

5.

Traditional epistemology sets itself the task of finding the foundations of our knowledge, and of showing how what we take for knowledge is related to these foundations. They were sought first in revelation, then in pure thought. Earlier forms of knowledge were *authoritarian*, they built on what a *person* said, be it a ruler or a god. His or her words were unconditionally true, immune from contradiction. The transition to pure thought changes *one* aspect of the situation – the proto-scientist no longer heeds one person, but an abstract principle. But this principle still ruled unconditionally and absolutely, as a sort of depersonalized tribal god. Theoretically speaking *empiricism* brings about, a change scarcely worth mentioning: the authority has changed – that's all. Seen in the light of this epistemology, science is a somewhat hypertrophic excrescence of traditional epistemology, nothing more. In *practice*, however, as we have seen, the difference is immense. Bacon's writings and the dogmas of the empiricists of the "Royal Society" gave rise to a form of knowledge radically different from [291]all previous forms, with the possible exception of the natural philosophy of the Presocratics. *Every* element of our knowledge can now be changed. Neither the existing results of experience nor the existing results of pure thought are excluded from the transformation. Empirical doctrines in the new theories stand equal to theoretical considerations – they are not shoved into an unimportant corner. But they are not sacrosanct. *Science is the first detailed example of nonauthoritarian knowledge.*

The difference with the *Presocratics*, who must rank as the inventors of nonauthoritarian knowledge, lies in two points. First, that the rise of science is accompanied by philosophy. Presocratic philosophers are epistemologically naïve: they theorize bravely, without taking much care about the basis of knowledge. The second difference lies in the fact, which we have already emphasized, that this accompanying philosophy remains authoritative, and proclaims experience as the unshakeable basis of knowledge. And it is not a mere philosophical appendix without a real function, as some scientists wish to persuade themselves. It has a strong and, by and large, hindering influence on the development of sciences. For small spirits, whose naïve instinct is not strong enough to keep up with the tension between a dogmatic epistemology and a critical activity in the sciences, it has become an unbearable straitjacket. But the leading scientists take advantage of the vagueness of the concept of experience. They introduce an "experience" found in the closest proximity to their own theoretical ideas, and try to draw these ideas from the experience. In this respect, Newton's proof of the law of gravity is exemplary. The same holds for his optical theories. Amazingly, he succeeded in convincing the radical empiricists of the Royal Society of the purely empirical character of his researches. The "experience" upon which his results are grounded, is, certainly, not everyday experience; it comes mathematically reinterpreted, in the sense of abstract physical concepts (such as the concept of the indices of refraction). Some members of Society, such as Hooke, had scruples, but their scruples disappeared as soon as the fruitfulness of the new theory became clear, and they became familiar with the new "experience". In his *Theory of Colors* Goethe showed that he understood the process and could criticize it. This development proves the extensive *elasticity* that characterizes empiricism in practice, and proves, at the same time, that *dogmatism*, the second element that marks empiricism, has not yet been abandoned.

I pointed out that the tension between dogmatic [292]epistemology and critical science, which marks classical physics, ends in the 20th century. The revolutionary discoveries of the theory of relativity and quantum theory show that the classical foundations were not only not solid, but did not exist at all. I also described two reactions to this situation: Einstein's admission that in physics we do not possess any fundamental basis and should not seek one. Einstein, and after him Popper, were the first epistemologists who intentionally introduced into physics the idea of a completely hypothetical knowledge, in which experience and theory play equal roles, but also in which both are capable of change in the light of new – empirical as well as theoretical – considerations. Neither the experience nor the theory provides us with absolute knowledge. But both together lead to the construction of a picture of the world that unifies sense and abstract elements in a coherent way. This picture of the world is subtler than a purely metaphysical view; and it is also much more difficult to construct. Indeed, it must satisfy a large number of completely new conditions. Besides, it takes in a larger number of reactions of the human spirit and hence leads to more satisfying explanations than those that sensory experience push aside as irrelevant. To be sure, knowledge offered by such a picture of the world is not *grounded* on experience, but *takes* experience *into consideration*. It is richer than purely abstract knowledge, but it is also richer than purely empirical knowledge, and it needs improvement to a larger degree. It is also *more human*, for it deals seriously with the statement that all knowledge derives from humans and, therefore, is full of human *errors*.

The second reaction, described more fully above, consists in the attempt to transfer to experience the authoritative role that classical physics gave it, at least in words, if not actually. It also sought to turn off the possibility of a constant reinterpretation of "experience" through more precise and explicit definitions. Now, it is very important to recognize that the *possibility* of such a process cannot be denied. It *is* possible to break down our knowledge into two components, "experience" on the one hand and "theory" on the other, and then require that "experience" is not changed. Theories are, after all, our own work, and if we do not want to give up certain propositions, we do not need to do so. But whether such a procedure is also *desirable* is very open to question. Certainty has always been regarded as a philosophical virtue. The insight, that it can be *generated*, should free us from this belief, and teach us that every system, however securely founded, is, in fact, only our own shadow. It makes no difference whether we attach this security to thinking or to experience. Only knowledge that is always [293] in need of improvement, a completely hypothetical knowledge, can guarantee that we will *sometimes* come into contact with something different from our own work.

9

Problems of Microphysics
(1964)

[1]It is frequently assumed that the development of physics consists of long periods of success which are interrupted by short periods of crisis. Successful periods are dominated either by a single theory, or by a few theories which complement each other. These theories are used for the explanation of known observational facts and for the prediction of new facts. Attempts to explain a particular fact may be difficult, and require great ingenuity. Still, there will be little doubt that the theory is correct, and that the difficulty will one day be overcome. Difficulties, therefore, are not regarded as being of any fundamental importance. Professor Kuhn who in his recent book *The Structure of Scientific Revolutions* has investigated the matter in some detail, expresses this attitude by calling them "puzzles", and not "problems". A puzzle may need great mental effort in order to be solved; however it does not necessitate the revision of basic theory.

Periods of crisis have a very different character. The optimism characterizing the successful periods has gone; problems have turned up which seem to indicate that a drastic change is needed. Numerous suggestions are made, many different theories are proposed, investigated, abandoned. Whereas the most conspicuous characteristic of a successful period of scientific research, or of a *normal* period as it has been called more recently, is the fact that a *single point of view* is used, and strictly adhered to, a crisis leads to the emergence of a great many theories. Normal science is *monistic*, crises are *pluralistic*.

This *factual* account of the history of the sciences is almost always supplemented, at least implicitly, by an *evaluation*. According to this evaluation normal science is desirable, and crises, or revolutions, are undesirable. Of course, a crisis leads to the discovery of fundamental weaknesses in the accepted theories, and it therefore precipitates progress to new, and better theories. It leads to improvement. But first of all, it is only regarded as a *stage* in a process leading to improvement and normal science. The latter is still the *aim*. And secondly, it is often added that a crisis would not have occurred had the previous theories been formulated with due circumspection. In short, it is believed that science is essentially normal science. Crises are embarrassments; periods of confusion which should be passed through as quickly as possible and which should not be unduly extended.

The general aversion toward periods of crisis has a marked influence upon what one could call the "ethics" of a scientific community; it also influences the appointed court historians of this community – the historians of science. Let me deal with the latter first. Historians of science are usually intent on stretching any successful period of normal [2]scientific activity until it almost covers the whole of the history of science from the so-called scientific revolution of the 16th and 17th centuries until about 1900. To present the revolution of the 16th and the 17th centuries as a crisis is regarded as quite legitimate: this, after all, was the emergence of science from myth and metaphysics. Nor is it possible to overlook the fundamental character of the more recent changes brought about by the discovery of the quantum of action. But the period in between these two dates is usually represented as a period of continued success; as a period where more and more *known* facts are subject to treatment by the received theories, and where more and more *new* facts are discovered.

The effect of the predilection for a theoretical monism upon the community of scientists is perhaps even more pronounced. Any procedure which *decreases* the impact of existing difficulties is welcome. Any procedure which *increases* this impact and which suggests that the difficulty completely undermines the received opinion, is frowned upon. Radical methods are used against those engaging in such subversive activity. It is of course no longer possible to eliminate opponents by recourse to *violence*. However they are still silenced most effectively by the refusal to publish their work, or else by the refusal to take such work seriously.

Some of the most extraordinary thinkers had to suffer from this unwillingness of the scientific community to change the delicate balance between difficulties and successes which characterizes a normal period. Of course, this conservatism is not the whole story. There exists hardly another domain of knowledge that is as full of daring innovations and of unusual ideas as is science. Yet the conservatism must not be overlooked as it will influence the *official* doctrine and will thereby contribute to the manner in which existing difficulties are evaluated. This has an immediate application to our topic – the problems of contemporary microphysical theory.

The history of the quantum theory is one of the most interesting and least explored periods of the history of science. It is a marvelous example of the way in which philosophical speculation, empirical research, and mathematical ingenuity can jointly contribute to the development of physical theory. In order properly to understand its interpretation, we must give a short account of the history of science from about 1600 until now. Of course, this account will have to be very superficial. Yet I hope that it will give some insight into principles which today are regarded as fundamental elements of the scientific method.

We must distinguish three periods. The *first period* is dominated by the Aristotelian philosophy. This philosophy contained a highly sophisticated physics which, in its turn, was based upon a somewhat radical empiricism. The Aristotelians at the time of Galileo are still being described by many historians as empty-headed dogmatists and the real force of their arguments against the motion of the earth as well as the [3]difficulties which these arguments created for the heliocentric system are hardly ever properly appreciated.

The *second period* is the classical physics of Galileo, Newton, Faraday, Maxwell. This is one of the most curious periods in the history of knowledge. The official philosophy is still empiricism – and indeed a very militant empiricism it is. Speculation is discouraged; hypotheses are frowned upon; experimentation and derivation from observational results are regarded as the only legitimate manner of obtaining knowledge. The domain of factual knowledge is considerably extended. The theories based upon this new material belong to the most subtle inventions of the human mind. At the same time, they seem to fit perfectly the empiricist ideal: most of them seemed to have been obtained by a strict mathematical derivation from experience. We know now, mainly through the work of Duhem and Einstein, that none of these alleged derivations is valid; that the defended theories go *beyond* existing observational results and that they are also *inconsistent* with them.

We therefore witness here the astonishing spectacle of men, who invent new and bold theories; who believe that these theories are nothing but a reflexion of observable facts; who support this belief by a procedure which is apparently a deduction from observations; who in this way deceive both themselves and their contemporaries and make them think that the empirical philosophy has been strictly adhered to. We have here a period of *schizophrenia*, characterized by a complete break between philosophical theory and scientific practice. This is an age, when the scientist does one thing, and insists that he is doing, and must do another.

To be sure, there were a few people who were aware of the difference between the philosophical ideal and the actual practice. Berkeley and Hume are examples. But the success of classical physics was taken to expose their argument as an exhibition of typical philosophical sophistry. This period of schizophrenia is terminated by the crisis connected with the invention of the theory of relativity and the discovery of the quantum of action.

It is hardly possible to overestimate the shock created by these changes. For almost 250 years, one had believed that one was in possession of the correct method; that one had applied it properly; and that one had in this way obtained valuable and trustworthy knowledge. It was of course necessary, now and then, to revise a theory, or a point of view – but such events were regarded as being of minor importance, or were perhaps even completely neglected. It now turned out that one had all the time based one's inquiry upon the wrong foundations. It is interesting to see how scientists reacted to this discovery.

Einstein drew the correct conclusion: science is incompatible with the empirical method, at least as envisaged by many classical physicists. A scientist intuitively invents theories which always go far beyond experience and which are therefore vulnerable by future considerations. The breakdown of a theory, or of a general point of view, is not an indication of [4]faulty method; its possibility is essential to science. Einstein also quite explicitly broke with the tradition of presenting a new theory as the result of a deduction from facts. His first paper on relativity, "Zur Elektrodynamik bewegter Körper", does not start with the enunciation of facts, but of principles such as the principle of the constancy of the velocity of light in all inertial systems. The development leading to the quantum theory was based upon a very different philosophy.

To start with, there was a period of experimentation when hardly any fundamental law remained unchallenged. "Every student of physics was trained in the art of overthrowing basic laws", writes the philosopher Hugo Dingler about this period (which Dingler himself regarded with a very critical eye). The investigations carried out during this period, mainly under the guidance of Bohr, led then to a curious result: there existed classical laws which *remained strictly* valid in the microphysical domain. This suggested that classical physics, though surely not adequate, was still not *completely* incorrect. It seemed to contain a factual core which had to be freed from nonfactual trappings. This suspicion was reinforced by the inclination of most physicists to retain the classical empiricism as the correct scientific method.

The breakdown of classical physics now proved to them that this empiricism had not been properly applied. Or, to put it in different words, it proved to them, that classical physics was physics only in part. It contained metaphysical constituents. Combining this *philosophical* conviction with the *factual* discovery of classical laws which were still strictly valid, they now set out, (1) to give an account of all the valuable parts of classical physics; (2) to add to this account the features dependent on the quantum of action; and (3) to find a coherent formal apparatus for the presentation of both (1) and (2). The principle of correspondence was used with the purpose to discover the parts of this formalism, and it was later on replaced by the method of quantization which transforms its more intuitive content into a mechanical procedure.

It is most important to realize that a theory obtained in the fashion just indicated is very different from universal theories such as, for example, Einstein's theory of relativity. The concepts of Einstein's theory can be applied to the world without qualification. They are relational concepts, true, but the relations asserted to hold are objective, and independent of the specific experimental arrangement used for ascertaining their presence.

The theory of relativity is therefore accessible to a *realistic interpretation* which turns it into a description of the objective (relational) features of the world we live in; a description that is correct whether or not experiments are actually carried out. The new quantum theory that was envisaged by Bohr and by his followers could not be a theory of this kind. For though its descriptive concepts are still those of classical physics, they have been stripped of what might be regarded as their metaphysical trappings. Thus, originally a concept such as the concept [5]of position was considered universally applicable – classical particles always have some well defined position.

Now, the conceptual spring-cleaning connected with the attempt to uncover the empirical core of classical physics and to utilize it for the purpose of prediction and explanation eliminates just this possibility of universal application. We may use the concept of position, and the corresponding more general concept of a particle only if certain experimental requirements are first satisfied. Many descriptive concepts are restricted in an analogous manner. The theory therefore cannot any longer be regarded as an account of a microlevel as it exists independently of observation and experiment; it can only say what will happen if certain experimental conditions are first satisfied. It is incapable of giving an account of what happens in between experiments. It is therefore *nothing but* a predictive device. Bohr's principle of complementarity gives a very striking intuitive account of the manner in which this predictive device works and thereby provides at least a partial picture of the microworld.

The interpretation which we have just sketched was used now for making some important predictions concerning the future development of the quantum theory. The basic structure of the quantum theory, the use of formalisms with noncommuting variables acting on a Hilbert-space or a suitable extension of it, the corresponding intuitive feature of complementary aspects in the world are the result of an analysis that has bared the observational core of classical physics. This structure and these features are at last firmly based upon experience, and they are therefore final and irrevocable. We are here not presented "... with a point of view which we may adopt, or reject, according to whether it agrees, or does not agree with some philosophical criterion. It is the unique result of an adaptation of our ideas to a new experimental situation", writes Professor Rosenfeld of Copenhagen.

It is of course admitted that the quantum theory will have to undergo some decisive changes in order to be able to cope with new phenomena and that physicists will be requested to introduce new concepts for the description of new facts. It is also considered legitimate to try different approaches and to develop different formalisms such as Chew's S-matrix theory for strong interactions, Heisenberg's unified field theories, and others. Nevertheless, it will be pointed out that however large these changes, and however different these formalisms, they will always leave unchanged the basic elements just mentioned, which are in any case needed to give them empirical content. The future development, therefore, can only be in the direction of greater indeterminism, and still further away from the point of view of classical physics. This attitude influences in an important way the evaluation of the existing theoretical problems of microphysics.

It is clear that such problems will now appear to be minor "puzzles" which do not necessitate a revision of the basic structure of the [6]quantum theory. It may be granted that new mathematical techniques will be needed for the purpose of bringing about a closer agreement between theory and experiment; yet it will be emphasized that these new techniques are still in need of interpretation, and that the interpretation must be carried out in the customary manner, by reference to the idea of complementarity. It is also admitted that some of the notions which had to be introduced in order to account for features going *beyond* complementarity may be faulty, and need qualification. (Nonconservation of parity in cases of weak interaction is a case in point.)

Still, such modifications will not affect the indeterministic framework. We have here therefore the beginning of a new period of *normal* science. The signs characteristic for a normal period such as aversion toward a different approach in fundamental matters are already present. Moreover this normal period may well be expected to last forever. Previous crises were due to the existence of metaphysical ingredients in the basis of science. These ingredients have been removed. The further development will therefore consist in the gradual addition of new facts and the gradual erection of a tower of facts on a solid basis. Minor crises may still occur, but they will modify only the upper layers of the tower. They will not necessitate the recasting of the fundaments on which the tower rests.

Now it should be obvious that the *reasons* given for such a prophecy are not acceptable. One of the main reasons given is that the indeterministic framework, the idea of complementarity and its corollaries are an immediate expression of experimental fact. This does not put them into a unique position. Experiment transfers upon the laws used for its description and approximate validity only. Given a set of experimental results, it is therefore always possible to devise alternative sets of experimental laws which agree with them within the margin of error characteristic for the equipment used.

The set of basic experimental laws, defended by the orthodox, is therefore only one set among many possible sets. Another argument given in defence of the point of view of the orthodox is that we need the descriptive terms of classical physics for expressing experimental results and that it is for this reason that alternative accounts are bound to fail. They might succeed in producing a *formalism*. However, in the attempt to give empirical meaning to this formalism the classical terms will again have to be used and with them the point of view of complementarity. As regards this argument, it must be noted that a new point of view will of course also have to provide new terms for the description of the observational level.

It is quite unreasonable to assume that observational terms possess greater stability than do theoretical terms. Observational terms usually are the terms of some theory which, having been applied under the most common circumstances, have become *familiar*. However familiarity is not a guarantee of adequacy. Quite the contrary, it has been found in the past that certain observational terms, such as the terms which were [7]used in the Aristotelian physics, were inadequate and in need of replacement. The transition to the classical physics of Galileo and Newton would have been quite impossible without such replacement.

It is therefore possible to abandon a given observational terminology and to replace it by a different terminology. This, by the way, shows the absurdity in the argument of some contemporary physicists, notably Heisenberg and von Weizsäcker, which maintains that since physicists are endowed with a classical observational vocabulary they will forever be unable to look at observable matters in a different manner. It is clear that an application of this principle at the time of Galileo would have given great comfort to the Aristotelians.

To repeat: It is possible to abandon a certain observational terminology and to replace it by a different terminology. But *should* we do such a thing? Clearly the decision will depend on whether we have convinced ourselves that the terminology in use is inadequate. Only very *fundamental* difficulties could have sufficient persuasive power. It is therefore necessary for us to find out whether the present quantum theory is indeed faced with fundamental difficulties.

It is clear that the distinction between periods of crisis and normal periods does not help us to answer this question. This distinction describes the situation *from the outside*, as it were, and *after* scientists have made up their mind. The individual scientist whose individual decision will contribute to the phenomenon which Professor Kuhn has described so clearly cannot base his decision on what is a *result* of the decisions made. At this point, Professor Kuhn's book seems to suggest, though perhaps not in accordance with the intentions of the author, that the *historical situation* might provide a valuable guide, a guide moreover, whose advice *should not be neglected*.

It suggests that a period of crisis *announces itself*, by an accumulation of unsolved problems, and by a feeling of general uneasiness and frustration, and that a separate analysis of the total theoretical situation by the individual scientist is not needed. This, it seems to me, means putting the cart before the horse. A mere accumulation of "puzzles" is neither necessary, nor sufficient for creating *genuine* problems and for suggesting that what is needed is a basic revision, and not only a minor adjustment. At the beginning, Newton's astronomy was faced with an ever increasing number of difficulties – but nobody, or only very few people saw in this an indication that Newton's theory was incorrect. The past triumphs of the theory, the new vision provided by it, the astonishing initial successes inspired sufficient confidence.

The theory of relativity, on the other hand, was not preceded by an *accumulation* of difficulties. Most of the problems which can now be cited in favor of its invention had already been solved by Lorentz's theory of the electron. What Einstein was after was a general principle, comparable to the second law of thermodynamics, that could be relied on in the changes which seemed to surround the discovery of the quantum [8]of action. He obtained this principle from fairly simple theoretical considerations, involving the symmetry properties of classical mechanics and of classical electrodynamics, and not under the pressure of adverse empirical material.

The further development leads to the general theory of relativity. It is this new point of view which raises certain annoying "puzzles", such as the excess movement of the perihelion of Mercury, into genuine problems, and thereby precipitates progress. And *it could not be otherwise*. After all, an accumulation of puzzles will cease to be regarded as a challenge to calculation on the basis of the old theory and will assume the appearance of a genuine new problem only if it is seen on the background of a new point of view that imposes upon it a coherence different from the coherence postulated by the established doctrine.

The recognition of a problem as fundamental therefore depends on the development, *in addition to the received theory*, of a new point of view. This new point of view may of course be refuted by future research; but without it, there is no chance of fundamental progress. Nor will it be possible to develop new points of view at once in a very detailed fashion – a vague metaphysical idea will in most cases be the starting point. The fruitfulness of new points of view is confirmed, apart from these more abstract considerations, by many episodes in the history of the sciences, and especially by the history of the atomic theory itself.

The exclusiveness of normal science; the strong moralizing against the use of new points of view which deviate from the established modes of thinking is therefore self-defeating. If successful, it will prevent the physicist from discovering the weaknesses of their favorite theory. It is also a little naive to suppose, as Professor Dirac did in a recent talk, that the more general, or "philosophical" difficulties of a theory are of no importance for a physicist and that the progress of physical theory will automatically dissolve them. The situation is just the other way around.

Philosophical difficulties emerge from the comparison of a physical theory with a more general point of view that has not yet been developed in detail and that is not in agreement with the basic principles of the theory. As we have pointed out above, such comparison is the motor which propels science, and the solution of the difficulties created by it is necessary if we want to achieve not only a few minor adaptations, but also fundamental progress. It is very important to repeat that the standard ideology of normal science, viewed in the light of these considerations, turns out to be a hindrance to such progress. Theoretical pluralism should be encouraged, and not frowned upon. It is only such pluralism, and not the accumulation of "puzzles" which advances knowledge and which prevents it from becoming the perfectly embalmed corpse of what at some time was an exciting discovery.

The consequence is, of course, that in science, unlike in politics, the battle cry should be – *revolution in permanence!*

10

Peculiarity and Change in Physical Knowledge
(1965)

[197] Physical knowledge has existed in nearly all peoples and at all times. For the most part it was intuitive, disorganized, and impregnated with superstition. Even the Egyptians, whose architectonic abilities still amaze us, and whose technique for moving heavy bodies is not yet completely understood, acted more by rules of thumb than according to comprehensive theory. Of course, they had, a picture of the world that allowed them to order individual events into a broader context and thereby make them understandable. But the assumptions on which their picture rested depended too much on emotion, were too much impregnated with religious feelings, to be regarded as objective descriptions of reality. Rational, objective thought about nature, and therefore theoretical physics, first begins with the Greeks, above all the Presocratics.

Only a few thinkers have appreciated the achievements of these brilliant men correctly. Everyone agrees that without them the development of philosophy and the natural sciences would not have been possible. But, to the question *in what* did the uniqueness and importance of the contribution of the Presocratics for our natural science consist, and in what way did and does *knowledge* of nature differ from a *myth* about nature, no single and satisfying answer can be found. Details are reported: Thales, Anaximander, Anaximenes tried to trace the great variety of appearances on Earth and in the starry heavens back to a few principles, in the ideal case to *one*. The principles chosen were materialistic. Heraclitus' principle is always in motion, and thereby enriched the Presocratic repertoire with an idea that has been very important for modern physics as well as for natural philosophy: the world does not consist of stable things, which are acted upon from outside and, therefore, either slowly [198]decay or change

211

into things of another kind, but: the world consists of *processes*. These processes can maintain their equilibrium in certain spatial areas, and thus bring about the appearance of stable things. The Pythagoreans acknowledged the central role of mathematics for the contemplation of nature. Democritus developed a primitive form of atomic theory, Aristarchus a primitive form of the Copernican picture of the world. The Greeks put forward cosmological assumptions of amazing boldness, studied the movements of the stars and explained them on general principles. It is impossible to give here a barely adequate inkling of the immeasurable quantity of ideas in all areas that Presocratic philosophers have left us. New things were constantly discovered, analyzed, criticized, modified, rejected; we look with amazement at this period of about 300 years, in which nearly all fundamental ideas of contemporary natural science, astronomy, physics, and biology were developed in embryo and conveyed with extraordinary clarity and simplicity.

But – and now comes a most decisive question – what is there to find in all those new things? In what does progress consist? What is the difference between the mythical pictures of the world characteristic of pre-Greek civilizations as well as contemporary myths, on the one hand, and the new ideas of the Ionians, on the other?

The wealth of inventions of mythical cultures cannot be denied. Often they esteem mathematics, as astrology and number magic show. And materialism is not as progressive as may appear at first blush: a myth does not stop being a myth when people disappear from it and are replaced by abstract principles (as in Aristotelian philosophy). Indeed, it is well known that materialism can be just as dogmatic, and exactly as unscientific, as any religion. If Presocratics really achieved progress, and if this progress made them the founders and the first trainees in theoretical astronomy and theoretical physics – that may have nothing to do with the content of their ideas. Progress must be connected rather with the role these ideas played in their thought. And this role was fundamentally different from the role a myth of nature, or a dogmatic philosophy, or a dogmatically anchored religion, in human thought.

THREE DIFFERENCES

The most important differences between myth and nonmyth in understanding physical knowledge are three. There is a difference first, in attitude; second, in the way of justifying assumptions; and third, in the logical structure of both thought constructions.

Attitude

Myth is accepted as correct, without any further questions. There is no doubt about its truth. The problem does not consist in finding the limits of application of the myth or its weaknesses; the problem is to understand the myth correctly and to apply it correctly. If difficulties appear, these are to [199]be traced back to a lack of understanding or to a false application, or to presumptuousness; they can never count as indicators of the weakness of the myth itself. This is perfect.

Clearly, an attitude such as this, a complete submission to a thought system, is encountered infrequently. There are doubters always, and everywhere. Our science still, or until recently, has sometimes shared in mythical construction, as in the physics and philosophy of the late 18th and early 19th centuries, and even still the early 20th century, when Newton's celestial mechanics was regarded as a fundamental and unchangeable truth. Difficulties were not viewed as difficulties with the theory itself, but as difficulties of its application to rough experience. Many physicists now believe that complementarity is the last, the absolutely last word in microphysical matters, and critics experience a supercilious and pitying treatment, similar to that granted to nonbelievers by defenders of certain religions. It is by no means true that physics always proceeds according to reason. Apart from physics, however, depth psychology is an excellent example of the thesis that so-called scientific progress has not necessarily moved us very far from our wildest ancestors. The content of myths has changed. Their *dogmatism*, and the *credulity* of their followers, are still with us.

The attitude of the Presocratics and, above all, of the Ionian natural philosophers, toward their own theories was, as K. R. Popper has most plausibly argued, completely different from that of mythographers. These thinkers were *rationalists*. They lived in relatively young cities that had become large through trade and commerce; they had consciously worked on the improvement of the constitutions of these cities and living conditions in them; they were practical people, thinkers, politicians, well acquainted with the imperfection of all human institutions but optimistic about the possibility of improvement. A theory, a picture of the world, was for them just like any other human product, afflicted with mistakes and in need – and capable – of improvement. The most important task, then, was detection of these mistakes. The appropriate attitude is critical: theories, pictures of the world, are faulty human attempts to understand it, and they must, therefore, be mercilessly criticized. Criticism may be directed against the form of the picture of the world, it may complain about its lack of simplicity or its inability to provide a satisfactory explanation of problematic phenomena; but it can also appeal to facts. If criticism is correctly applied, the picture of the world does not last long. Instead of the intellectual (although not always acoustic) silence that surrounds an old (or modern) myth, rapid progress to increasingly better theories occurs.

Justification

The questions "How do you know that?", "What is the reason for your belief?", or "How do you justify your assumption?" are often – and also fully justifiably – put to a myth or a dogmatic philosophical system. We do not want to be taken in by some trick. If criticism is out, we should at least have grounds for accepting one theory and not another. Requests for reassurance are answered by referring to some authority, to divine decision, to what wise men say – and this authority must be accepted unconditionally. It is important to notice that this authority does not have to be a person. Philosophical foundations of thought, such as that of "clear and distinct ideas", suggested by Descartes, or the experience of the empiricists have a similar authoritative function: one builds on them, but never doubts their word, or does so only a little.

By contrast, a rational theory of the sort Ionian natural philosophers suggested, or a theory like relativity, is not capable of justification. They originate from human brains and make statements about the world as a whole, or at least about all things with a given property. They go far beyond what is available to a single individual. Of course, we might appeal to authority to bridge over the finiteness and limitation of human knowledge. But a rationalist does not take the statement of an authority unexamined. The words of the authority, such as, the "lesson of experience", must be critically examined, and might survive to serve as a solid foundation on which to build the rest. This position gives thought and cosmological speculation great freedom. One is not fearful, does not constantly ask "How can I justify that?", "Are my reasons for this hypothesis good enough?", but first draws up a theory and then looks for what must be rejected from it. Einstein has brilliantly expressed this position of critical rationalism when he characterized theories as free creations of the human spirit. Perhaps it is already clear that not all physicists share Einstein's position, and that some appeal to the goddess "experience" with the same full faith as shamans or medicine men turn to the inspiration provided by their tribal gods. But more on that later.

Logical Structure

Above I said that myths are believed unconditionally. This appears to suggest that all humans who follow myths are subject to an illusion, and that they need only to be awakened to see how the world really is. This quite natural thought overlooks the fact that a myth justifies the attitude of unconditional belief not only by referring to authority, but also because it has an explanation for almost everything at hand. There is no event that could embarrass an ideal myth. Whatever happens, it is always possible to find an explanation that shows how the event follows from the fundamental assumptions of the myth. We can express this feature of myths by saying that they are absolutely true.

The believer already knows that he or she possesses a tool, which, with a bit of adroitness, can be applied to everything. This characteristic has always counted as the triumph of a dogmatic system. Such a system is not simply naïve or obstinate. It is not merely founded on its own dictum, [201]come what may. It tries to show, in each individual case, how the apparent difficulty arose from fundamental principles unconditionally assumed, and so boasts an omniscience that impresses clever people in all times, distracting them from rational, critical research. Even in certain fields of physics people try to find a theory, or at least some very general assumptions, to hold on to as certain boundary posts in times of crisis. Now, it is very easy to show that this quest is not worth the effort that is made and will always be made to achieve it. First, because the certainty is still merely human: it is attained by describing the experiences that contradict it so as to make the clash with principles purely verbal. Second, because an absolutely certain theory is empty: it is true whatever happens in the world and, therefore, it is unable to distinguish what actually happens from what fantasy can imagine.

INVENTION OF SCIENTIFIC METHODS

Ionian natural philosophers intuitively knew the facts about understanding nature. As a consequence, they refused to protect their theories against attacks, and developed them in a way that gives criticism and nature the best possible chance to indicate possible mistakes. From this point of view, they have to be regarded as the inventors of scientific methodology.

Theories, pictures of the world, are not unchangeable and absolute truths; rather, they are attempts by thinking but imperfect human beings to understand the world surrounding them. The expectation that the pictures possess a kernel of truth cannot be justified. There is no authority whose communications can replace our lack of knowledge. Experience, too, is ultimately only a human event, and, as such, subject to mistakes and criticism. Theories must be developed in a form that makes them easily open to criticism, and must then be criticized relentlessly. Certainty, "sure results", are neither achievable nor desirable. The proof that a result is "secure" is at the same time a proof it is *not* objectively relevant. Even long acceptance is no sign of truth; an obvious weakness might have been overlooked due to prejudice or lack of multifaceted criticism. This is, basically, the scientific method. There are, of course, many details to be added, but these details are unimportant with respect to the thesis: physics, cosmology, astronomy, require critical thinking, that is, speculation, governed by criticism, applied to nature. The inventors of this method were the Ionians. Einstein and Popper described its features with great clarity in the 20th century.

This thesis, and some comments about its history, could be the end of my exposition. But the matter is by no means so simple. It is not the case that once the Ionians thought in a critical way, mythical thought, the demand for certainty, for justification, for final and enduring results, was once and for all overcome. [202]Philosophy, to name only one discipline, has made it its task since Plato to reintroduce mythical thought in a rational form, that is, with greater cleverness. In physics, however, ever since the so-called "scientific revolution" of the 16th and 17th centuries, the situation became very complicated. In words one says *one* thing, and in practice does something *completely different*, trying to reinterpret facts according to the words, which does not happens without violence and distortion. To put it in a more concrete way: just as the Ionians did, men of the present invent bold theories for which they have no justification and which are by no means appropriate to give dogmatic certainty. But they present these theories as if they had derived them from authority. They are inventors, geniuses with bad consciences, and this bad consciousness comes from the fact that mythological forms of knowledge, the quest for certainty, for a secure foundation, for an authority, still count ideals of knowledge worth striving for.

Let us take the example of Newton. He presents the discovery of his law of gravitation as if directly derived from observed facts, namely, from Kepler's laws. Many thinkers, among them Max Born, concluded from this claim that the theory of gravitation does not include very much that is new and, in fact, is only an abbreviated description of observed facts. This is by no means correct. In fact, a detailed examination shows that

Newton's theory contradicts Kepler's laws and, therefore, cannot refer to them as a basis. Newton's theory was therefore regarded as a speculative system in his time; a theory that criticized the so-called facts and that, therefore, has no "solid foundation" in them. It is a theory of exactly the same kind as Thales' idea that everything is made of water, or at least comes from one substance; an idea that is outright refuted by unbiased observation. Hence we may think of theory distinct from a myth. *But it is presented as a myth.* By way of a mistaken mathematical derivation physicists try to link it to the solid foundation of experience, and, in so doing, to justify it. The belief that this justification is successful leads to a highly dogmatic attitude, which identifies scientificity with Newtonianism, and which made it so difficult for Einstein to obtain acceptance for his own, completely different ideas. We can quite generally say that most physicists from 1700 until about 1920 found themselves in a sort of schizophrenia: they did one thing, and they tried to persuade both the world and themselves that they were doing something completely different. They followed the tracks of the Ionians, they speculated, even if somewhat fearfully at times, but very often regardless of prejudices and even of experience. They tried to give the impression that, beginning with secure facts, they slowly built a solid system of thought that could, and should, be nothing other than a modern myth.

THE SITUATION TODAY

The 20th century marked the end of this schizophrenia in physics. The revolutionary discoveries [203]bound up in the theory of relativity and quantum theory exposed the pretension of the physicists of the 19th century, that they worked on a construction securely founded, and that they very gradually added to it, stone by stone. These discoveries showed that the alleged foundations were not only not solid, but did not even exist. Confronted with this situation, two possible positions were open.

The first consisted in the decisive admission that in physics we possess no foundation, and should not seek one. This was, from time to time, Einstein's position. The second position consisted in the attempt to avoid any conscious or unconscious attempt to move away from experience. This is the stance of the majority of contemporary physicists. The first position accepts the "classical" practice of building theories, and rejects its philosophy of radical empiricism. The second position accepts this philosophy and rejects speculative building of physical knowledge. Let us not be confused by the fact that today many things are in flux, that formalisms are readily developed and as easily rejected. Formalisms are, indeed, explicitly held free from any interpretation. They say nothing about the world, they are only means to predict events whose nature experience is to determine. And, as we have seen earlier, such a determination is by no means difficult: as long as we decide to explain away difficulties wherever they appear, and are willing to complicate a theory formally so much that it can accommodate any problem by adding variables – precisely so long can we be certain that modifications will be hardly necessary. So it is not too much to say that today we find in physics a repetition of the division prompting the rise of Ionian natural philosophy, that is, the division between a mythical way of thinking, which was abandoned after having acquired certain results, and critical rationalism, which invites speculation and can accept no result as final and certain.

This situation calls for a clear decision. Should we join Einstein's tradition, should we propose theories (*full* theories, interpreted with respect to their content and not only their formalisms) bravely, modify them through criticism and exempt no element of physical knowledge from criticism? Or should we participate in the gradual construction of a thought system that promises security but is nothing other than a new, grand myth? This is a fundamental decision, which today a physicist must make. The very need to make this decision shows that physics is far from reaching, a system of objective knowledge that towers high over the quarrels of the day; rather, it stands in the middle of these quarrels. It is closely linked to ideology. And although the appearance of a new myth, rich with formulas, is a very disturbing thing, we can still draw comfort from the knowledge that it is up to us to overcome it, since physics – just as every part of knowledge – is not something that is forced upon us from outside, but is completely our own creation.

Dialectical Materialism and the Quantum Theory
(1966)

[414]The discussion, or even the mere description, of procedures and events in a society whose basic principles are not generally accepted is always a touchy matter. Too often the difficulties of the subject described and the quite natural errors made by those developing it further are projected onto the unloved ideology and are regarded as clear proofs of its weakness and insufficiency. Studies of the historical development of dialectical materialism and attempts to evaluate the influence of this philosophy invariably suffer from this effect. They also suffer from the ignorance of their authors in scientific matters and from their rather primitive philosophy of science. There are not many writers in the field today who are as well acquainted with contemporary science as was Lenin with the science of his time, and no one can match the philosophical intuition of that astounding author.[1] Wetter's book[1] is full of interesting information; it is a good starting point. But it fails completely because of Wetter's rather primitive ideas concerning the manner in which science might influence ideology and might in turn be influenced by it. The same applies to Bocheński. Occasionally an author admits his shortcomings, as does Lukacs in the preface to his *Zerstörung der Vernunft*. The need for a presentation that takes the sciences into account is only emphasized thereby. Mr. Graham's interesting, well informed, clear, and

---

[1] I am here thinking mainly of Lenin's comments on Hegel's *Logik* and *Geschichte der Philosophie*. *Materialism and Empiriocriticism* is a different story.

[1] Gustav A. Wetter, *Der dialektische Materialismus: seine Geschichte und sein System in der Sowjetunion*, Vienna: Herder, 1952; English translation by Peter Heath, *Dialectical Materialism: A Historical and Systematic Survey of Philosophy in the Soviet Union*, London: Routledge & Kegan Paul and New York: Praeger, 1958.

straightforward narration[2] of an important episode in the history of science goes a long way toward fulfilling this need. However, a purely descriptive account[2] such as his cannot suffice. It may be read as a further example of the restriction of freedom which scientists must suffer in some countries, of the irrelevant and time-consuming debates they are involved in, and as an indirect plea for leaving pure science entirely to itself. It may lead to overlooking[3] the subtle relation that exists between what one is pleased to call pure science and ideology, on the one hand, and between science and politics, on the other. For one might want to say that the ideas introduced by modern science, especially by the founders of the quantum theory, are too important and too far-reaching to be left in the hands of a few specialists and that a more [415]general, and "ideological", discussion is urgently required. One might also want to say that these ideas, because of their generality, must also be taken seriously by politicians, must be examined and criticized by them. Party discussions have, of course, a tendency to become one-sided and dogmatic. Yet one cannot disregard the fact that party lines are not restricted to politics but occur right in the center of science.[4] However, enough of general remarks! Let us get down to specifics.

It is not easy to judge the concrete work of a scientist according to the standards of dialectical materialism. The reason is that the philosophy of dialectical materialism has until now failed to develop a methodology that might guide scientists in their research.[5] Of course, one frequently hears that good scientists have proceeded in accordance with dialectical principles, but just what these principles are and how a person who has not yet achieved greatness is supposed to proceed – this is left undetermined. However, even if there is not a systematic methodology, there are yet pieces which one may try to unite into a coherent point of view. The pieces are: (1) the emphasis on the fact that in nature there are no isolated

---

[2] Loren R. Graham, "Quantum Mechanics and Dialectical Materialism", *Slavic Review*, 25, 3, 1966, pp. 381–410.

[2] *Almost* purely descriptive. For there is sufficient insinuation to make one acquainted with the author's own point of view.

[3] Despite the author's own *caveat* at the end of his essay.

[4] They are there, but concealed as "experimental facts". The principle that philosophy must be partial, rightly understood, brings them to the fore and thereby makes them accessible to criticism.

[5] Empiricism has not yet developed such standards either, but it has at least progressed on this path a little further, as is evident from Newton's rules and the research of Carnap and his school. Of course, Carnap's principles are all highly questionable, but at least here we have some material which we can examine and criticize.

elements but that everything is related to everything else; (2) the emphasis on the existence of discontinuities, indicating essential limitations of our knowledge; (3) connected with this, the emphasis on the approximate character of knowledge; (4) the demand to unite practice with theory, so that there is neither unreflected practice nor empty theory; (5) the change (motion) of concepts in the course of the development of our knowledge. Items (2), (3), and (4) are usually united in the (inductivistic) demand to take the results of experimental practice seriously, to generalize from them, but to be aware of the limits of all generalization.[6] Item (1) again emphasizes the essential incompleteness of our knowledge, which can never take into account all the relations by which a particular object is characterized. From (5), together with (4), it follows that no part of our knowledge is ever exempt from change and that it is futile to base eternal truths on conceptual considerations.

Now it seems to me that Bohr's point of view fits exactly into this methodological frame and that most of the principles enunciated above are also accepted by him. All the research done in Copenhagen before 1924 was aimed at finding a new theory of atomic processes (realism). In this attempt one tried to locate, and to isolate, those parts of the practice of classical physics that still led to correct predictions, and it was hoped that a new and coherent account of the atomic level might one day [416]emerge from the mass of experimental material, if combined with these remnants of classical physics. The relation to concrete physical practice was therefore always very close, as was the realization of the essential limitations inherent in the classical framework, or, for that matter, in any other framework, of thought.[7] The need for a more elastic handling of the classical concepts was emphasized again and again. In this respect the method practiced in Copenhagen differed decisively from that of Einstein, who was much more prepared to invent extreme views and to make an isolated fact the starting point of a new world view. And note that we are here talking not about the end products but about the manner in which theories were developed out of problems and theoretical results. It cannot be denied that after 1924, after the theory of Bohr, Kramers, and

---

[6] There may be a more "idealistic" elaboration of (2), (3), and (4) which allows for an initial clash between praxis and idea and expects that a revision of experience will lead to progress. These versions are not very widespread, however, as is seen from the earlier objections to Einstein's theory of relativity.

[7] Bohr's favorite example, which he used again and again in discussions, was the refutation of the original Pythagorean creed by the discovery of the square root of 2, and the subsequent extension of the concept of number.

Slater had been refuted by experiment, Bohr gave up hope of finding a new coherent theory of the classical kind, and it must also be admitted that from 1925 on his point of view ceased to affect actual research but became more and more an apology for the basic features of the commonly accepted quantum theory. But even then the charge of idealism was unjustified; (5) and especially (1) were still adhered to, but (1) was now extended to include the human observer and his peculiarities. And this extension is not simply a repetition of Kant, as one might be inclined to think (and as I was once inclined to think), but provides objective criteria for the implications of the fact that in theorizing we are restricted by our own (material) organization.[8] We know now[9] that Bohr was much influenced by the philosophy of William James, whose ideas admit of a straightforward materialistic interpretation (as do so many other idealistic ideas). We have to conclude, then, that those philosophers who criticize Bohr for being an idealist or a positivist (a charge which he himself always emphatically denied) are not acquainted with his ideas and his work. Where, then, do they obtain the material for their criticism?

They obtain this material from what one might call secondary philosophies, that is, from philosophical points of view which, though not effective in research, are used afterward for explaining the results in general terms. The philosophical perorations of Jordan and Jeans and the dramatic announcements concerning the disappearance of the boundary between observer and object all belong to this category. (Heisenberg, who actively participated in the early investigations and who himself made important contributions, must unfortunately be reckoned in this group also, for he was quick to establish a relation between his own activity and positivism and to report the results of this research in positivistic terms).[10] Parasitic [417]philosophies of this kind are not at all a new phenomenon. Usually they are the remnants of a once efficient point of view which have outlived their usefulness but are still adhered to dogmatically.[11] Expert philosophers should

---

[8] Cf. Abner Shimony, "Role of the Observer in Quantum Theory", *American Journal of Physics*, XXXI, No. 10 (Oct. 1963), esp. 768 ff., as well as G. Ludwig, *Die Grundlagen der Quantenmechanik* (Berlin, 1954), Chap. VI, esp. p. 171.

[9] Klaus Meyer-Abich, *Korrespondenz, Individualität, und Komplementarität* (Wiesbaden, 1965).

[10] *Physical Principles of the Quantum Theory* (Chicago [1930?]), a book Bohr was not too happy about.

[11] That idealism can have a positive function at certain periods of the history of our thought was recognized by Lenin: "It is the crude, metaphysical, simplistic materialist who regards

have been able to distinguish between them and the ideas which led to the quantum theory in the first place. This did not happen; the theory was confounded with its parasitic interpretations and was occasionally criticized even because of some uninformed popularization that had come to the author's notice. It is to be welcomed when ideas are taken seriously and when their general effects are carefully examined and criticized. But it is regrettable when really interesting ideas are not allowed to come to the fore and when attention is paid only to the pious afterthoughts.

It is very different with physicists such as Fock and Blokhintsev, however. But now one must ask oneself whether the interference of philosophy and, more generally, of ideology in matters of pure science is to be condoned or whether science is not better left to its own resources.

Is science to be left to its own resources? Should pure science be protected from all ideological interference? Is the progress of science impeded by such interference? More especially, can ideologies make a positive contribution toward the development of the sciences? It seems that the majority of Western scientists and scientific philosophers would give a negative reply to this last question. So great is the awe in which science is held today by philosophers, and so small is confidence in the power of philosophy to advance our state of knowledge, that all a philosopher is now prepared to do is either to "analyze" science, that is, to present its results and methods in a language of his own, or to make it the basis of philosophical generalizations. But all this overlooks the fact that science is full of ideology, although its ideological commitments are usually masqueraded as "obvious methodological rules" or as "well established experimental facts".[12] Now obviously such commitments can be examined only by a philosophy that is bold enough to oppose the sciences. Mach provided such a philosophy.[13] The existence of such opposition in some eastern countries, the importance ascribed to philosophy, to human thought, even by politicians, the optimistic belief that philosophy can advance

---

philosophical idealism as being merely nonsense" ("Concerning Dialectics", in Lenin, *Aus dem philosophischen Nachlass* (Berlin, 1949), page 288). He did not apply this principle to Mach, however. And yet it was Mach who by his opposition to the supremacy of mechanics opened the way to a more dialectical point of view.

[12] See my "Problems of Empiricism", in *Beyond the Edge of Certainty*, ed. R. G. Colodny (Englewood Cliffs, 1965).

[13] See his debate with Planck in *Zwei Aufsätze* (Leipzig, 1916).

our knowledge – all this is a most welcome sight compared with what goes on in the West. All we need in order to make these features still more effective is a further democratization of institutions. But once the power of philosophy is admitted, this will follow almost as a matter of course.

12

Remarks about the Application of Non-Classical Logics in Quantum Theory (1966)

1. ONTOLOGICAL INTERPRETATIONS

[351]Consider a certain formalism of quantum theory, for example that of von Neumann, and a certain connection of this formalism with experience, for example Born's rules in the form in which they appear in von Neumann's representation. We designate the interpretation of a theory that arises from connecting its formalism with experience its *empirical interpretation*, and the resulting structure of theoretical and empirical statements its *prediction scheme*. A thinker who sees the function of a theory only in its predictions is merely interested in its empirical interpretation.

But the history of science contains many attempts to pull more than predictions out of a theory. Again and again people have tried to understand theories as pictures of the world that make general accounts of the structure of the world. That requires that further elements be added to the formalism and the empirical rules of correspondence. These elements allow conclusions about relationships that are relatively independent of observations, and permit the theoretician to see the world with the help of an empirically interpreted formal expression in a unified way relatively independent of formal peculiarities. I call a totality of elements of this kind an *ontological interpretation*.

A doctrine received by nearly all empiricist philosophers is that science and, indeed, every kind of useful knowledge must make do with empirical interpretations only. Whatever goes beyond that cannot be controlled by experience and must be rejected. Ever since Kant, however, general pictures of the world, whether grounded in experience or not, are held suspect by a much wider circle of philosophers. We will not deny these pictures [352]a psychological function – like alcohol or coffee or sexual adventures they can encourage or impede a thinker. But this has nothing to do with the *content* of the examined theory.

In the following remarks I shall attempt to refute the basis of this widespread suspicion. The refutation makes use of the proof that it is possible to *argue* for or against various ontological interpretations of the same scheme of prediction, as well as of examples of the arguments employed. The proof is not compelling, but it virtually amounts to the remark that there is more in an ontological interpretation than in a cup of coffee.

As examples, I shall offer three different interpretations of quantum theory, namely, the so-called Copenhagen interpretation, Einstein's objective-statistical interpretation, and interpretations based on nonclassical logic. I will try to show that Einstein's interpretation is preferable to the other two.

2. THE COPENHAGEN INTERPRETATION

Behind the name "Copenhagen interpretation" hide various ideas that are hard to bring over a common denominator. These ideas were developed in the years between 1913 and 1926 on the basis of many discussions of concrete problems, systematically exposed for the first time in 1926, and brought into final form after Einstein's fundamental criticism in 1935. As to what this final form is, opinions differ. Heisenberg, Jordan, and Pauli endorse a version close to positivism; von Weizsäcker, and recently also Heisenberg, add elements of Kantian philosophy. The Aristotelian concept of potentiality has been mobilized (by Heisenberg and Bohm, and also by Havemann in his well received, but philosophically inadequate small book, *Dialektik ohne Dogma*). Rosenfeld sees relationships with dialectical materialism, whereas Bohr explicitly distances himself from links with philosophical schools. Only with some difficulty can Bohr's own point of view be extracted from the literature. In this situation, one must try to construct a consistent picture that is as close as possible to the literature without being able to guarantee historical correctness.[1] This procedure is especially relevant to Bohr's writings.

---

[1] For a more detailed analysis see my essay, "Problems of Microphysics", in *Frontiers of Science and Philosophy*, ed. R. Colodny, Pittsburgh 1962, London 1964; as well as my article "Quantum Theory, Philosophical Problems of", in *Encyclopedia of Philosophy*, ed. Paul Edwards {the former is reprinted as Chapter 7 of the present collection; the latter was never published, and is printed here as Chapter 25 for the first time}.

In any such construction Bohr's concept of phenomenon must play an important role. A *phenomenon* is an experimental result together with a description of the measuring apparatus with which [353]it was made. It is a complex macroscopic event, which can be described in everyday language or, when details and precision are needed, in the language of classical physics. The phenomenon referred to as the Stern-Gerlach experiment would also include a description of the source of the particle radiation waves and the method of its focusing, indications of the magnet and the strength of field forces, and indications of the final distribution of the particle hits on the photographic plate (a particle's coordinates after it passes through the magnet are here the classical indicator of the spins). Now it is claimed that the task of quantum theory lies exclusively in the correlation of phenomena in this sense, that is, in the correlation of certain classical events, *and that it does nothing beyond that*. This means that we must be careful not to interpret well-known aspects of the calculation (such as the appearance of wave functions) as signs of real and objective processes (the existence of objective real waves). In analyzing the Stern-Gerlach experiment, for example, one assigns a wave function to the incoming stream of particles, and then develops their spin eigenfunctions; this leads to a separation of the eigenfunctions and correlation with the position coordinates of the particles through interaction, and, finally, as they hit plate, reduces this locally fanned out wave function to a wave packet. The formalism, which can be imagined in this way, leads to correct predictions, for example, of the macroscopic determinable position on the photographic plate. But the assumption that the scheme of the graphic representation pictures real processes occurring between the source and the plate – this assumption leads to difficulties, and must be given up, according to the Copenhagen interpretation. This is not the place to go into details about these difficulties; suffice it to indicate that contradictions arise with laws that are also valid in quantum theory, such as the laws of the conservation of energy and momentum, as well as the laws of interference. These are the *physical* reasons why the formalism is not burdened with more than the task of producing correlations between phenomena. Clearly, these physical reasons agree well with a *philosophical* inclination for prediction schemes. (This has sometimes caused the false impression that quantum theory is simply the offspring of a certain philosophy, and that, beyond this genetic connection, there are no further reasons for assuming the Copenhagen interpretation.) Hence, from the ontological point of view, microobjects are nothing other than a bundle of macroscopic situations, they are mere evident subsidiary constructions that make our treatment of the formalism easier. The principle of *comple-mentarity* approximately provides the conditions under which certain elements of the bundles appear.

3. EINSTEIN'S INTERPRETATION

[354]Einstein's interpretation of quantum theory allows differentiation between assumptions with differing grades of generality. I will deal here only with the most general assumption, the assumption of the existence of hidden parameters. The special deterministic models that Einstein con-sidered from time to time are not affected by the arguments, and, for that reason, are not defended.

According to Einstein, there are events describable by so-called hidden parameters, that is, unknown quantities not taken into account by quan-tum theory that probably do not follow the laws of quantum mechanics. On this interpretation the usual theory is incomplete and must be extended with a dynamics of individual processes. In this respect, it is similar to statistical mechanics. The quantities of statistical mechanics give average values and additional quantities (that is, hidden parameters) are needed to describe the individual process completely. Neither the interconnection of these quantities nor their connection with the quan-tities of quantum theory is precisely known yet, although already there are various very interesting attempts in this direction. The following arguments do not deal so much with an existing theory, but with the question whether it is scientifically fruitful, whether it furthers scientific progress, if theories with hidden parameters are developed.

The advocates of the Copenhagen interpretation answer this question with a plain and simple no. Quantum theory, so goes their argument, is a complete theory in the sense that every question whose answer can be experimentally decided can in principle be answered. In principle – that is, either with the help of the basic formalism of Hermitian operators in a Hilbert space, or with the help of new variables, which add to the formalism without removing their built-in indeterminacy. But the hidden variables will either repeat what the theory says, and amount only to a useless formal cleaning up; or else they will make predictions that contra-dict the theory, and thus experience, which has confirmed the theory to a high degree. In neither case, therefore, are they scientifically fruitful.

This widespread argument, which is sometimes extended with a detailed proof of the contradiction – or alleged contradiction – between the idea of hidden variables and the usual theory, cannot be maintained.

First, here "complete" means only this: complete with respect to the predictions of phenomena (in Bohr's sense). But we have no guarantee that phenomena exhaust all the facts of the world. Second, a theory is not refuted because it contradicts another theory, even if this other theory is empirically confirmed to a high degree. There is always only a finite number of obser[355]vations, and even these are only determined within a finite interval of vagueness. Theories go far beyond this kind of finite information. It is therefore perfectly possible that assumptions like the existence of hidden parameters, which contradict a highly confirmed theory such as quantum theory, are nevertheless empirically unobjectionable. But – and here is a practical objection – is it not a pure waste of time to contemplate fantastic possibilities, if we already have reached the only goal of science, namely, a useful theory that corresponds to the facts? Should we not concentrate all our efforts on this theory, improve it, extend it to new fields, and investigate its philosophical consequences? And does not the fresh and happy invention of new and always newer ideas stand in the way of this desirable concentration?

No doubt a theory capable of producing results must occupy the center of interest and be open to constant scrutiny of its correctness. Theories express more than the facts they are grounded upon; hence, they can be refuted by the discovery of new facts. Now, it turns out that one cannot avoid a quest for critical facts, following from new ideas, that ultimately clash with the theory under examination.[2] The greater the number of these ideas, the richer the empirical content of the theory. The use of ideas that contradict contemporary quantum theory is thus very much in the sense of empirical methods. With this, we have proven the superiority of Einstein's interpretation, which recommends the construction of alternatives, over the Copenhagen interpretation, which rejects them.

4. REICHENBACH'S INTERPRETATION

We begin now to examine interpretations that assume that quantum theory proves the inadequacy of classical two-valued logic, and hence modify logic to provide a better agreement between facts and theory. I do not intend to fight for a definitive demonstration of the impossibility of such an interpretation. But I believe that the cases I will discuss lead to undesirable consequences that clearly show its inferiority to both

---

[2] For details, see my "A Note on the Problem of Induction", *Journal of Philosophy*, Vol. LXI (1964), pp. 349ff.

the interpretations just discussed. I will first discuss the simpler, but still formally unsatisfying, interpretation of Reichenbach, and then Mittelstaedt's interpretation, which is far better developed.

The cancerous damage of Reichenbach's interpretation can be illustrated best via a suggestion that Prof. Putnam, a student of Reichenbach, made a few years ago.[3] Putnam observed [356]that quantum theory (he had in mind here *elementary* quantum theory) is incompatible with the principle of locality. He wanted to retain the principle of locality and so suggested that the logic be changed to make the contradiction disappear. Let us assume, for the moment, that such a change can be carried out without fatally affecting other areas, and consider the consequences.

The principle of locality is confirmed to a high degree. On the basis of the contradiction just described, its confirming instances are (indirectly) refuting instances of quantum theory (and, vice versa, the confirming instances of quantum theory are also, on the basis of the contradiction, indirectly refuting instances of the principle of locality). Putnam's move eliminates these instances and thereby lessens the empirical content of quantum theory and of the principle of locality. It contradicts the basic principle that a theory with high empirical content is better than a theory with low empirical content. It is therefore to be rejected.

Let us assume now that the content of a theory only depends on its directly refuting instances. Furthermore, the contradiction between two very fundamental principles always demands an *experimentum crucis* that directly excludes one principle or the other, and forces the scientist to do better. This is the way things went in the case discussed by Putnam. One developed relativistic theories such as Dirac's relativistic theory of the electron and the various relativistic field theories; drew from them predictions, such as the existence of the positron; and then, using the confirmation of its existence, refuted nonrelativistic theories, and increased their empirical content. Putnam's method removes the appeal to such a procedure. Generally applied, it would lead science to stagnation. It is now quite possible to limit a fundamental discovery to one narrow field, and leave the rest of the sciences as alone. For example, it is possible to introduce the postulate of the constancy of the speed of light without presupposing corresponding changes in the transformation equations of mechanics. But this reduces the degree of falsifiability of our knowledge, and thus its empirical content, and eliminates the impulse

---

[3] "Three-Valued Logic", *Philosoph. Studies*, Vol. VIII (1957) pp. 73ff. – See my criticism in *Philosoph. Studies*, Vol. IX (1958), pp. 49ff.

toward the development of new and more general theories. Putnam's method stops progress of the sciences.

Exactly the same must be said about Reichenbach's more complicated investigations.[4] Reichenbach notices that in quantum theory there are so-called anomalies that differentiate quantum theory from classical physics. Sometimes his portrayal gives the impression that anomalies [357]are unusual, previously unnoticed, physical processes. In order to examine the case more closely, let us consider a light wave hurrying toward a photomultiplier, and let us reduce the intensity of the light so much that in a macroscopic period of time only a single photon is expected to arrive. The photomultiplier reacts. What happens to the wave? Let us assume that it continues undisturbed. Then, in order to satisfy the conservation laws, we must demand that it no longer carries any energy. Let us assume that the wave shrinks together to a point: an event that cannot be described by the wave equation and also cannot be followed in detail in any way whatever. Waves without energy, inaccessible, and suddenly shrinking: those are examples of the processes Reichenbach calls anomalies.

A further anomaly is the strange behavior of microparticles in interference processes: it seems that fields of force totally unrelated to any expenditure of energy steer the particles into the right paths. These examples make clear that anomalies are simply processes with a strange description, which refute certain ideas about microparticles. The photoelectric effect, for example, refutes the assumption that light is a physical wave phenomenon. The introduction of ghost waves without energy, or of unobservable changes in the conditions of waves, masks this situation by describing it in a strange way (one could call a white crow a sick black crow, and talk about an anomaly of the law according to which "all crows are black"). Similarly, the occurrence of interference refutes the assumption that light consists simply of a particle, which is once again masked by speaking of interactions without energy. To sum up: descriptions of anomalies are completely artificial, ad hoc hypotheses, which must conceal refutations of certain assumptions (light consists of particles; light consists of waves). From this point of view, anomalies are not special problems. They disappear by removing ad hoc hypotheses, by taking the refutations seriously, and by trying to construct a theory that is more successful than the simple particle theory or the simple wave theory.

---

[4] *Philosophic Foundations of Quantum Mechanics*, University of California Press, 1946 {Feyerabend's review of this book is reprinted as Chapter 23 of the present collection}. For details and further literature see also my criticism mentioned above, in footnote 3.

Reichenbach chooses another way. He retains the ad hoc hypotheses and defuses them by ascribing to them the truth value "indefinite". The empirical content of classical ideas, whose mistakes hide behind these hypotheses, is thereby reduced without replacing these ideas with anything better. The reduction of empirical content in itself [358] aims at retaining classical ideas, in spite of the difficulties that might arise. And besides all this, Reichenbach's method suffers from pure logical difficulties. It is therefore to be rejected.

5. MITTELSTAEDT

Peter Mittelstaedt, of the Max Planck Institute for physics and astrophysics in Munich, has suggested another change in classical logic, which lies along the course of thought followed by von Neumann and Birkhoff, and stands out for clarity and lucidity.[5] It is therefore possible to quote and criticize his results in a few words.

In an interference experiment we consider the fate of particles passing through the openings S and S'. Let A be the statement that a certain particle turns up upon a certain place on the photographic plate; let B be the statement that the particle has passed through the slit S. The statement that the particle has passed through the slit S' is then $\overline{B}$, the negation of B. From the basic statements of the calculus of probability it follows that

$$P(A) = P(AB) + P(A\overline{B}). \tag{1}$$

As a matter of fact, however, due to the interference, the situation is:

$$P(A) = P(AB) + P(A\overline{B}) + I \tag{2}$$

where I is the interference term. Mittelstaedt explains the contradiction between (1) and (2) by claiming that the statements A, B, etc., no longer form a Boolean lattice. In this special case, the equation

$$AB \vee A\overline{B} = A \tag{3}$$

is no longer valid; now, only

$$AB \vee A\overline{B} \rightarrow A \tag{4}$$

[5] *Philosophische Probleme der Modernen Physik*, Mannheim, 1963.

holds, but not

$$A \to AB \vee A\overline{B}. \tag{5}$$

Now it is clear, that by constructing (1) and (2), we have adopted a description in terms of particles. Only if we speak of particles do B and $\overline{B}$ have a meaning. The interference experiments, that is result (2), contradict the usual particle picture. Hence, they belong to the empirical content of the particle picture. The change in the logic by the elimination of (5) and the retention of (4) alone (we see very clearly how the change in the logic is closely related to the *weakening* of possible attacks on a theory: conclusions valid earlier are now eliminated) removes the interference experiments from the empirical content of the particle picture, reduces the empirical content of this picture without compensations anywhere else, and, consequently, is to be rejected, as it clashes with the empirical method.

6. SUMMARY

[359]Three ontological interpretations of quantum theory have been examined. It has been shown that certain ways of changing the logic reduce the empirical content of our knowledge without bringing other advantages (apart of course from the "advantage" of the retention of old and dear ideas); as a consequence, they should be rejected. It has also been shown that the detailed implementation of Einstein's interpretation raises the empirical content of our knowledge and, therefore, is to be favored over the Copenhagen interpretation. Both conclusions demonstrate that we can argue about ontological or "metaphysical" interpretations, as they might well be called; hence, these interpretations cannot be relegated to the field of psychology.

13

On the Possibility of a Perpetuum Mobile
of the Second Kind
(1966)

I.

[409]In one of his by now classic papers on the kinetic theory of matter von Smoluchowski admits that the second law "in the usual formulation of Clausius and Thomson certainly is in need of revision". Yet he denies that the fluctuations might constitute "a perpetual source of income" (in terms of readily available work). His argument is that one-way valves, lids, and, for that matter, *all* mechanical contraptions designed for transforming fluctuations into such a "perpetual source of income" are themselves subject to fluctuations. "These machines work in normal circumstances because they must remain in a position of equilibrium which corresponds to a minimum of potential energy. Yet in the case of molecular fluctuations all other positions are possible in addition to the position of minimal energy and they are distributed in accordance with the magnitude of total work. The valve has its own tendency of fluctuation; either the spring is so strong that it does not open at all, or it is so weak that it fluctuates all the time and is bound to remain ineffective. A perpetuum mobile would therefore seem to be possible only if one could construct a valve of a quite different kind, and without the tendency to fluctuate, and to do this we see not the slightest possibility today." [4, page 248; translation by P. K. F.]

Smoluchowski adds that a human observer who like a *deus ex machina* "always knows the exact state of nature and who can start or stop macroscopic processes at any arbitrary moment without having to spend any work" might perhaps be capable of establishing those correlations be[410]tween microphenomena and macrophenomena which are needed for a perpetual and systematic violation of the second law [4, page 3961].

2.

It is well known how the last remark has led, via the investigations of Szilard, Brillouin, and others, to a rather subjectivistic account of thermodynamic properties. The steps may be briefly characterized as follows: (i) Szilard tried to show that the process of obtaining the information necessary for establishing the needed correlations was accompanied by an increase of entropy that exactly balanced the decrease brought about by utilizing that information. A general proof of such equivalence was not given by Szilard, who restricted himself to the analysis of a particular and rather complicated example. (ii) A general "proof" appeared only much later. It was based on a *definition* of information that ascribed to it exactly the value needed for bringing about the compensation. It is hardly possible to regard such a procedure as the required general proof. It is *not* shown, on the basis of a physical analysis of the process of measurement, that it will always lead to entropy changes which guarantee the validity of the second law. Quite the contrary; the validity of this law is *taken for granted* and used for defining the entropy changes connected with the information obtained. (iii) This circular definition was itself misunderstood in a subjectivistic sense. *Originally* the "information" that a particle P dwells in part V' of a larger volume means that it is *physically restricted* to V'. "Loss of information" after removal of the wall W means that this physical restriction is no longer present, and that new possibilities of behavior have been introduced. *Now* "information" means *knowledge* where the particle is and not what occupational possibilities are open to it. In the original version the entropy of a system may increase although our knowledge concerning its ingredients remains unchanged (P moves over a larger volume, though we still know where it is). In the more recent version such knowledge prevents the entropy from increasing. It is interesting to consider some of the double-talk resulting from the mixture of both versions: "*Entropy*", writes Leon Brillouin, "*measures the lack of information* about the actual structure of the system. This lack of information introduces the possibility of a great variety of microscopically distinct structures, which we are ... unable to distinguish from one an[411]other" [I, page 160]. This means that there are states of the system which we cannot distinguish from one another (second version). These states can therefore be equally occupied by the system (first version). Lack of knowledge on our part conveniently removes all physical barriers in the world! (And therefore is not really lack of knowledge at all.)

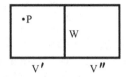

3.

Continuation of this kind of attack is more entertaining than enlightening. A much more fruitful line of approach would seem to consist in pointing out that the general proof in the form in which it was originally envisaged by Szilard cannot be given and that its circular substitutes are therefore pointless. It is possible to correlate fluctuations with devices in such a fashion that a "perpetual source of income" is obtained. We shall of course have to assume frictionless devices. They are used by von Smoluchowski and Szilard and may therefore also be used here. The *Höllenmaschine* I am going to introduce is a modification of a machine used by J. R. Pierce [2, pages 200–201] to establish the exact opposite, namely, that "we use up all the output of the machine in transmitting enough information to make [it] run". A cylindrical vessel Z contains a piston P, whose center piece O can be removed, and a single molecule M. The piston is held in balance by a string S leading via two wheels, L and M, to the pans C and D. On top of each pan a stick, A and B respectively, is kept suspended on two forks (see top view below Z). The whole arrangement is kept at temperature T. We start the machine by opening O and allowing M to wander freely through Z. We then close O. Now if M is to the left of O, then it will press to the right, lift C, lift A off the fork, and transport it to a higher level. If it is to the right of O, then the same will happen to D and B. The process can be repeated indefinitely without loss of entropy from observing M. We have here a "perpetual source of income" of the kind von Smoluchowski did not think to be possible. Where lies the weakness of his argument?

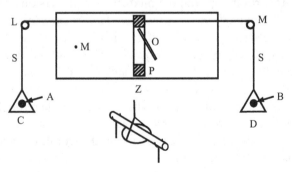

4.

Consider for that purpose a one-way valve V whose upward move[412]-ment is connected with a mechanism similar to the one mentioned in Section 1. According to von Smoluchowski the arrangement cannot work because it is bound to fluctuate between 1 and 2, and because correlation between 2 and activity of the machine would presuppose knowledge accessible to a *deus ex machina* only. (It is *this* feature of the case, the random motion of V between 1 and 2, which led to the intrusion of knowledge into physics and started the trend toward subjectivism.) But if V can be fastened to A at all, then it can also be prevented, say by a ring AB, from moving down to 1. Of course, it will still fluctuate back and forth between 0 and 2, but it will not fluctuate in this manner because of its "own tendency" as von Smoluchowski seems to believe,[1] but because of the impact of fast moving particles.[2] It will move in this way because it takes over the motion of particles moving in a privileged direction, because it is acting as a "perpetual source of income".

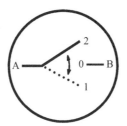

5.

To sum up: the attempt to save the second law from systematic deviations, apart from being circular and based upon an ambiguous use of the term "information", is also ill-conceived, for such deviations are in principle possible.[3]

---

1 [4], p. 395. Cf. the quotation in Section 1 above.
2 Cf. Smoluchowski's account of a similar mistake in the case of diffusion, in [4], p. 536.
3 The present note is to be read as an addendum to Karl R. Popper's [3], and especially to Section 4 of that paper.

References

[1] Brillouin, Leon. *Science and Information Theory*. New York: Academic Press, 1956.
[2] Pierce, John R. *Symbols, Signals and Noise*. New York: Harper, 1961.
[3] Popper, Karl R. "Irreversibility; or, Entropy since 1905", *British Journal for the Philosophy of Science*, 8: 151–155 (1957).
[4] Smoluchowski, Maryan von. *Oeuvres*, Vol. II. Cracow, 1928.

14

In Defence of Classical Physics
(1970)

[59]In 1965, at the International Conference on the Philosophy of Science in London, K. R. Popper delivered an opening address entitled "Rationality and the Search for Invariants".[*] In this address Popper criticized "the doctrine that science is limited to the search for invariants" and proposed "the conjecture that though a search for invariants is undoubtedly one of the most important of all scientific enterprises, it does not constitute or determine the limits of rationality, or of the scientific enterprise" {WP, p. 154}. I am of a different opinion. I believe that restricting oneself to a search for invariants may be an advantage and that loosening the framework may have some very unfortunate consequences. More especially I want to argue that the classical framework has been given up prematurely and that one should continue using and exploiting it. Popper's views as expressed in his address will be presented in greater detail in this paper. (Quotations or paraphrases from the address are marked by the initials KRP.)[1]

---

[*] Popper's paper was published posthumously, as Essay 7 in Karl R. Popper, *The World of Parmenides: Essays on the Presocratic Enlightenment*, edited by Arne F. Petersen and Jørgen Mejer, London-New York: Routledge, 1998, pp. 146–250, with the title "Beyond the Search for Invariants"; page references to Feyerabend's quotations will be indicated with *WP*, followed by the relevant page(s).

[79]1 Earlier versions of this paper were criticized by Imre Lakatos and Alan Musgrave. In the present version I have gratefully made use of their advice. For support of research I am again indebted to the National Science Foundation.

I. THREE VIEWS OF HUMAN KNOWLEDGE

The Parmenidean world view which Professor Popper presents and criticizes in his address distinguishes between a realm of perfection that is characterized by simple, unchangeable, reversible causal laws, and a very different domain, our world, "the way of appearance and illusion", where irregularities abound, and where approximation, imperfection, variation, emergence, chance arc the order of the day. Considering these "two ways" we may adopt the one or the other of the following three procedures:

(*i*) We may retain the perfect world and abandon our world, the world which we explore with the help of our senses, our instruments and our theories, by declaring it to be an illusion. This would seem to be the point of view of Parmenides. It certainly is the point of view of those physicists who use the subject, or his ignorance, in order to remove some glaring discrepancies between their theories and the facts. For it is the [60]characteristic feature of "Parmenidean apologies" (KRP {*WP*, pp. 147, 150 and *passim*}) of this kind that the subject is introduced in a purely verbal fashion, as another way of describing the discrepancy, not as an explanation of it that is capable of being tested independently. There is no attempt to frame hypotheses which can be examined in a reasonable way and to use them in order to explain why a perfect and unchangeable world should look so very different when viewed with a human eye. No such attempt is made, it is simply asserted that the discrepancy is the fault of the observer.

Certain versions of the quantum theory of measurement are an excellent case in point. Assume that we want to measure the position of an elementary particle known to be in a situation characterized by a well-defined momentum.[2] In such a situation particle states associated with different positions interfere with each other, entailing physical consequences not present when the particle is at a definite place. We therefore cannot assume that the particle, while dwelling in this situation, has a well-defined though unknown position. Yet a measurement always produces a particle in a more or less well-defined place, whose behavior is moreover independent of the behavior of other particles from the same source: the interference between the states of different position has disappeared. This disappearance of interference effects cannot be explained as

---

[2] We now neglect the complications due to relativistic considerations and square integrability.

the result of an interaction between the particle and the measuring instrument. Quite the contrary, any interaction that leads to correlations between macrostates (pointer readings) and microstates will just transform the original state $\sum c_i \psi_i$, where the ψ_i are the eigenstates of the magnitude to be measured (position in our case) into $\sum A_i \psi_i$ where the A_i are those macroscopic states which the measurement correlates with the ψ_i. The interferences therefore do not disappear in the theory. They disappear as a matter of fact. The theory is therefore not strong enough to bring about agreement with facts unless a further element is added. This further element is the famous reduction of the wave-packet.

Now the reduction of the wave-packet as it is conceived by some physicists (but not, for example, by Bohr) is a very curious process. It does not take place by itself, when the system is left alone; nor is it without physical consequences (as is, for example, the replacement of a vague description by a more precise one). The first point suggests that something nonphysical is involved. The second point suggests that the action of this nonphysical thing has physical consequences. Hence, the dictum that the observer has intruded into the world. But of course this dictum has no explanatory power whatsoever. We do not bring in the observer because [61]we have some independent knowledge of him that would help us to understand the reduction of the wave-packet.[3] Quite the contrary – the talk about observers is merely a concise repetition of the two points just made. In other words, we are dealing here with a classic example of an ad hoc hypothesis.

Another case in point that has been discussed by Popper himself is the information-theoretical version of the second law of thermodynamics. Roughly speaking, the procedure is as follows. One finds situations in which the second law seems to be violated. Now one inquires whether one has perhaps overlooked some sources of entropy. This is an entirely natural procedure. It is still natural to ask whether the observer is perhaps the guilty party. Observers are complex thermodynamic systems and may well influence the objects they inspect. However, no account is given of

---

[3] This applies only to one specific form of the theory of measurement, not to all such theories. Günther Ludwig, for example, and the members of the Italian school, try to explain the reduction of the wave-packet by showing that it is an approximation involving certain assumptions about the measuring instruments. Their account has other drawbacks, however. Bohr's account of the reduction is also unobjectionable and free from subjective elements. Cf. section 3 of my "On a Recent Critique of Complementarity", *Philosophy of Science*, 35, 1968 and 36, 1969, 82–105 {reprinted in *PP1*, ch. 16, with the title "Niels Bohr's World View"}, where the difficulties of Ludwig's approach are discussed as well.

their thermodynamic properties that could explain the occasional loss of entropy. It is simply postulated that any loss of entropy that cannot be accounted for in a straightforward way and that has some mysterious properties should be ascribed to the observer, and it is said in addition that the observer provides exactly the amount of entropy necessary for preservation of the second law of thermodynamics. The talk about the observer does not increase our knowledge, it just describes the existence of a clash between theory and fact in a manner that makes the clash invisible. It is but another ad hoc hypothesis.[4]

The first procedure, then, consists in the formulation of ad hoc hypotheses which remove the clash between the "true way" and the "way of appearance" in a purely verbal fashion.

(*ii*) The second procedure is distinct from the first in one point only. It is admitted that the "true world" and "the world of appearance" are very different from each other. It is also conjectured that this difference may be connected with the observer, with his instruments, and with the peculiar (physical) position from which he views the universe. But such conjectures, though often inarticulate and primitive, are no longer ad hoc. Or at least, one is not satisfied with them as long as they are ad hoc, but tries to find independent tests.

Galileo's attitude toward Copernicus is a perfect example of the second procedure. He praises him for having persisted despite the glaring conflict between his theory and the facts of observation. He frames hypotheses designed to bridge the gulf between the idea of the motion of the earth and the familiar everyday world of the senses; and he tries in this way [62]to save the former. He subjects the saving hypotheses to a relentless examination that produces new and apparently independent evidence.[5] This evidence eventually[6] leads to a content-increasing[7]

---

[4] Newton's assumption that God restores the momentum lost in inelastic collisions is of exactly the same kind. Today it is the word "observer" and not the word "God" that occurs in the most favorite ad hoc hypotheses. For thermodynamics, cf. also my brief note "On the Possibility of a Perpetuum Mobile of the Second Kind" in *Mind, Matter and Method: Essays in Honor of Herbert Feigl* (Minneapolis, 1966) {reprinted as Chapter 13 of the present collection}.

[5] For a more detailed account including some surprising aspects of seventeenth-century science the reader is invited to turn to sections 2–7 of "Problems of Empiricism, Part II" in *The Nature and Function of Scientific Theories*, ed. Colodny (Pittsburgh, 1970).

[6] For the meaning of "eventually" (which word strictly speaking refers to an infinite process) cf. the text immediately below.

[7] I have adopted the very apt phrase from Lakatos' contribution to *Criticism and the Growth of Knowledge* (Cambridge, 1970).

reconciliation between the "true world" and the world of experience. Indeed, it seems that this is the most decisive difference between the Aristotelian point of view and the "new astronomy" of Galileo, Bruno, Kepler, and Newton: for while Aristotle eliminates whatever is contradicted by experience, and while he tries to build his physics on assumptions that mirror our world in a more or less direct way,[8] the thinkers of the sixteenth and seventeenth centuries introduce laws which *prima facie* have nothing whatever to do with the world in which we live but describe a perfect and "Parmenidean" world without friction, refraction, atmospheric disturbances, tremors, illusions of the senses, and so forth. It is surprising to see how quickly the resistance against these strange and unfamiliar laws is overcome and how soon they are used as a new framework for the explanation of everyday events. (We disregard here, of course, the attitude of the Holy Roman Church which rejects the Copernican view but accepts Galileo's telescopic discoveries and his mechanics.) In the course of such explanations one tries to show the manner in which the complex and often erratic processes of our world can be understood as results which emerge when such laws are first made to work under specific, highly complex, and "imperfect" conditions (on the surface of the earth, for example, where winds, clouds, resistance of material and irregularity disturb every experiment),[9] and are then examined with instruments (the eye, eyeglasses, magnifying glasses, telescopes, thermometers, and so on) exhibiting idiosyncrasies of their own, many of them too complex to be immediately understood. Reference to the observer then does occur, but in a content-increasing way. For the intention is

---

[80][8] Cf., for example, his theory of elements as described in ch. 17 of H. Solmsen, *Aristotle's System of the Physical World* (New York, 1960). For a more general account cf. G. E. L. Owen, "ΤΙΘΈΝΑΙ ΤᾺ ΦΑΙΝΌΜΕΝΑ", *Aristote et les Problèmes de la Méthode* (Louvain, 1961), 83–103. For the development of Aristotelian thought in the middle ages, cf. A. C. Crombie, *Robert Grosseteste and the Origins of Experimental Science* (Oxford, 1953). The relevant works of Aristotle are *Anal. Post., De Anima, De Sensu.* Concerning the motion of the earth, cf. *De Coelo*, 293a 28f: "But there are many others who would agree that it is wrong to give the earth the central position, *looking for confirmation rather to theory than to the facts of observation*" [my italics].

[9] Referring to Galileo's account of the floating of metal chips, Ludovico Geymonat (*Galileo Galilei*, trans. Stillman Drake [New York, 1965], 64) describes his procedure as follows: "It was his merit to perceive that here he was dealing with a special phenomenon, where the exceptional behavior was rooted in the special situation, and that the exception could not be invoked to overthrow general laws, like that of Archimedes ...". Cf. also Galileo Galilei, *Discourse on Bodies in Water*, trans. Salusbury, ed. Drake (Urbana, Ill., 1960), 55ff. One should note that surface tension is here also accounted for, but in a more or less ad hoc fashion.

that whatever is said about the observer and his surroundings should be subjected to independent tests, if not at once then at least some time in the future. This finishes the description of the *intended* difference between the second procedure and the first.

Now just as the laws of the world do not appear to us in a pure form, but are obscured by disturbances, so the intended difference between the second method and the first often disappears in practice and may remain invisible over long periods of time. The reason is easy to explain. What counts and what does not count as a relevant test depends on the theory tested as well as on the current views concerning the process of cognition. In the Aristotelian philosophy such views contained the assumption that a normal observer who is in possession of his faculties is in harmony with [63]the universe and capable of giving adequate reports on all aspects of it. "The traditional conception of nature", writes Hans Blumenberg, commenting on this feature,[10] "was connected with a kind of *postulate of visibility* which corresponds both to the finite extension of the universe and to the idea that it was related to man as its centre. That there should be things in the world which are inaccessible to man not only now, or for the time being, but in principle, and because of his natural endowment and which could therefore never be seen by him – this was quite inconceivable for later antiquity as well as for the Middle Ages". However, precisely such a view was implied by the new Copernican cosmology, especially in the radical form given to it by Bruno (though not by Galileo; and certainly not by Kepler). The observer is no longer situated at the centre; he is separated from the true laws of the universe both by the special physical conditions of his observation post, the moving earth, and by the idiosyncrasies of his main instrument of observation, the human eye. What is needed for a test of the Copernican cosmology is therefore not just a simple-minded and direct comparison of its predictions with what is seen, but the interpolation, between the "perfect world" and "our world", of a well-developed meteorology (in the good old sense of the word, as dealing with the things below the moon) and of an equally well-developed science of physiological optics dealing both with the subjective

---

[10] Galileo Galilei, *Sidereus Nuncius, Nachricht von neuen Sternen* (Sammlung Insel, vol. i, Frankfurt, 1965), 13. Aristotle himself was more open-minded: "The evidence [concerning celestial phenomena] is furnished but scantily by sensation; whereas respecting perishable plants and animals we have abundant information, living as we do in their midst, and ample data may be collected concerning all their various kinds, if only we are willing to take sufficient pains. Both departments however have their special charms ...". *De part. anim.*, 644b26ff.

(brain) and the objective (light, lenses) aspects of vision, telescopic vision included. One can say at once that the evidence obtained in accordance with the older Aristotelian view of vision was bound to clash with the new astronomy; and one is also perfectly justified, for the reasons just given, to regard the clash as irrelevant.

The most important element in this peculiar and complex situation, an element which is disregarded by all those methodologies which do not, as Popper does, consider our background knowledge, is, however, the *time factor.*[11]

The Copernican view conflicts with the evidence that has been assembled in accordance with the older views of cognition and is directly compared with it. This is not a relevant test. Relevant tests must interpolate meteorological and physiological disturbances between the basic laws and the perceptions of the observer. Now there is no guarantee that the supplementary sciences providing these disturbances and their laws will be available immediately after the discovery of the new astronomy. Quite the contrary – such a sequence of events is extremely unlikely, because of the vast complexity of the phenomena one has to consider. [64]It may take hundreds of years before the first reasonable hypotheses appear. But in the meantime the new astronomy and the new Parmenidean laws it contains must not be abandoned. We must preserve it, and we must even try to develop it in order to keep alive, and to further articulate the motive for the invention and the improvement of the supplementary sciences. One of the best methods to effect such postponement is, of course, propaganda and the use of ad hoc hypotheses. Galileo fully exploits these methods – but he gives them a peculiar twist of his own.

Galileo's first publication of his telescopic findings, the *Sidereus Nuncius* (1610), insinuates that they prove the Copernican view. In the *Dialogues Concerning the Two Chief World Systems* the insinuation has become outright assertion. But when we look at the situation a little more closely we find that Galileo is unable to support this assertion by independent evidence concerning the relation of telescopic vision to the real world. (See note 5 above.) All we are given is a new and strange source of perception which occasionally shows things the way Copernicus

---

[11] The existence of "phase-differences" such as I am about to explain is well known to dialectical materialists. Thus Trotsky writes in *The First Four Years of the Communist International* (New York, 1953), vol. II, 5, "the gist of the matter lies in this, that the different respects of the historical process ... do not develop simultaneously along parallel lines". Cf. Lenin, *Left Wing Communism, an Infantile Disorder* (Peking, Foreign Languages Press, 1965), 59.

says they are and which on other occasions produces puzzling and obviously illusory images. Using late sixteenth-century evidence and theories of cognition one would have to say, strictly speaking, that both the Copernican view and the idea that the telescope, taken separately, gives a better account of the real world than the naked eye are refuted, but that these two ideas, while undermined by the evidence, are able to support each other. It is this rather peculiar situation that Galileo exploits, using it to prevent the elimination of either idea.[12] Exactly the same procedure preserves his new dynamics. This science, too, is only vaguely connected with the elements in our natural surroundings[13] – but the support it lends to the idea of the motion of the earth now seems to make them both more acceptable, although the relation to actual phenomena (of "our" world) is established in a perfectly ad hoc fashion, by treating friction and other disturbances as tendencies defined by the observed discrepancies rather than as physical events explained by a theory of friction for which new and independent evidence can be obtained.[14]

---

[12] "The *Nuncius*", says Franz Hammer in the best and most concise account I have seen of this matter, "... contains two unknowns, the one being solved with the help of the other". (Johannes Kepler, *Gesammelte Werke*, hgg. M. Caspar and F. Hammer (Munich, 1941), vol. IV, 447.)

[13] "Simplicio: So you have not made a hundred tests, or even one, and yet you so freely declare it to be certain? ... Salviati: Without experiment I am sure the effect [that the falling stone will arrive at the bottom of the mast even when the ship moves] will happen as I tell you, because it must happen that way ...": Galileo Galilei, *Dialogue Concerning the Two Chief World Systems*, trans. Stillman Drake (University of California Press, 1953), 145. Einstein's attitude was in many respects very similar to that of Galileo. Thus he wrote to his friend Besso that the test "through little effects" of his general theory was not too important, even if it led to a negative result. The "sense of the thing [die Vernunft der Sache]", he said, "is too obvious" (cf. Gerald Holton, "Influences on Einstein's Early Work", *Organon*, 3 (x966), 242). The same attitude is expressed in a letter to Karl Seelig of January 1953. Commenting on the great precision of the measurements concerning the deflection of light near the sun he writes: "For someone who knows these matters [*für einen Kenner*] it is not very important, for the main point of the theory is not that it is confirmed by little effects; its main importance lies in the great simplification of the whole basis of physics it provides. However, only few will be able to appreciate this properly". (C. Seelig, *Albert Einstein* (Europa Verlag, Zürich, 1960), 271) Einstein also shares with Galileo (and Bohr) an astounding intuition concerning the nature of secondary effects. Cf. Shankland, *American Journal of Physics*, 31, 1963, 46ff. For Galileo, cf. note 9 above.

[14] As Truesdell observes: "While the ancients had been ready to make precise, mathematical statements, right or wrong, about the doings of the planets, their approach to motions on the earth had been largely qualitative, using the language of cause and effect, and tendency as in biology and medicine. This separation of heavenly geometry from earthly mechanics had been maintained by Galileo who, while he had done much to publicize the kinematical properties of uniformly accelerated motion, derived correctly by the schoolmen 300 years before, had widened the gap between theoretical mechanics and practical

The reader will realize that a more detailed study of historical phenomena like these may create considerable difficulties for the view that the transition from the pre-Copernican cosmology to Galileo consisted in the replacement of a refuted theory by a more general conjecture which explains the refuting instances, makes new predictions, and is corroborated by the observations carried out to test these new predictions. And he will [65]perhaps see the merits of a different view which asserts that while the pre-Copernican astronomy was in trouble (was confronted by a series of refuting instances), the Copernican theory was in even greater trouble (was confronted by even more drastic refuting instances);[15] but that being

mechanical phenomena, first by refusal to connect celestial and terrestrial motions in any way, and secondly by setting up as governing presumably near and familiar motions laws valid only in an ideal or vacuous medium, [81]the unfortunate practicer who had to work on the real earth being put off with effects recognized by name as being due to resistance and friction but explained only as qualities, or tendencies in the style of the Aristotelian physics so contemned by Galileo ..." (C. Truesdell, "A Program Towards Rediscovering the Rational Mechanics of the Age of Reason", *Archive for the History of Exact Sciences*, 1, 190, 6). Truesdell's article contains an excellent account of the difficulties encountered in the attempt to construct a "meteorology" of the kind referred to in the text above and he points out that these difficulties were gradually solved only by the invention of still another Parmenidean frame: "Euler's success in this most difficult matter lay in his *analysis of concepts*. After years of trial, sometimes adopting a semi-empirical compromise with experimental data [of the kind now again recommended by Popper in his address], Euler saw that the *experiments had to be set aside for a time*; some remain not fully understood today. By creating a simple field model for fluids, defined by a set of partial differential equations, Euler opened to us a new range of vision in physical science". *Op. cit.*, 28. Cf. also Truesdell's introductions to vols. 13 and 13 of *L. Euleri Opera Omnia*, 2 (1954, 1956 respectively). For a detailed examination of the conceptual changes entailed in Galileo's procedure as well as of his peculiar technique of persuasion, cf. my paper "Bemerkungen zur Geschichte und Systematik des Empirismus", *Metaphysik und Wissenschaft*, ed. Paul Weingartner (Salzburg, Anton Pustet, 1968) as well as "Problems of Empiricism, II", *loc. cit.*, and "Against Method" in vol. IV of the *Minnesota Studies in the Philosophy of Science*.

[15] For an excellent account, cf. Derek J. de S. Price "Contra-Copernicus: A Critical Re-Estimation of the Mathematical Planetary Theory of Ptolemy, Copernicus, and Kepler" in *Critical Problems in the History of Science*, ed. M. Clagett (Madison, 1959), 197–218. Price deals only with the kinematic and the optical difficulties of the new views. Consideration of the dynamical difficulties of the Copernican system further strengthens his case (as Galileo himself would have admitted). Cf. my "Problems of Empiricism, II", *loc. cit.*

Bohr has expressed a very similar situation in the quantum theory by saying that his objective was not "to propose an explanation of the spectral laws", but rather to "indicate a way in which it appears possible to bring the spectral laws into closer connection with other properties of the elements *which appear to be equally inexplicable* on the basis of the present state of science" (*The Theory of Spectra and Atomic Constitution*, Cambridge, 1922, 115). The procedures of Bohr and of Galileo are exactly the same. The difference is that Bohr does not try to conceal the actual structure of the situation and of his contribution to it by the use of phrases such as "search for the truth"

in harmony with still further inadequate theories it gained strength, and was retained, the refutations being made ineffective by ad hoc hypotheses and clever techniques of persuasion.[16] Moreover, he will perhaps also see that a situation like this is anything but undesirable.

As was pointed out above, the test of a new cosmology describing a Parmenidean world involves theories of cognition and "meteorologies" whose development may take considerable time. One of the main problems to be solved by such theories is to find the correct relation between the Parmenidean world as it exists in itself and the same world as it presents itself to the senses. The new cosmology must be kept alive until satisfactory solutions of this problem are obtained, for it is only then that a proper test can be carried out. Moreover, it is wise to further develop the cosmology in order to make the problem more definite. All these desiderata are satisfied by the procedure just described which allows inadequate

---

which have over-impressed some admirers of Galileo. For this seventeenth-century split between procedure and verbal description, cf. my essay "Classical Empiricism", in *The Methodological Heritage of Newton*, ed. R. E. Butts and J. W. Davis (Toronto, 1970) which analyzes the most extreme example, Newton. Galileo was much more open, and much more aware of the real nature of the moves he thought necessary for the defence of Copernicus. Cf. also the next note.

[16] Cf., for example, note 13. Galileo says here that he can predict what will happen in a specific case "without the help of experiment". But he does not proceed without the help of experience. Quite the contrary – the whole rich reservoir of the everyday experience of his readers is utilized in the argument, but the facts which they are invited to recall are arranged in a new way; approximations are made, different conceptual lines are drawn so that a new kind of experience arises, manufactured almost out of thin air. This new experience is then solidified by insinuating that the reader has been familiar with it all the time. Thus a characteristic argument (establishing the rough surface of the moon) concludes: "you can see [now] Simplicio how you yourself really knew that the earth shone no less than the moon and that not my instruction but merely the recollection of certain things already known to you have made you sure of it ..." (*Dialogue*, trans. Drake, 89f). Again Salviati, who "act[s] the part of Copernicus" (ibid., 131), says to Simplicio, the Aristotelian: "The unravelling [of the problems posed by the centrifugal force of the rotating earth] depends upon some data well known and believed by you just as much as me, but because they do not strike you, you do not see the solution. Without teaching them to you then, since you already know them, I shall cause you to resolve the objection by merely recalling them". Simplicio: "I have frequently studied your manner of arguing, which gives me the impression that you lean towards Plato's opinion that *nostrum scire sit quoddam reminisci* ..." Koyré (*Etudes Galiléennes*, vol. iii (Paris, 1939), 53ff.) [82]is perfectly right in calling the work a "polemic, ... a pedagogical, ... a propagandistic ... book" – but one must add that in the peculiar historical situation in which Galileo found himself such propaganda was essential for the advancement of knowledge. Considering that every interesting problem situation has similar features one must also admit that "honest" argument without propaganda will often be ineffective, or will lead to the premature elimination of excellent views. For this cf. my essay "Against Method", to appear in vol. IV of the *Minnesota Studies in the Philosophy of Science*.

views (that is, inadequate *vis-à-vis* the evidence) to become crystallization points for the aggregation of other inadequate views and protects the steadily growing bundle by propaganda tricks and ad hoc hypotheses (which latter are really nothing but temporary place holders for the not yet existing epistemological theories).[17] Such place-holding can go on for centuries – for why should a reasonable conjecture be condemned just because human ingenuity has so far failed to provide the necessary content-increasing connections? In the case of Copernicus the ad hoc hypotheses of Galileo and others have served an excellent purpose indeed, for it is only now that theories of telescopic vision are being gradually developed[18] and that we have hope of getting a little more insight into the behavior of *continua*.[19] The moral to be learned from such episodes is that the tension between a "true" world and "our" world should always be maintained, even if this means cheating and specious argumentation. Not being prophets we must also admit that any inadequate view, however often refuted and however implausible, is capable of becoming the crystallization point of a new world picture which, taken in conjunction with a new (and perhaps utopian) theory of knowledge might in the end prove more effective than the most attractive and most highly corroborated element of the *status quo* (no method can guarantee an approach to the "truth").

[66](*iii*) The third position with respect to the "two ways" is closely connected with the more recent development of the sciences, especially of

---

[17] A marvellous example of this very principle, though taken from the domain of individual history, is reported in the *Autobiography* of John Stuart Mill (quoted from *Essential Works of John Stuart Mill*, ed. Max Lerner (Bantam Books, New York, 1965), 21): "The explanations [which Mill's father gave him on logical matters] did not make the matter at all clear to me at the time; but they were not therefore useless; *they remained a nucleus* for my observations and reflections to crystallize upon; the import of his general remarks being interpreted to me by the particular instances which came under my notice *afterwards* ..." (my italics). The Copernican theory formed precisely a nucleus of this kind for later generations to let their "observations and reflections ... crystallize upon" and it would have been fatal had it been rejected because such "observations and reflections" were as yet absent.

[18] Cf. V. Ronchi, *Optics, The Science of Vision*, trans. E. Rosen (New York, 1957). To a certain extent such theories have been made superfluous by the invention of the photographic plate which, however, has led to perfectly analogous problems. Cf. Mees, *The Theory of the Photographic Process* (New York, 1954). For the similarity of certain photographic "phantoms" with visual illusions, cf. the marvellous discussion in F. Ratcliff, *Mach Bands* (Holden Inc., San Francisco, 1965), 235ff.

[19] Cf. C. Truesdell and R. Toupin in *Encyclopaedia of Physics*, vol. iii/1 (Springer, Berlin, 1960), as well as C. Truesdell, *The Elements of Continuum Mechanics* (New York, 1966).

biology, of thermodynamics, and of cosmology, as well as of atomic physics, and with the strong empiricist tendencies of our times. According to this position we must give up at least some features of the perfect world as abortive attempts to account for our world which have not stood up to certain tests. The conjecture that the probabilistic aspects of the quantum theory will not admit explanation on the basis of deterministic laws is of exactly this kind. It entails the renunciation of a very important feature of the Parmenidean framework. We may also mention here the further conjecture that irreversibility will have to play a more essential role than is granted it by classical mechanics. We may say that this third procedure leads from the classical empiricism outlined under (*ii*) forward into a broader viewpoint that also allows for the use, without any attempt at a reduction, of more "empirical" statements and laws which *prima facie* have nothing to do with classical physics.

Now, if I understand Popper correctly, he would seem to say that physicists have now generally recognized the need for a revision even of very fundamental "Parmenidean" assumptions. They have also started to carry out the necessary changes. But while trying to make their physics fit the facts they have also tried (with a few notable exceptions) to leave it as Parmenidean as possible. They have done this in the belief that "the search for invariants ... constitute[s] or determine[s] the limits ... of the scientific enterprise" (KRP {*WP*, p. 154}). As a result they had to fall back upon procedure (*i*) and had to use ad hoc hypotheses in order to remove the discrepancies between their physical theories and the new findings of scientific research. In these ad hoc hypotheses they referred to the observer, thus turning them into Parmenidean apologies. Such apologies are dearly unsatisfactory. But – so Popper would seem to say – they are also entirely unnecessary. They are unnecessary because science and rationality are not bound to the Parmenidean framework. Science can, and occasionally already does, proceed in a non-Parmenidean fashion. Such tentative attempts must be strengthened by a philosophy which frees them from their bad conscience, which shows that leaving an entirely Parmenidean framework means widening and not abandoning rationality, and which also makes concrete suggestions for research. It is such a philosophy that Popper sets out to develop.

Now I quite agree with Popper that more recent developments within the sciences, if faced squarely, have decidedly non-Parmenidean and [67]rather Aristotelian features. Science has indeed left the Parmenidean framework, at least in part, without ceasing to exist (or, better, without loss of recognition and without loss of the label "scientific"). But it is also my conviction that this is an undesirable change. Abandoning even a tiny part of the Parmenidean framework offers no advantage (except increased agreement with the empiricist and anti-metaphysical tendencies of today). Quite the contrary, I shall try to show that it is bound to do much harm. Considering the drawbacks, I am convinced that upholding a physics that is fully Parmenidean is a good policy at any time. One should uphold such a physics, not because the truth of the Parmenidean point of view can "be established by rational argument" (KRP {see *WP*, pp. 152–153}), but because a fully Parmenidean point of view can be shown to possess some quite definite advantages. Moreover, I think that the attempt to keep science fully Parmenidean is not quite as hopeless as Popper seems to think it is. As a matter of fact it seems to me that one particular Parmenidean framework, namely the framework of classical physics, has been abandoned prematurely, that its resources are not exhausted, and that its theories and the philosophy of mechanicism and materialism which forms their background constitute, and always will constitute, a most powerful instrument of explanation, and of the criticism of alternative ideas.[20] The ad hoc hypotheses, however, which may arise in our attempt to retain the fruitful tension between an ideal Parmenidean world and "our world",

---

[20] Interest in classical physics has been considerably revived by Professor R. H. Dicke's investigations. These investigations make it clear that it was premature to believe that the explanation of, say, the advance of the perihelion of Mercury was beyond the reach of classical physics. Cf. his "The Observational Basis of General Relativity" in Chiu-Hoffmann (eds.), *Gravitation and Relativity* (New York, 1964), 1–16. The experimental basis of Dicke's investigations was reported at the January 1967 meeting of the American Physical Society and made public in the press. For predecessors of Dicke's hypothesis, cf. S. Newcomb, *The Elements of The Four Inner Planets and the Fundamental Constants of Astronomy* (Washington, 1895).

The foregoing is not supposed to imply that Professor Dicke's own positive theory is classical. All I want to point out is that he is using classical physics when criticizing Einstein's theory of relativity. We have here therefore a very interesting relation between the theory to be criticized, the theory that does the criticizing, and the theory that is supposed to take over after the criticism has been effective. The theory that does the criticizing is not thought to be adequate as a description of reality. Still, having played a decisive role in the acceptance of general relativity (the 43″ which support general relativity were, after all, obtained by observation in conjunction with classical perturbation theory) it can now play an equally decisive role in its rejection. The reader should also be aware of the fact that classical mechanics has recently undergone a "renascence [which] broadened and deepened it but rather elevated than annulled the older parts of the subject" (Truesdell, *The Elements of Continuum Mechanics*, I: the reference is to continuum mechanics as well as to the science of general mechanics, which was founded by Birkhoff). For the role of classical mechanics in the criticism of the Quantum Theory, cf. section 7 of my "Problems of Empiricism", in *Beyond the Edge of Certainty*, ed. Colodny (Prentice Hall, N. Y., 1965).

need not be at all vicious but may be regarded as temporary substitutes, keeping the place open for a view of man and his natural surroundings that transcends whatever highly corroborated laws we already possess.

I shall now give a brief statement of the general arguments which have convinced me of the advantages of the Parmenidean framework and especially of classical mechanics.

II. GENERAL ARGUMENTS FOR A PARMENIDEAN APPROACH

The Parmenidean framework, when combined with procedure (*ii*), is attractive chiefly because of its versatility. The laws of classical physics and especially the laws of classical mechanics are not directly related to the processes in this world, and they are not refuted when such processes change in a decisive manner. A major disturbance of the planetary system by a passing fixed star will create serious difficulties for Aristotelian [68]physics which describes how the undisturbed planets (undisturbed in Newton's sense) look when viewed from an undisturbed earth. No such difficulty arises in classical mechanics. Of course, the predictions made by this theory under the new conditions may turn out to be false (as will be the case when the passing star is very heavy, or moves very quickly). But the point is that such predictions can be produced, whereas the Aristotelian point of view is stuck with the undisturbed case.[21] They can be produced because classical mechanics is not confined to our world but can deal with different and even with entirely imaginary situations. Thus we have on the one hand considerable freedom to construct and critically examine many different models of a given natural process; yet on the other hand this freedom is not too great a burden upon our imagination, for we are not entirely left to our own resources, as all the models follow from the same basic formalism. Any attempt to tolerate on the theoretical level a closer fit with "our world" will remove the versatility, and it will make the invention of alternative hypotheses a much more difficult business. It should therefore be resisted, though it may of course eliminate at one stroke the difficult problem of bringing about a content-increasing reconciliation with the known facts.

---

[21] The same remark applies to the different phases of the world process which Boltzmann invokes in order to explain the (local) direction of time. Cf. para. 4 of his second reply to Zermelo, *Ann. Phys.*, 60 (1897), 392ff. (The paper has been translated and published by S. Brush in *Kinetic Theory*, vol. 2 (Pergamon Press, London, 1966). The relevant passage occurs here on page 242).

The last remark is quite topical today. Close agreement between theory and fact is one of the most basic demands of almost all contemporary science: "things which cannot be ascertained in principle should not appear in the theory".[22] This attitude is closely connected with Mach's view "that the universe is given to us only *once*"[23] and that our theories have the task of describing this one existing universe and nothing else. Considering that the universe is nowadays often regarded as a developing entity whose main features are subjected to temporal changes, one has inferred that the laws of nature must be essentially time dependent, too. Landé's attempt to introduce probabilities into the basic postulates of kinematics is guided by the very same ideology: he wants to eliminate discrepancies between the ideas of classical physics and certain properties of the world we live in by modifying the former (in accordance with (*iii*) above), and he does not even consider the possibility that the statistical features he observes and describes so dramatically[24] might be the result, not of our basic laws, but of the special conditions under which the nonstatistical laws of classical physics work in our particular universe. Popper feels that the belief that "gas molecules show an inherent tendency to get mixed up" is very natural (KRP {*WP*, p. 184}). This feeling is still another instance of the non-Galilean[9] habit of at once projecting obvious and well-known [69]empirical facts on to our basic laws rather than trying to explain or even replace them by more comprehensive views which *prima facie* tell a very different story. It is of course true, and no one would try to deny it, that "the universe is given to us only once". But it is not at all certain that abandoning Parmenidean laws and resting content with obvious empirical regularities will help us to find interesting descriptions of this one and only universe. After all, even the most advanced theories and the most advanced observations may deceive us by reflecting the properties of our close surroundings, measuring instruments, and senses included, rather than the laws of the world at large. In this situation it is of paramount importance not to be tied too closely to one particular conceptual system, but to have alternatives available. And this need becomes the greater the smaller the distance between theory and fact has become. It is for this reason that I would

---

[22] G. Ludwig, "Gelöste und ungelöste Probleme des Messprozesses in der Quantenmechanik", in *Werner Heisenberg und die Physik unserer Zeit* (Braunschweig, 1961), 157. Ludwig calls this principle "Heisenberg's principle".

[83][23] E. Mach, *Die Mechanik in ihrer Entwicklung* (Leipzig, 1933), 222.

[24] *Foundations of Quantum Theory* (New Haven, 1955), 3ff.

look very critically at a point of view that regards time-dependent laws, or laws containing individual constants, or purely statistical laws, as being on a par with, or even preferable to, laws which do not have this feature.[25] And my first objection to this habit is, I repeat, my belief that it reduces the versatility of our physics, making it much more dependent on experience, and putting the task of considering alternatives into our own hands entirely. Classical physics is capable of doing part of this task itself.

The versatility of classical mechanics is closely connected with a second feature which I also find most valuable: it enables us to adopt a more critical attitude toward experience and it does so automatically, without any effort on our part, by the sheer wealth of imaginary worlds it provides.[26] Experimental laws such as "metals expand when heated" are dominated by experience to an extent which severely restricts the possibilities of research. We may repeat our observations; we may try to uncover new experimental evidence (e.g., we may look for another piece of metal); but we are entirely lacking in means which enable us not only to doubt, but to further examine the validity or the relevance of the *kind* of evidence we are assembling. Classical mechanics, on the other hand, gives us a rich reservoir of consequences, of possible worlds which may be used for the examination even of experience itself. It is precisely in this spirit that Boltzmann inquires what the "fact" of irreversibility implies about mechanics.[27] *Prima facie* we are faced with a "contradiction between experimental laws and theoretical laws". But it would be very rash indeed to infer that this "amount[s] to a definite condemnation of

---

[25] Considerations like these would seem to lead to new arguments in favor of Maxwell's conditions for natural laws and in favor of Eddington's "fundamental theory".

[26] This was already seen by Descartes, at the very beginning of the history of mechanistic doctrines.

[27] Boltzmann's philosophy is (1) pluralistic (he recommends the use of rival theories for the advancement of knowledge and defends on this basis the right of Mach and of others to develop their own point of view); (2) critical (he once suggested a journal where all misfired experiments should be recorded); (3) non-gradualist (science advances by revolutions, not by accumulation); (4) anti-observationalist ("almost all experience is theory"). For details, also concerning the relation to Popper's philosophy, cf. my article "Ludwig Boltzmann" in the *Encyclopedia of Philosophy*, ed. Paul Edwards {reprinted as Chapter 26 of the present collection}. Boltzmann's philosophy is expressed most clearly in his *Popularwissenschaftliche Vorlesungen* (Leipzig, 1906). Unfortunately there does not yet exist an English translation of this remarkable book. It would also be very interesting to examine to what extent Einstein was influenced by Boltzmann's general outlook. Ehrenfest and Schrödinger certainly were influenced.

mechanicism"[28] unless we have made certain that there is not a single consequence of the [70]classical point of view which either approximates to the said experimental laws or produces them when tested with the instruments that were used to establish them in the first place.[29] Conversely, all those consequences of mechanics which are vaguely similar to what we obtain in thermodynamics give rise to further tests of the latter discipline as well as of the "facts" which seem to endanger a Parmenidean world view. All this is in perfect agreement with Popper's own philosophy and has been said by him, long ago, especially in his classic paper "The Aim of Science".[30] However, while he seems to think that such attempts at a mechanical criticism of irreversibility (and indeterminism) have already taken place and have failed,[31] I am persuaded by the two features just described (namely, versatility and criticism) that giving up (ii) and the classical point of view would mean a tremendous loss of rationality. I would therefore be prepared to try a little harder even in the face of overwhelming adverse evidence. And I am somewhat encouraged in my attitude by the progress that has been made in our knowledge of mechanics since Boltzmann, especially by Birkhoff and his school. A new trial therefore seems worth while.

Such a new trial will of course have to use what Popper calls "Parmenidean apologies". For Popper these are questionable procedures. For me the need to employ them yields a third argument in favor of the Parmenidean outlook.

[28] Cf. the last line of H. Poincaré, "Le Mécanisme et l'Expérience", *Revue de Metaphysique et de Morale*, 1, 1893, pp. 534–537. The article is reprinted in an English version in Brush, *op. cit.*, and the quotation occurs there on page 207. Popper seems to accept Poincaré's argument.

[29] Cf. Boltzmann's first reply to Zermelo, *Ann. Phys.*, 57 (1896), 777–84. I am quoting from Brush, *op. cit.*, 223: "Thus, when Zermelo concludes from the theoretical fact that the initial state in a gas must recur – *without having calculated how long a time this will take* – that the hypotheses of gas theory must be rejected, or else fundamentally changed, he is just like a dice player who has calculated that the probability of a sequence of 1000 ones is not zero, and then concludes that his dice must be loaded since he has not yet observed such a sequence!" (my italics).

[30] *Ratio*, 1, 1952.

[31] The Ehrenfests seem to have been a little more optimistic than Popper makes them out to be (KRP {see *WP*, p. 171}). "Our discussion" they say (p. 3 of *The Conceptual Foundations of the Statistical Approach in Mechanics*, Cornell University Press, Ithaca, 1959) "will be guided by the conviction that such inconsistencies do not exist". What is meant is the apparent inconsistency between "reversibility in the premises and irreversibility in the conclusion" as Poincaré expresses the matter (see above, note 28). "We have to agree", they repeat at the end of their analysis (p. 24), "with Boltzmann that we can find no inner inconsistencies".

To start with, it is clear that reference to the subject cannot be indiscriminately condemned. There are many things which an observer does not notice and many other things which he notices but which depend on his own organization and have no correlate in the objective world. The magnificent blue dome, the "brazen bowl"[32] of the sky, to take but one example, is entirely his own creation and must be explained as such, as the illusion of erring mortals. As far as I know there exists as yet no content-increasing explanation of this phenomenon that makes it compatible with our general astronomical outlook (as opposed to the Aristotelian astronomy which regarded the starry sphere as a real physical thing[33]). At the present state of our knowledge the assumption that the dome of the sky is the work of the observer is a Parmenidean apology of the worst kind, entirely in line with procedure (*i*) above. Would Popper want to eliminate this assumption? And if he does not want to eliminate this account of the sky, then why does he attack the various attempts to explain irreversibility in a similar manner? It is entirely conceivable to me that mortal organisms living in an eternal and unchangeable universe [71]project upon it their own feeling of getting older. Life, after all, need not be the result of an evolution of matter (although as an unrepentant materialist I believe it is); it may be an extraneous phenomenon, superimposed upon an otherwise static world. This is entirely possible.[34] But how should we examine conjectures of this kind? Not, I suggest – and with this I come to my second point – by turning our attention upon ourselves, but by testing hypotheses about what observers of a certain structure

---

[32] *Iliad* 17, 425; *Od.* 3, 2; 15, 329; 17, 565; Pindar N. 6, 3–4. Similar expressions can be found in Egyptian, Babylonian, and Assyric myths. For more refined phenomenological descriptions which, however, give us the same result, cf. D. Katz, *Die Erscheinungsweise der Farben* (Leipzig, 1911), Abschnitt 2. Cf. also Ronchi, *op. cit.*, sections 102, 119f. (the latter dealing with the appearance of sun and moon).

[33] For the arguments, cf. Ptolemy's *Syntaxis*, Book One, Chapter 3 (Heiberg), 11ff.: "... hence perceptions of this kind were bound to suggest that the heavens are spherical". And "perceptions of this kind" are the path of the stars, the regularity of their revolutions, the constancy of the size of the constellations as well as the constancy of the brightness of all the stars. These arguments are compressed in the first astronomical textbook of the West, Sacrobosco's *De Spera* (I am quoting from pp. 120f. of Thorndike's translation). Here Alfraganus is quoted as saying that "if the sky were flat one part of it would be nearer to us than another, namely that which is directly overhead", which indicates that the idea of the sky as a physical surface, "which is so clearly suggested by observation", had already become well established. It is still defended, partly on metaphysical grounds, partly by reference to observation, in Galileo's early *Trattato della Sfera* (*Opere, edizione nazionale*, vol. ii, esp. 215f).

[34] Our world, after all, might be entirely our dream.

experience when viewing a fully Parmenidean world; or, to express it in different terms (see again the description of procedure (*ii*) above), by testing content-increasing or potentially content-increasing "Parmenidean apologies". This is how phenomena such as the sky or the moon illusion were first examined[35] and this is how we can hope to further increase our knowledge of the observer. I entirely agree with Popper that "it is for the philosopher of science ... to combat the lapses into subjectivism and irrationalism which ..., are due to an obviously unwilling and therefore repressed abrogation of an unconsciously held belief" (KRP {*WP*, p. 199}). But I would like to restrict this remark to observer-hypotheses that are quite explicitly ad hoc and have no other intention than to create a verbal bridge between an "unconsciously held belief" and contradictory evidence. Observer-hypotheses that either make new assumptions about ourselves or expect such assumptions to become available in the future, and therefore aim at explaining the discrepancies between observation and theory while at the same time contributing to our knowledge of ourselves – such hypotheses should be encouraged.[36] But the hypotheses also need a Parmenidean background for testing. Hence, it would seem to be advisable to adopt the following methodological principle:

Assume that two different accounts are available of (some spatio-temporal part of) the world, one using laws that describe its most prominent features directly, another one involving laws that *prima facie* seem to describe an entirely different world plus suitable Parmenidean apologies; then always prefer the latter to the former.

---

[35] Aristotle, characteristically enough, tries to explain the moon illusion objectively, by reference to the magnifying power of the terrestrial atmosphere. Similarly Tycho interpreted the observed variations of the diameter of the moon during a synodic month as being due to real pulsations while Kepler, framing an ingenious Parmenidean apology, attributed them to the methods of observation. Cf. *Astronomiae Pars Optica* in Joh. Kepler, *Gesammelte Werke*, ed. [84]Franz Hammer (Munich, 1939), Bd. II 48 (concerning Tycho's conjecture) and 60f. (containing Kepler's "apology"). All of Kepler's optics is built upon the "apology" that "the object seen is a function of vision" (*imago sit visus opus*), *op. cit.*, 64.

[36] This applies for example to Professor Ludwig's version of the quantum theory of measurement which characterizes the observer by certain operators and shows that given these operators and Schrödinger's equation the reduction of the wave packet follows to a high degree of approximation. Cf. G. Ludwig, *Grundlagen der Quantenmechanik* (Berlin, 1954), ch. 5. For an earlier attempt along the same lines, cf. D. Bohm, *Quantum Theory* (Prentice-Hall, New York, 1951), ch. 21. Note, however, that new assumptions need not be immediately forthcoming and that it may take a considerable time to bring about the required increase of content. The phenomenon of the sky is an example of a Parmenidean apology which so far has not led to any noticeable increase of content.

Considering how little we know about thinking and perception, and considering also the surprising discoveries made by more recent research (by Michotte, by the school of transformational psychology and by others), this would seem to be a very reasonable principle to adopt. I am a little afraid that an acceptance, into our basic laws, of emergence, irreversibility, time-dependence, variation and of all the other non-Parmenidean categories will severely retard further progress in these [2]fields. Having been projected on to the world, these properties can never again return to the observer, who will therefore remain a very primitive and unstructured entity. (The most important and fascinating problem – whether time is our own way of seeing things or whether it is an objective feature of the world – may now completely disappear from view). Conversely we must say that our self-understanding depends to a large extent on the further development of extremely Parmenidean frameworks.

III. SURVIVAL OF CLASSICAL IDEAS

I would like to conclude these arguments with some general considerations which have convinced me that a particular Parmenidean framework, viz., classical physics and especially classical mechanics, should be taken more seriously than one is inclined to take them today. Classical mechanics, certainly, is vastly better than anything that has gone on before.[37] But can it be superseded? And has it, as a matter of fact, been superseded? Is it true, for example, that the quantum theory or the general theory of relativity has superseded it both in generality and in adequacy? It seems to me that the question must be answered negatively on both counts.

To start with, there are excellent reasons to suppose that the quantum theory does not give us classical mechanics as a special case unless it is supplemented both by classical concepts and by classical assumptions. In the last few years it has become increasingly clear to what extent the theory still depends on the correspondence principle and how little it can stand on its own feet.[38] Apparently we have here not a fully fledged new

---

[37] Classical mechanics is now undergoing a new renascence which has considerably enriched it and led to quite surprising developments. Cf. note 20 above.

[38] On this point cf. section 9 of my "Problems of Empiricism, II", *loc. cit.*, as well as section 3 of "On a Recent Critique of Complementarity", *loc. cit.*, where Popper's ideas on quantum theory are examined in detail.

view of the microcosm and, therefore, of the world, but a half-way house in between classical physics and some utopian microcosmology of the future.

It is surprising to see that the theory of relativity shows a very similar incompleteness, at least partly for the very same reasons. To start with, the adequacy of the general theory of relativity is now seriously endangered by the results of Professor Dicke, which use the resources of classical physics in an essential way (cf. above, note 20). And as regards its generality we must remember that any actual calculation employs both the theory and classical mechanics in its premises. Classical ideas turn up first in the calculation of planetary orbits, which rests on classical perturbation theory, adding but a small relativistic correction.[39] We have no clue what a purely relativistic calculation would yield (the problem [73]of stability, for example, is entirely open). This is a practical difficulty, but one wonders, in view of certain features of relativity, whether it does not reflect matters of principle also. Secondly, we need the classical ideas when dealing with extended solid objects. The reason is that we do not possess a concise relativistic description of those phenomena which were described (approximately) in classical physics with the help of the notion of a rigid body.[40] Classical ideas, then, occur essentially in all those predictions which are today regarded as decisive tests of the theory of relativity. It is therefore quite mistaken to assign such predictions to the content of relativity exclusively and to declare that an increase in content and in explanatory power has taken place.[41] So far the actual situation.

---

[39] "The ephemerides are calculated in accordance with the Newtonian law of gravitation, modified by the theory of general relativity", *Explanatory Support to the Astronomical Ephemeris and the American Ephemeris and Nautical Almanack* (London, 1961), 11. Cf. also the detailed account in J. Chazy, *Théorie de la Relativité et la Mécanique Céleste* (Paris, 1928), vol. i, 61, 87, 135, 228. "This mixture of the theories of Newton and Einstein is intellectually repellent, since the two theories are based on such different fundamental concepts. The situation will be made clear only when the many body problem has been handled relativistically in a rational and mathematically satisfactory way". J. L. Synge, *Relativity, the General Theory* (North Holland, 1964), 296f.

[40] Born, *Ann. Phys.*, 30, 1909, 1ff. has given a definition of rigid motion which demands that the distance between adjacent world lines of the motion remain constant. However, a rigid object in this sense cannot be made to rotate (Ehrenfest, *Phys. Z.* 10, 1909, 918) and has only three degrees of freedom (Herglotz, *Ann. Phys.*, 31, 1910, 393 and Noether, *Ann. Phys.*, 31, 1909, 919ff.). Rotating solid objects must be dealt with by relativistic fluid mechanics which has not yet led to a simple and concise characterization of isolated bodies.

[41] Dicke's investigations (cf. note 20 above) have done much to separate the classical and the "relativistic" elements of the allegedly purely relativistic predictions. One might be

However, I suspect that the general theory of relativity is also unable in principle to produce many empirical predictions. My reasons are, roughly, as follows: the theory is very simple. It is also coherent to a high degree. This means that distant points of any curve expressing a law of the theory are strongly dependent upon each other. Hence, if large parts of the curve move in an arbitrarily selected domain with certain simple topological properties (such as the domain of theoretical entities), then it is very unlikely that intermediate parts will leave the domain (and move, say, into the domain of observable entities). This last consideration is of course highly conjectural. But taken together with the preceding description of the actual situation it shows that there may be something seriously wrong with the customary philosophical evaluation of the general theory of relativity.[42] (Einstein's own ideas are untouched by these considerations, for he was interested in "unification" and not in "verification by little effects". "This however", he continues in the letter from which this quotation is taken, "is properly appreciated only by few".[43])

All these arguments seem to show that classical physics represents a plateau of success in our attempts to cope with the world and that we should not abandon it too rashly. (One might regard this as an argument in favor of a modified Kantianism.) In the remaining sections I shall

---

inclined to deny the conclusion in the text by referring to the numerous "derivations" of classical mechanics from the general theory of relativity. However, such derivations are but formal exercises unless it is shown that not only momentary effects but also long-term effects are excluded, and this for the whole period for which useful astronomical observations are available (more than 3000 years): one would have to show that the minute deviations neglected in the usual approximations have no cumulative effects which might endanger the stability of the planetary system. No such proof is available, as far as I know. (J. Chazy, *op. cit.*, vol. ii, ch. ix–xi, gives an approximate derivation of the Schwarzschild solution from the basic equations of relativity, but without taking such cumulation effects into account. His statement that the method outlined at the beginning of note 39 "has thereby been shown to be legitimate" (182) must therefore be taken with a grain of salt.) Considering the difficulties of the relativistic many-body problem (note 39), it is not very likely that it will soon be found. But without it the "derivations" referred to are useless for the purpose of unified prediction. That they are useless for an even wider purpose, viz., for showing the conceptual continuity of the theories of Einstein and Newton, is demonstrated by more detailed examinations, such as those by Havas, *Rev. Mod. Phys.*, 36, 1964, 938ff. As regards the Havas paper the reader is invited not to remain content with the introduction which promises to show continuity between Newton and Einstein, but to consult also the remaining sections, which tell a very different story.

[42] This is of course realized in physics where the so-called evidence for relativity is now subjected to a critical re-examination that separates the classical and the relativistic elements in the derivation. Cf. notes 20 and 41 and the literature given there.

[43] See reference in note 13 above.

argue that it is not too unreasonable to hope that two of Popper's anti-Parmenidean phenomena – irreversibility and chance – can be accommodated in it.

IV. CLASSICAL STATISTICS

Do molecules possess an inherent tendency to get mixed up? The answer seems to be no: left alone in a part of the world without forces a cloud of [74]molecules will soon get unmixed. Enclosed in a container the molecules will speedily get mixed up again. This suggests that the mixing is due not to an inherent tendency of the molecules, but rather to the special conditions in which they have been put. Whatever inherent tendencies the molecules possess are determined by the laws of mechanics, and these favor neither reversibility, nor irreversibility. Take another example. A molecule enters a box with perfectly reflecting walls of height h and length l, under an angle a at A.

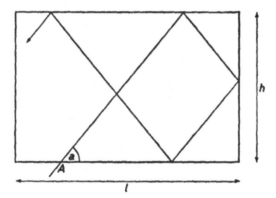

The molecule will leave the opening A after a definite time only if there are integers M and N such that $2h \cdot \tan a \cdot N = Ml$. If $2h/l \cdot \tan a$ cannot be represented by a finite decimal fraction, then reversal will never occur.

A modification of the model, making it only slightly more complex, explains how coherence between different wave trains may get destroyed and never recur. Yet we have not introduced a single statistical element, and lack of knowledge has not played any role either. The laws used are of the most simple reversible type imaginable. Again it seems possible to obtain an irreversible process by letting reversible laws work under conditions that need not even be particularly complex (note that reversal can be obtained simply by changing the length of the container).

A more realistic example is this. A molecule enters an irregularly shaped vessel. It is subjected to reflections from the walls only and is not otherwise disturbed. Under what conditions can a reversal be expected? If the vessel does not exhibit any obvious symmetries then the only way in which reversal can be brought about is that the molecule hits the wall perpendicularly and is reflected back into its own path. Conversely, reversal can be forbidden by postulating that for no molecular path does there exist a situation such as the one just described. If one fears that this postulate may lead (*via* the recurrence theorem) to a contradiction with the laws of mechanics, then we just introduce arbitrary disturbances of the paths of the molecules which are justified by the fact that "every [75] [molecule] is really interacting with the entire universe".[44] Again, irreversibility results not from the laws, nor from the complexity of the moving system (the single molecule), but from the conditions to which the system is subjected and from the surroundings in which it moves.

We make a further step toward the interesting example, viz., the mixing of two different gases, the "red" one and the "blue" one, which are initially separated by a wall. It is often asserted that we obtain irreversibility only if we first introduce some coarse-grained magnitudes and then follow their development. Such a procedure does not seem to be necessary in the present case, where both the macro-observer and the micro-observer notice irreversible changes. For the macro-observer the color changes from red left, blue right, to purple all over. The micro-observer (in the right part of the container) first sees only blue stars on his "molecular sky", then more and more red stars turn up, until the sky is almost equally sprinkled with red and blue molecular stars. Why don't these stars disappear after some finite time? They stay for the same reason as given in the second example, the only difference being that the complexity of the arrangement has been considerably increased. One can now try to construct or to imagine containers which effect reversal after one, two, three minutes, given certain initial conditions which must then be realized with a remarkable degree of precision (by an expert billiard player, for example). These thought experiments at once drive home the lesson that no craftsman will ever be able to work out the details and that no billiard player will ever be able to realize the corresponding initial conditions. We assume that our universe has not succeeded in working

---

[44] Boltzmann, first reply to Zermelo, Brush, *op. cit.*, 225.

them out either, for any initial state of the separate gases that is going to be realized. And if this seems too bold an assumption to make then one can always add, what is in fact true, that even a situation perfectly fitted for reversal would soon be destroyed by the (outer or inner) disturbances due to the nonexistence of closed systems.

To sum up: it seems that the existence of irreversible processes (of imperfect resonances as well as of other types of imperfection) can be explained without assuming inherent tendencies toward equilibrium, simply by giving an account of the conditions under which certain reversible laws are working. Nor would it seem to be necessary to learn "to think better in ... 'imperfect' ways" as Bondi has expressed it in his comment on Popper's address – at least it is not necessary to let such modes of thinking invade the domain of our laws. This domain can still remain entirely Parmenidean.

V. PROBABILITIES

[76]Probabilities, then, seem to enter not when we consider the existence of irreversible processes, but when we wish to examine the details of the motion toward equilibrium. But even here mechanics seems to be able to achieve a good deal, and perhaps everything without addition of probability assumptions.

Consider for this purpose the dynamical (not the statistical) behavior of a large number N of billiard balls (about one hundred) which shall constitute our molecules. We may ask how long it takes for the Maxwell distribution to appear to a reasonable degree of approximation where the Maxwell distribution is characterized by the behavior of the dynamical function $v_i - \bar{v}^2$ for all molecules. This is a purely mechanical problem. One may of course take the occurrence of average velocities as an invitation to talk about probabilities, but I think it is preferable to interpret them as dynamical magnitudes on a par with the total momentum: $\bar{v} = \frac{1}{N} \sum v_i$. This manner of speaking is preferable as it makes clear at once that no new assumptions are involved and that our calculations are based on the usual initial conditions (position and momentum for each single molecule) and the laws of mechanics only.[45] Probability theory proper enters when we decide to calculate the averages in a different way, not by constructing

[85]45 Dynamical predictions of the kind discussed have been made by Alden and Wainwright. Cf. their paper "Molecular Dynamics by Electronic Computers", *Proc. Int. Symp. on Transport Processes in Statistical Mechanics*, ed. Prigogine (Interscience, 1958), 97–131.

the path of each single molecule by adding the obtained values and by then dividing their sum by their total number, but by regarding the v_i as random variables using certain assumptions of independence as well as the one or the other form of the central limit theorem. These are indeed genuinely novel assumptions which we make sometimes in order to see where they lead, but which are often forced upon us by our ignorance *vis-à-vis* the complexity of the problem. (Who would refuse to use information about the behavior of each single molecule of a gas, if he could obtain, handle, digest it and could calculate dynamical averages with its help in a reasonably short time?) Unlike Popper, I do not see any harm in this particular reference to ignorance, especially as we can envisage conditions in which it could be removed and also know how we would proceed then. Of course, we may decide to stick to our probabilistic framework even if complete information about all molecules should be given to us. In this case, however, an examination [77]of the relation between the dynamical and the probabilistic averages would seem to be especially important. After all, both these averages are supposed to describe one thing: gases (in containers and in open space). And the fact that the probabilistic approach assumes independence where the dynamical approach suggests dependence makes such an examination even more urgent.[46] What, then, is the relation of the usual probability assumptions to the body of mechanical theory? I venture to suggest that it is reasonable to attempt to *derive* them from mechanics.[47]

VI. BIRKHOFF'S THEOREM

In statistical mechanics we start from *a priori* probabilities and we then aim to show that certain physical situations have a probability close to

---

[46] Popper's idea that probabilities enter because of the problems we want to solve must therefore be taken with a grain of salt. This is so especially in the quantum theory where we learned that certain definitely nonstatistical problems ("when, and where on its orbit does the electron commence to rise to a higher orbit?") admitted only of a statistical answer. And the role of probabilities in the calculation of the intensity of spectral lines emerged only after a revision of the classical picture which forced us to replace certain problems by very different problems. The decisive turning point is Einstein's theory of 1916 (published 1917) where Planck's law is derived from purely statistical assumptions concerning transitions (this phase of the development seems to be overlooked by Professor Bondi in his comments): in the classical picture the intensity was simply given by the amplitude of the classical wave field.

[47] This seems also to be Einstein's opinion. Cf. the penultimate paragraph of his letter to Popper as published in *The Logic of Scientific Discovery*, 459f.

one. The disadvantage of this procedure is that the systems considered are in addition supposed to obey the laws of mechanics (I am now dealing with classical statistics only), and we have to examine whether the *a priori* assignments are in agreement with these laws. It is of course possible to regard statistical mechanics as an entirely new discipline and to forget about its relation to already existing theories. However, "such an approach, ready to *assume* any result for which some difficulty of proof turns up, breaks physics into disconnected fragmentary rules of thumb, forgetting that in physics it is no less important to understand than it is to calculate useful numbers".[48] We therefore start with magnitudes that can be calculated dynamically, for example with time averages. If it is possible to calculate time averages without any assistance from ensemble theory, then statistical physics has been reduced to mechanics proper and this particular part of the anti-Parmenidean argument disappears. But this is exactly what is suggested by Birkhoff's theorem! According to this theorem the time average $<f>_t$ of an arbitrary phase function f (which has but to satisfy certain criteria of differentiability and summability) is identical with the phase average $<f>_p$ almost everywhere in a metrically transitive part of phase space. Let us now examine the various ideas of this theorem.

Time averages are here calculated as dynamical magnitudes. Averages, of course, also occur in probability theory, but they are then connected with random variables. In our present case the variables do not at all change in a random fashion, but obey the laws of mechanics.

[78]*Phase averages* are the averages usually calculated in statistical mechanics on the basis of some assumption of equiprobability (microcanonical ensemble; Tolman's approach). No such assumption is made here. The interpretation of the formula containing the average is given rather by the theorem itself, which asserts that the phase average expresses some long-range dynamical property of the systems concerned. On the other hand we now have a mechanical explanation of the success of the statistical procedure: this procedure happens to lead to the correct value of long-range mechanical properties and must also therefore give a correct account of equilibrium properties.[49]

---

[48] C. Truesdell, in *Ergodic Theories*, Course 14 of the Proceedings of the Enrico Fermi International School of Physics, Varenna (Academic Press, New York and London, 1961), 41.

[49] It is quite possible that some theorems of mechanics can be readily expressed in probabilistic terms. However, that would show just the opposite of what Popper wants to establish! It would show that probability theorems follow from mechanics and not that

The introduction of the phrase "almost everywhere" is the first point where assumptions different from those of mechanics may be needed. It has to be assumed that the phases for which $<f>_t \neq <f>_p$ are of no physical consequence. I am not so sure that this assumption should be made lightly, especially in view of the fact that the most minute deviations may lead to noticeable macroscopic effects (avalanches). However, first of all the assumption is much weaker than the traditional postulate of equiprobability. And, secondly, one should not forget that the same difficulty also occurs in probability theory proper.[50] Besides, it may be possible to remove the "almost" by restricting phase functions in a suitable manner so that the theorem no longer applies to all such functions, but only to the functions of a select class K. If the functions determined by our thermodynamic measurements turn out to form a subclass of K, then this part of the problem would seem to be solved. As far as I can see there is no reason to suspect that the suggestion will lead to a dead end.

Metrical transitivity, too, is a real stumbling block as far as the calculation of actual time averages is concerned, for there is no simple way to identify the metrically transitive parts of phase space.[51] However, these difficulties do not seem to have any relation at all to our problem, which is the question whether probabilities can be obtained from purely mechanical considerations. For the very nature of the troublesome conditions suggests that it is a further examination of mechanics that will provide the answer.[52]

---

the mechanics of complex systems cannot work unless it is supplemented with, or partially replaced by, statistical elements. That the science of general mechanics has itself statistical features has been asserted by A. I. Khinchin, *Mathematical Foundations of Statistical Mechanics* (Dover, N.Y., 1949), 10.

[50] It occurs in two different forms. First as the problem whether $p(A) = 0$ should be interpreted as meaning that A will not occur (this is difficult to reconcile with the frequency interpretation). Secondly, as the problem of small probabilities: should one put $p(A) = 0$ whenever $p(A)$ is smaller than some very small number?

[51] The same applies to the procedure suggested by Lewis's first theorem, according to which certain restricted time-averages can be calculated by averaging over a vanishingly small shell around a surface that is determined by a complete invariant of the motion. Here again one does not seem to have even an inkling of what phase functions are contained in such an invariant.

[52] It must be mentioned that Maxwell's original suggestion, according to which the encounter of the molecules with the wall of the container guarantees (for irregularly shaped vessels) that the energy will be the only invariant, can be given a statistical twist. It need not be interpreted statistically, however, but can be looked at in the way suggested in Section IV above.

So far we have examined some common features of all mechanical systems. We have not introduced coarse-grained densities. Nor have we made use of the special character of gases (large number of molecules, Hamiltonian separable into parts containing one degree of freedom only, *etc.*). Yet we have already come very far. These are excellent reasons for further examining the power of the mechanical point of view.

VII. CONCLUSION

[79]I have given some general methodological arguments in favor of a wholly Parmenidean outlook, including vague and ad hoc Parmenidean apologies. I have tried to show in a special case that the situation may not be quite as desperate as Popper presents it and that there is reason to expect that the resources of mechanics and of its Parmenidean background are not yet exhausted. This encourages one to support the classical view also in fields where it has been definitely abandoned, for example in the quantum theory. One must do this, for it is only in this way, by such "swimming against the stream" (KRP {*WP*, p. 147}) that a more critical attitude can be obtained both toward classical physics, and toward the more recent theories which claim to be able to replace it.

And if it should be true that classical physics represents a high point in the development of our knowledge, then this would also be the only way to prevent us from slipping, and from losing what we have already achieved.

PART TWO

REVIEWS AND COMMENTS
(1957–1967)

15

Review of Alfred Landé, *Foundations of Quantum-Mechanics: A Study in Continuity and Symmetry*, New Haven: Yale University Press, 1955 (1957)

[354]In various articles[1] as well as in a book Professor Landé has suggested a new approach to quantum mechanics. Whereas most presentations of the subject rest content with the assertion that one has to accept the peculiar features of the atomic world (as expressed, e.g., by the principle of complementarity, or by the commutation-rules, or by the calculus of the ψ-functions) as fundamental, however difficult it may be to understand them properly, Landé makes the attempt to elucidate those features by showing how they can be understood on the basis of some very simple and plausible assumptions. The assumptions are (*a*) Leibniz's principle of continuity (an infinitely small change of a cause cannot lead to a finite change of the effect) and (*b*) some further principles, to be mentioned presently.

Applying (*a*) to the case of two different diffusing gases (or else of two different states of one and the same gas) leads to the postulate that the maximal increase of entropy resulting from the mixture should be dependent on the degree of likeness of the components. This postulate, which Landé calls the "postulate of entropy-continuity", removes a very curious paradox of statistical mechanics which was first discussed by Gibbs, and leads at the same time to some well known features of quantum mechanics: ordering of states into sets of mutually orthogonal states, splitting effect, jumps, the Born interpretation.

---

[1] I only mention: "The Logic of Quanta", *The British Journal for the Philosophy of Science*, 6, 1956, pp. 300ff.; "Quantum-Indeterminacy, A Consequence of Cause-Effect-Continuity", *Dialectica*, 8, 1954, pp. 199ff.; "Continuity, a Key to Quantum Mechanics", *Philosophy of Science*, 20, 1953, pp. 101ff. (cf. also the literature given in this last article).

For example, let us start with the class K of all possible states of a certain system (gas) and choose a state A_1 then a state A_2 completely unlike A_1 (i.e., a state which on mixture with A_1 gives the classical value for the maximal increase of entropy), then a state A_3 which is completely unlike A_1 as well as A_2 and so on until no further state is left which is completely unlike all the preceding states. In this way a class (A_i) of "mutually orthogonal" states is obtained. Take from the remaining states a state B_1 and proceed as above until there is no further state left which is orthogonal to all the B_i already selected, which gives the class (B_i) and continue in this way until all members of K have been sorted out in one way or another. This method [355](which is not necessarily unambiguous) terminates in a classification of the elements of K into sets of mutually orthogonal states.

Again, within thermodynamics the relations between states are usually defined by ideal experiments. Two states are unequal if and only if there is a semi-permeable membrane which allows us to separate them. Introducing likeness-fractions means admitting that there are intermediate cases between complete separation and complete absence of separation. Thus, if an A-filter is a device which passes the state A but completely rejects any state which is orthogonal to A, it is to be expected that a state $B \neq A$ will neither pass completely, nor be completely rejected. B is split into two parts, one passing, one rejected. (One mol of B-gas is split into q (A, B) mols passing and $1-q$ mols rejected. This leads to an operational definition of the q's as passing fractions.)

Further, according to the definition of an A-filter given above, the part of the B-gas which has passed the filter will be in the state A: the B-gas reacts to the filter by jumping partly into the state A, partly into the state $\bar{A}$. And as in the case of a B-gas exposed to an A-filter the only splitting is into A and $\bar{A}$, it follows that $q(B, A) + q(B, \bar{A}) = 1$, or, elaborating, $\sum_i q(B, A_i) = 1$ the summation being extended over all elements of (A_i).

The symmetry of the q follows from very simple thermodynamic considerations. Adding to this account the assumption that any kind of substance consists of indivisible units we arrive at the following concerning probability: $q(A, B)$ is the probability that a particle belonging to a gas in the state B will pass an A-filter when exposed to it: "The correct statistical interpretation (Born) follows immediately from the thermodynamic origin of quantum-theory" (p. 56).

Apart from the principle of entropy-continuity, some of whose consequences we have just been discussing Landé uses (b_1) the principle that there is a general law which allows us to derive $\{q(A, B)\}$ (i.e., the matrix with the elements $q(A_i, B_k))$ from $\{q(A, C)\}$ and $\{q(C, B)\}$, as well as (b_2) the postulate of a constant probability-density in phase-space (a postulate which Landé erroneously identifies with Liouville's theorem). The simplest, if not the only possible way, to satisfy (5_X) consists in assuming that $\psi(A, B) = \psi(A, C) \cdot \psi(C, B)$ where $\psi = \sqrt{q} \, \exp(i\varphi)$ – that is, in assuming a "law of superposition" for probability-amplitudes. (b_2) leads directly to Born's commutation-rules. And this finishes Landés account of general quantum-mechanics.

Trying to evaluate this account we must at once admit its methodological importance. We have here a well-ordered presentation which makes it easier to understand the principles on which the whole theory rests. But the specific principles employed suggest also a new way of looking at this theory which is much nearer to classical thought than the favorite house-philosophy of many physicists would have it nowadays. For it is commonly [356]assumed that these matters have once and forever been settled by what is known as the "Copenhagen-Interpretation".

As regards the Copenhagen-Interpretation, one ought to realize that it is not an interpretation in the usual sense in which one speaks, for example, of the Born "interpretation". The Born interpretation is necessary and sufficient for applying what would otherwise be a purely mathematical formalism to reality, or (if the word "reality" is disliked because of its apparently "metaphysical" character) for tackling physical problems in terms of quantum theory. Any additional element, however interesting and fruitful it may be, is not thus necessary and it is therefore not forced upon us by physics, or by "experience", or by any similar source, it is a bit of speculation which we have to judge according to its fruitfulness, its inner consistence as well as according to its elucidatory power. This speculative, or, if you like, metaphysical character of Bohr's interpretation is often overlooked by people who are in the habit of already seeing quantum theory in its light and who therefore think that either physics, or simple analysis of physical theories forces them to adopt Bohr's views. Admitting this implies that we are to a certain extent free to invent and to consider other "metaphysical" interpretations. Of all the interpretations which have been suggested so far, Landé's attempt seems to be the one "next to reality", that is, it seems to be the attempt which goes least beyond the Born interpretation and which therefore should be most suitable to throw light upon quantum mechanics *as it is*. And whatever one may think about minor points, one must certainly admit that the curious and unsatisfactory features of the Bohr picture are completely absent from Landé's presentation. Quantum theory is much nearer to classical physics than one is usually inclined to assume. Having shown this is another important asset of Landé's undertaking. (As it has been doubted whether Landé's approach is tenable on thermo-dynamic grounds it may be useful to point out that the partial separability of states, advocated by Landé, corresponds to the well known fact that, within quantum mechanics, states are completely separable by semi-permeable walls if and only if they are orthogonal. It was von Neumann (*Mathematical Foundations of Quantum Mechanics*, Princeton, 1955, p. 370) who first connected this theorem with Gibbs' paradox, but without making this fact the starting point of a new presentation of the theory.)

Two critical points ought to be mentioned, one general and one specific. First of all the connection between the postulate of entropy continuity and the general principle of continuity itself is hardly as close as Landé would have it (viz. direct entailment). For there are numerous cases in which a straightforward application of the more general principle leads to undesirable consequences. For example, a mass-point is moving on a straight line l with constant speed. A point divides l into two parts, l_1 and l_2. The mass-point's passing from l_1 to l_2 involves a discontinuity. [357]Now in this case it would hardly be advisable to introduce degrees of "being in l_1" and "being in l_2" and thus to abandon the concept of a continuous movement altogether. For the undesirable discontinuity can easily be removed by using a different presentation of the process and by treating the empirical aspect within a theory of errors. This shows that, although the principle of continuity may be adopted as a guiding principle, it is not sufficient for deriving the entropy-continuity from classical statistical mechanics. Another remark is about the role which the wave-function plays in Landé's presentation. Landé thinks that there exists a form of quantum theory, admittedly very impracticable, which does not contain any probability-amplitudes but only probabilities and which is equivalent to the usual presentation by means of probability-amplitudes; that is, he thinks that the probability amplitudes have no independent physical meaning. Considering the importance of the symmetry-properties of the wave-function as well as of Schrödinger's equation I doubt whether this account of the matter is correct. But the value of Landé's undertaking for those who want to become clear about physics – and this should include also the physicists – remains unimpaired by those minor difficulties.

16

Excerpts from Discussions with Léon Rosenfeld and David Bohm (and Others)
(1957)

I. ON DAVID BOHM, "A PROPOSED EXPLANATION OF
QUANTUM THEORY IN TERMS OF HIDDEN VARIABLES AT A
SUB-QUANTUM-MECHANICAL LEVEL" AND LÉON ROSENFELD,
"MISUNDERSTANDINGS ABOUT THE FOUNDATIONS OF
QUANTUM THEORY"

(with David Bohm, Ernest H. Hutten, Maurice H. L. Pryce, Léon Rosenfeld,
and Jean-Pierre Vigier)

[48]*FEYERABEND*. The discussion and the remarks which Professor
Bohm has made in his lecture have greatly clarified the import of von
Neumann's proof. He has rightly emphasized that the statement that
hidden variables do not exist is a theorem of quantum mechanics rather
than a statement whose truth has been proved absolutely and once for
ever. What, then, is the reason which made Bohm's opponents say that his
theory is very unlikely to succeed? It is a belief in the truth of some
principles of present-day quantum mechanics which is based on an
assumption, common to all inductivistic interpretations of science,
namely on the assumption (a) that present-day quantum mechanics is a
rational summary of facts, (b) that any future theory must also be a
rational summary of facts and (c) that, therefore, no future theory can
possibly contradict quantum mechanics in its present form. It is true – if
challenged, hardly any physicist will subscribe to this view and Professor
Rosenfeld obviously does not subscribe to it. But analysis of various more
philosophical writings of physicists, such as of Born's recent book, shows
that this view is the implicit reason of the conviction of many physicists,
that present-day quantum mechanics, or at least the quantum of action

will have to be an essential and unmodified part of any future theory. This, I think, is a very arbitrary view. Another arbitrary view which is part of the Copenhagen-Interpretation is the view that fundamentally the theory is about nothing but classically describable events, experience is to be described in classical terms; quantum theory proper is nothing but a means of predicting classical states of affairs. It is this idea which has led to the famous intrusion of the observer into physical theory. This should show that, far from being "forced upon us" as Professor Rosenfeld expressed himself (this is again an inductivistic idea) the Copenhagen-Interpretation contains as elements many things which are arbitrary and perhaps even false. I am also not impressed by the fact, mentioned by Professor Rosenfeld, that the law of the quantum of action has always and in all domains been found to be correct. For nothing is easier than preserving a law which one wants to preserve; and I wonder whether in the case of the quantum of action this preservation is not due to the efforts of various physicists to have it preserved rather than to the fact that it is true. One question to Professor Bohm; he has said that he was able to derive the Bohr-Sommerfeld conditions, the Dirac equation, as well as the symmetry-properties of the wave-function, from his model. This is not sufficient for showing the adequacy of this model on the level of quantum mechanics. For there exist many formal analogies between present-day quantum mechanics and, say the theory of diffusion, analogies which are very seductive but which always break down at one point or another. Hence it is possible that Bohm has done nothing but derive parts of a theory which possesses formal analogies to quantum mechanics rather than quantum mechanics itself. Only the derivation of the whole formal apparatus of the present theory *together with its proper interpretation* will show that his model is adequate on the level of quantum mechanics. – A final remark about the relation between present-day quantum mechanics and Bohm's new model. The first thing [49]that comes to one's mind when one hears the objections of Bohm's opponents, are objections which were raised, in the nineteenth century, against the atomic theory. Those objections were much more formidable, as they seemed to reveal a contradiction in the idea of a reduction of thermodynamics to the mechanics of a collection of atoms. There was the reversibility-objection of Loschmidt and the recurrence objection of Zermelo and Poincaré – and nevertheless the atomic theory was finally successful and could even be used for showing that the phenomenological account could not be universally correct. At that time the situation was much clearer, classical mechanics and classical thermodynamics were better understood and less surrounded by a philosophical fog than quantum mechanics is today; and that quantum mechanics, at least in the form in which it is presented by the Copenhagen school is surrounded by a philosophical fog should be clear to anybody who has tread, for example, von Neumann's account which rests upon nothing but the formalism together with the Born-Interpretation, and who compares this account with the orthodox view which contains, apart from elements which follow from the theory in the Born-Interpretation, many other arbitrary assumptions. What we need before any discussion on that matter can be successful is a "minimum-presentation" of quantum mechanics, a presentation which does not contain anything but what is absolutely necessary. That has not yet been achieved and the Copenhagen-Interpretation is far from having achieved it....

[52]*ROSENFELD*. ... Now with regard to Dr. Feyerabend's remarks: I think it was not quite fair to quote a single sentence from my paper and not to quote a sentence two or three pages later in which I said explicitly that certainly quantum mechanics as it stands is not sufficient to cope with those new phenomena. Nobody thinks of attributing an absolute validity to the principles of quantum theory. Everybody agrees that something has to be done about it; whether it has to be done in the Bohm way or in any other way remains to be seen; but whatever is done certainly would imply a limitation of the validity of quantum theory in its present form. Now Bohm says that this limitation would be a limitation of the principle of the quantum of action. What I have tried to point out was that hitherto we have not had any induction from experience of such a limitation. On the contrary all the new phenomena have been found to satisfy this principle. Why should one just suspect that it will be this principle which will be at fault? Now Dr. Feyerabend said that there is an arbitrary assumption in quantum theory, namely the assumption that the concepts must be defined in a classical way. Well, it is an assumption, if you like; but isn't that an obvious thing? How could you do it otherwise?

PRYCE. Well, you don't do it that way anyway.

ROSENFELD. Of course we do, when we want to be careful. We often forget about it. The only way to define position or momentum is to refer to some operation to which we attribute this property and the reason why we do this in a classical way and by classical means is because classical physics is the kind of physics which is adapted to our scale of observation; the reason why we are forced to do so is that we are after all beings consisting of many atoms and endowed with certain senses, and we see the external world in a certain way. Hence, if we want to describe our experience in a communicable way so that other people like us can understand what we are saying, we must use concepts which refer to these possibilities of observation. Isn't that absolutely obvious to everybody?

PRYCE. No. The point behind these interjections, which I did not make explicit, is that we have to reduce our description to "everyday" concepts, that is, to what we actually do or can do, and *not* to "classical mechanical" concepts, which are themselves already abstractions therefrom.

ROSENFELD. That is very surprising.

FEYERABEND. As soon as we have a new theory we have to use it throughout, for the description of experience as well as of anything else.

[53]ROSENFELD. Yes, but what does it mean? I am not suggesting that we must refer every single concept to observation. But we must do that ultimately; the way may be a quite roundabout one and in most cases we will not do it explicitly – we can thus forget about it – but ultimately, surely, whatever you do in order to communicate what you think or what you observe to other people, you must use a language which they are capable of understanding; you must put them in such a situation that they can repeat your own observations, so as to understand what you mean. Isn't that a quite obvious thing?

FEYERABEND. May I try to explain what I mean by way of an example. Take the special theory of relativity. According to this theory the concept of length is a relative concept in the sense that you cannot attribute length to a given object without at the same time referring to some co-ordinate system. Now before the special theory was introduced, length was regarded as a property of an object. Following the procedure which the Copenhagen school adopts in the case of quantum mechanics – that is, describing experience not in terms of the best theory available but in terms of some previous theory which has become so intimately connected with our ways of experiencing things that we can hardly imagine things being different – we would be forced, in the case of special relativity, to describe the results of our measurements in a nonrelativistic way, that is, we would be forced to describe them without referring to the co-ordinate system used. Now this would be a very odd procedure and it would also be very artificial to say that the reference to the co-ordinate system comes in only in the theoretical account of the experiment made. The natural procedure is to say that so far our experiences have misled us into believing that length is a property rather than a relation involving a co-ordinate system, and to try and improve the situation by describing even the results of experiments in terms of the new theory of relativity rather than in terms of a previous theory (e.g., Newtonian theory) which did not know anything about the relative character of length.

ROSENFELD. This is a misunderstanding about classical physics.

HUTTEN. I think that logic can help here quite a lot. We must always have a known language in which to speak about things which are known. If you have a new language and introduce new ideas you must first express them in a language you know – in other words we use classical physics as a metalanguage.

ROSENFELD. But in relativity the situation is exactly the same. Relativity is part of classical physics – by classical physics I mean physics in our scale, including relativity – so, therefore, it is true as you say that as long as we remain within the domain of classical phenomena we attribute the length to the object; that is an approximation we are allowed to make at that level. In connection with this I would say to Professor Bopp that it seems to me that the kind of analysis he wants has been made by Bohr. He just uses classical language for this kind of experience because classical language is our only tool for describing this kind of experience although its application is limited by laws of nature which are independent of us. This brings me to the question of Vigier who absolutely wants to put the accusation of positivism [54]upon me. I think there is a great difference between the early positivists for whom I have much respect – even for Mach, in spite of his aversion against atomism – and the later positivists. I have not the same respect for the later positivists because I think that where positivism went wrong was when it became a system in which those nice points which I think were made already by Mach, were fossilized into principles of absolute validity; it was maintained that science was only a language and that one could decide about statements only by reducing them to sense data and so on. Mach himself was not very consistent and he made many mistakes which encouraged that direction. However, as far as quantum mechanics is concerned, I would say that it is impossible to understand it without assuming that there is an external world which is independent of what we think and which is the ultimate origin of all our ideas. In that sense I absolutely reject the suggestion of present-day positivists about the subjectivity of our statements. As regards the atoms and the atomic theory which has been invoked by Bohm and also by other people as an example of a theory with hidden parameters which has been successful, I have to say this: of course there are atoms and therefore people who spoke of atoms before it was possible for us to observe them were very lucky; that corresponds also to what Fierz said about the possibility of hitting upon the truth, although one has really no right to do that if one goes the wrong way about it. The latter was certainly the case with Francis Bacon who did not know much about science but who happened to put out certain ideas which proved later to be right, together with many others which turned out to be wrong; so I don't think that argument is very convincing. Now as regards the atomic theory in the nineteenth century it is necessary to go a bit closer into the historical development and to follow up the transformation of this idea from a speculation which might be true or not into a scientific theory. I think that the fact that individual atomic processes were only observed at the end of the nineteenth century is, from our present point of view, more or less an accident. The atomists like Boltzmann were convinced of the reality of atoms for very cogent reasons although they had never seen individual atomic processes. For Maxwell the connection between the various parameters which are undetermined on the phenomenological level, was a very strong argument in favor of the reality of the underlying mechanism. Another very important point was that it was possible to make predictions, and to find relations between phenomenological parameters without going into details about the structure of atoms; in this way the degree of arbitrariness in the atomic theory was very much reduced. It is true that Maxwell invented his law of distribution but that was not essential. The most beautiful application of atomic theory was the earliest one, namely, Laplace's theory of capillarity. In this theory he only assumed that the forces between molecules are forces of short range and he could deduce all the macroscopic laws of capillarity from that assumption alone. This shows that there was very little arbitrariness in that conception and it is only that feature of early atomic theory which in my view conferred upon it the quality of a scientific theory as opposed to a speculation. When Mach criticized the atomic theory he of course bet upon the wrong horse and he has to bear the blame for it. But apart from that his paper on the *Erhaltung der Arbeit* was just a very sound warning against the danger of introducing arbitrary elements into atomic theory. His whole point was that one has no right to introduce and to apply to atoms the mechanical concepts which have [55]been derived from experiences about macroscopic bodies, unless one has cogent experimental reasons for doing so. I therefore think that it is unfair to criticize Mach just because he happened to draw the wrong conclusion, as his criticisms were completely sound. Now Vigier has presented us with an example where the same physical laws are interpreted in two different ways. I would not describe this example as he does. I would rather say that we have two different theories which are both self contained – a microscopic description and a phenomenological description: the latter is a definite system of concepts, connected with mathematical symbols; and the former is another system of concepts, namely the properties of systems of a large number of particles assumed to move according to the laws of mechanics; and we define also quantities in that theory which are uniquely connected with the other mode of description and which, of course, satisfy the same relationships.

VIGIER. I was talking of the time when the kinetic theory of gases had not yet covered the whole range of macroscopic laws. As soon as it did this – as soon as it covered all the experiments together with things which could not be foreseen on the basis of the macroscopic laws the situation was different.

ROSENFELD. Now with regard to Dr. Hutten's emphasis on the fact that Bohm made a suggestion how to treat the new processes which require new interactions, I should like perhaps to point out that there are other people who are doing the same thing and have gone a long way already to find, on the basis of quantum theory, descriptions of those couplings and even to reduce the number of independent types of interaction. So there is the same search for simplicity in both cases.

BOHM. I agree with what Dr. Feyerabend has said about using a new theory to modify the definition of older concepts. With regard to the derivation of the Dirac equation, this was meant in the sense of an approximation or a limiting case. I always supposed that at a deeper level there will be a fundamental contradiction between this theory and quantum theory. Also the question was raised as to how we got discontinuity out of continuity. This is all very well known from de Broglie's idea, in which he explained the discrete energy levels as discrete frequencies of vibration of a continuous wave, with the aid of Einstein's relation $E = hv$. My next problem would then be to explain Einstein's relation and this is what I have begun to do here. I can show that if there is a continuous distribution of phases, then the requirement of continuity, applied to the integral $\oint pdq$, leads precisely to the Bohr-Sommerfeld relation which implies the Einstein-relation. Such cases are already familiar from classical mechanics where it turns out that nonlinear oscillations have stable modes of oscillation with certain energies; and the actual oscillation fluctuates around this stable energy. This one can generalize – it is a very common phenomenon in physics and mathematics that from one conception you can deduce the opposite conception as an approximation, so that in a certain sense one conception implies the other. This is what happens here: from the continuity of the phases of the particles that are coming out of the background of the subquantum-mechanical level I deduce precisely the Einstein-relation and the de Broglie-relation for energy and momentum. [56]I do not wish to say by this that only continuity is a correct concept and that discontinuity is wrong; I want to say that both conceptions are correct but that each one reflects only a certain side of a phenomenon. Each one is reflected into the other, so that you can deduce the one from the other as an approximation. Next the question was raised about symmetry and anti-symmetry – it is true that by starting with the field-theory you can deduce that the wave-functions are either symmetric or anti-symmetric – but this is done in a rather arbitrary way. In the theory that I am developing, one does not begin by assuming a single field for each kind of particle, with the further stipulation that each field leads either to symmetric or anti-symmetric wave functions. Rather, one begins with a single subquantum-mechanical level, out of which come the various kinds of particles with symmetrical or anti-symmetrical wave functions as different sides of one and the same thing. Moreover, the interactions of those particles are implicit in the theory and not postulated. Indeed, two particles are seen to interact, not because they are first assumed in separate existence and then put into interaction, but rather because they are different sides of the same thing. Hence the characteristics of their interaction are deduced from the fundamental theory rather than assumed. This theory therefore opens up the possibility of showing why there are different kinds of particles of different masses, why they interact in the way they do, and why they have such and such properties. The question of epicycles was raised. I believe our theory is more like the Copernican theory and that the usual quantum mechanics is closer to the epicycle point of view. For at present there are over twenty kinds of particles. To each particle is ascribed a field, and between each two fields a certain interaction is assumed. Nobody can deny the value of that picture; but the situation still remains that you must put in as many fields as there are particles, and that you must assume the corresponding interactions. On the other hand from our point of view there is a deeper level where all this will come out as an approximation. It is explained why there are those particles and why they interact in the way that they do. This raises another point, namely that the basic object of a theory is not just to obtain simplicity. The stress on simplicity is partially right but it is not the best way to say it. The purpose of a theory is to obtain the essence or unity behind the diversity of phenomena When one has found this essence, then generally one can apply the associated theory in a much broader domain than the original domain of facts. In other words a law which one has discovered holds not only for the facts that one had before; but it also holds for new facts not only in that same domain, but very often in new domains. The world is so built that it has laws of this kind, so that one can extract the essence from a limited number of facts and obtain a theory which contains more than the facts on which it is based. Hence the main point of science is not so much to make a summary of the things one knows, but rather to find those laws which are essential and which therefore hold in new domains and predict new facts. It is an objective fact that the world is so constructed that generally speaking, by finding the unity behind the diversity, one will get laws which contain more than the original facts. And here the infinity of nature comes in: the whole scientific method implies that no theory is final. It is always possible there is something that one has missed. *At least as a working hypothesis* science assumes the infinity of nature; and this assumption fits the facts much better than any other point of view that we know. The [57]next question was raised by Hutten about the determinate theories which are now obtained as limits of statistical theories. But the opposite is also true. I discussed this point briefly yesterday. If you have a large number of variables undergoing sufficiently complex motions then you can show that most of the essential consequences of a statistical theory, such as randomness and the Gaussian distribution, can be approximated to an arbitrary degree from a determinate theory. Similarly you have the opposite, that any determinate theory can be approximated to an arbitrary degree as the limit of a statistical theory. This again is characteristic of our conceptions. Each particular concept finds its reflection in its opposite and can be approximated to an arbitrary degree in its opposite. Then I want to bring up this question of the discovery of atoms. Again I believe this was not a question of luck. For the foundation of the atomic theory consisted in the fact that it was necessary to explain certain difficulties of the large scale theory. Demokritos tried to explain the difficulties of Zeno's paradoxes and this is really a fundamental way by which progress in science is made: difficulties of conceptions at a certain level are resolved by suggesting a new conception. We go beyond the phenomena and try to learn more than the phenomena suggest. Finally I want to raise the question of how we are to define a thing or the conception of a thing, such as position and so on. This was raised by Dr. Feyerabend. I believe that we must define a property such as position in terms of all the concepts that are available. During the Middle Ages or before that time people were not very successful in defining positions accurately; then, when geometry developed, they were better able to define positions. Then with the development of mechanics and wave optics, it was possible to do this still better, for one could use mechanical and optical parameters and so on. And now if we should come to a subquantum level then we could define position in terms of the parameters of the subquantum level. In other words, the conception of position like all conceptions, may have infinite possibilities as we go deeper and deeper into nature and find out more and more of what is there. We cannot measure position apart from the actual physical properties of the things that we use for measuring. This is one of the lessons of the theory of relativity. In the theory of relativity we use measuring rods and clocks which behave in a certain way, and we must change our conception of position in accordance.

2. ON MARKUS FIERZ, "DOES PHYSICAL THEORY
COMPREHEND AN 'OBJECTIVE, REAL, SINGLE PROCESS'?",
STEPHAN KÖRNER, "ON PHILOSOPHICAL ARGUMENTS
IN PHYSICS", AND MICHAEL POLANYI, "BEAUTY,
ELEGANCE, AND REALITY IN SCIENCE"

[112]*FEYERABEND*. Professor Fierz has said that people like Einstein and others want to give a description of an individual single process, and he has tried to point out that the [113]attempt to give such a description would lead to difficulties already in classical mechanics. In trying to show this, he has argued (1) that we always make idealizations and (2) that any assertion about an individual can be tested only by comparing it with other objects. I think that in saying this he has forgotten that in different situations these two methods may lead to different results. Take classical physics and quantum mechanics. In both cases we make idealizations. In both cases we base our assertions or our theory on comparisons between different objects. But in the first case the idealization made is of a completely different kind than the second case, and what Einstein wants to arrive at is the classical kind of idealization rather than a complete description in Professor Fierz's sense. As to the controversy, between Professor Körner and Professor Rosenfeld – Professor Körner has said that we can argue about the regulative principles we introduce and that there is a certain freedom. Professor Rosenfeld has objected; he said there is not very much freedom, at least not in the case of quantum mechanics, and he has given two arguments which I think are both invalid. The first argument is: if we try to use a different "meta-" interpretation we arrive at epistemological contradictions. My point is we arrive at epistemological contradictions with a presupposition which the orthodox hold and which

I think must be attacked – it is the presupposition that experience is to be described in terms of classical mechanics rather than in terms of the best theory available. This presupposition, Professor Rosenfeld has said yesterday, is a triviality, a matter of course. I do not think it is a triviality as so far physicists have never yet adopted this principle, they have always described experience in terms of the best theory available, and I think it is this principle which leads to the epistemological contradictions. The second argument used by Professor Rosenfeld is the argument of Einstein-Podolsky-Rosen. But this is not really an argument as long as there is no experimental evidence showing that the result predicted by quantum mechanics is in fact correct. A final brief remark about Professor Bohm's infinity of nature. He seems to oscillate between two interpretations which should be kept separated; in one interpretation the principle of infinity of nature would be a methodological principle saying that every statement of physical theory is not final, but may turn out to be false at some time. In the other interpretation, this principle may be called an empirical statement, a statement describing the world and saying that there are infinitely many individuals in the world, and that each individual has infinitely many properties.

3. ON PAUL K. FEYERABEND, "ON THE QUANTUM-THEORY OF MEASUREMENT" AND G. SÜSSMANN, "AN ANALYSIS OF MEASUREMENT"

(with David Bohm and Jean-Pierre Vigier)

[138]*BOHM.* I would like to call attention to an important problem concerning the theory of measurements, first raised by Schrödinger. He supposed a hypothetical experiment, in which a single photon passed through a half-silvered mirror. This photon is either reflected or transmitted. If it is reflected, nothing happens. If it is transmitted then it activates a photo-electric cell, which fires a gun that kills a cat which was placed inside a small box.

After the experiment is over, but before any person has looked, the wave-function for the combined system is a linear combination of functions which represent a dead cat and functions which represent a live cat. But at this time it is not only impossible to say whether the cat is dead or alive; it is impossible even to say that the cat is either dead or alive. Of course, when somebody looks he then either sees that the cat is alive or that it is dead – and adjusts the wave-function accordingly. Now if we wish to have an objective treatment of the whole process, that is to say, a treatment where the observer plays no essential role, then there must be something more in the theory than the wave-function. For the wave-function does not give the most complete possible description of reality, as is usually assumed. This suggests that quantum mechanics is incomplete, that there should be in addition to the wave-function some further parameters which tell you what the actual state of the system is after the interaction. After the apparatus has functioned but before anybody has looked, the system is already in some state – and then the observer looks and he finds what it is. The reduction of the wave-packet takes place objectively, that is without the aid of the observer, but with the aid of those additional parameters.

[139]*FEYERABEND.* I don't think that this radical procedure is justified, and I think that the paradox can be solved within quantum mechanics in its present form and interpretation. One has only to consider that, when the interaction is over the combined wave-function is of such a kind that a macroscopic observer cannot discover any difference between the statistical results which it implies and the statistical results implied by a mixture in which the measuring apparatus (or the cat) can be said to be in one of various different states. Hence, if one takes into account the fact that the observer who looks at the cat (or the device which is used for ascertaining the state of the cat) can only distinguish macroscopic events, states, and so forth, one is allowed to identify the wave-function and the mixture. And if one is allowed to make this identification, then the assumption that the cat is in a definite state even before anybody has looked at it, will not lead to any contradiction.

BOHM. That is very similar to the idea that I was proposing – in the large scale there are some classical variables not satisfying quantum theory.

FEYERABEND. That is a misunderstanding. On the large scale there are all sorts of variables. Some of them can be measured by the macroscopic observer, some can't be measured by the macroscopic observer or by a macroscopic measuring device. But if the macroscopic observer takes into account only those variables which he can measure then he will find no difference between the combined wave-function which forbids him to say that the cat is in a well defined, though perhaps unknown state (dead or alive) and the mixture which does not forbid him to say this.

BOHM. But then the observer is being given a fundamental role and the fact that it is in a definite state seems to depend on the observer once again.

VIGIER. The Schrödinger paradox points to the fact that there is something which has happened at the macroscopic level and which we have not been able to describe. There is the fact that the cat is either dead or alive and this fact does not come into the description of the process as this description is purely statistical. But this brings us back to the argument that the only knowledge we have at the macroscopic level is a statistical knowledge.. . .

[146]*FEYERABEND*. We have a measuring device and the outcome of the measurement is indicated by a pointer which has only two possible positions, either pointing up or down. Now when the interaction with the system to be measured is over two different questions may be asked. The first question is whether from the combined state of the system and the measuring device one can derive a statement to the effect that the pointer will be in a well-defined, though perhaps unknown, position. This is the question which Bohm and Vigier have been asking and this question must be answered in the negative. The second question which is of importance for the theory of measurement is: is the statement that the combined system is in a certain combined state, compatible with the statement that the position of the pointer is already well defined? According to the strict account those two statements are not compatible (which shows, by the way, that the negative answer to the first question does not imply, as Bohm and Vigier seem to believe, that quantum mechanics is incomplete. For the combined wave-function is not too "poor" to say anything about the position of the pointer, it strictly *forbids* us to make a definite statement about the position of the pointer). Therefore the strict account forces us to introduce such things as "quantum-jumps" which lead us from the combined state to the classical level where the cat is definitely either dead or alive and where the pointer has a definite, though unknown, position. My point is that this is due to the incompleteness of the strict account which does not consider (a) that the measuring apparatus is a macroscopic system, and (b) that the observer who looks at it or the device (photographic camera) which is used for ascertaining its state cannot determine all its finer properties. (a) and (b) allow us to replace the combined wave-function by a mixture, not because a quantum-jump has happened in nature but because the error involved is negligible. Hence, a complete theory of measurement which takes into account the special (viz. macroscopic) properties of the systems concerned is no longer faced by the cat paradox and it does not contain quantum-jumps either. It contains nothing but the equations of motion and some special assumptions about the character of the systems involved in the measurement. My objection to Dr. Süssmann is therefore simply that his account, although correct, is incomplete. Of course, he is not content with

the combined wave-function; he wants to have this mixture in order to be able to describe the measuring apparatus in a more or less classical way. But instead of taking into consideration the approximations which this transition involves and justifying them on the basis of the special properties of the measuring-device, he introduces, as many physicists before him (e.g., Heisenberg), the jumps as unanalysed wholes. Therefore his strict account is not at all as strict as it looks, as it leaves out the interferences anyway – but without in any way justifying the procedure. This is the reason why I cannot consider his account, or any similar account, [147]to be a satisfactory one. This is also the answer to Dr. Landsberg's first question. For when introducing macroscopic observers we must at the same time introduce coarse-grained densities. In fact a macroscopic observer is defined by certain (coarse-grained) possibilities of subdividing phase-space.

4. ON WILLIAM C. KNEALE, "WHAT CAN WE SEE?" AND WALTER B. GALLIE, "THE LIMITS OF PREDICTION"

(with Alfred Jules Ayer, David Bohm, Richard B. Braithwaite, Walter B. Gallie, Léon Rosenfeld, Jean-Pierre Vigier)

[182]*FEYERABEND*. I would like to make a contribution to the discussion between what has been called the extensionalists and the Spinozists or the Laplacians. The Extensionalist, as I understood him, analyses determinism in terms of predictability whereas the Spinozist would try to analyse determinism as a property oft he universe. Now I suggest the following formulation of determinism which is independent of start with a syntactical definition: *a theory is deterministic with respect to a class K of variables*, if and only if, a conjunction C of statements to the effect that some of the variables of K have a certain value at the time t can he derived from the theory together with another conjunction of statements which again contains only variables of K – not necessarily the same as in C – and which asserts that those variables had some other value at some other time $t' < t$. On the basis of this definition I propose to formulate *determinism* thus: The world is deterministic if and only if the class of all variables can be sub-divided into sub-classes in such a way that for every sub-class there exists a theory which is deterministic with respect to that sub-class and which is true. Two remarks about this formulation of determinism: first it is a statement about the world because it involves the concept of truth. It does not involve anything about observers. Secondly, it can be tested and can also be refuted. For assume that quantum theory in its present form is true and that the Neumann-proof is correct.

This would imply that no true and deterministic theory exists with respect to the class of variables of elementary quantum mechanics, hence, determinism in the form which I have just suggested would be refuted.

GALLIE. I admit that in this form determinism could be refuted, but I do not think it could be confirmed.

FEYERABEND. But this does not matter, perhaps it can be confirmed.

GALLIE. But if it is all one-sided then it does matter.

FEYERABEND. It can even be confirmed: within classical mechanics we could confirm it because classical mechanics was deterministic and we had also confirmation to the effect that this theory was universal in the sense that all variables could be defined in its terms; so it could be confirmed; but I am not keen on confirming it – it can be refuted.

ROSENFELD. May I ask how you can speak of truth without observer. I thought that the truth of a statement relates on its being possible to have a meaning for an observer.

FEYERABEND. I do not define this statement of determinism in such a way that the defining sequence contains anything about observers, about possibilities of prediction, and so on. It only contains the notion of a theory being deterministic which is a syntactical notion, together with the concept of truth – how we test a true theory is an altogether different matter.

GALLIE. If you only take one test referring to one of these sub-classes of variables – just one little exception would refute your determinism. Hence you could never improve or strengthen the case for determinism.

[183]FEYERABEND. This only shows that the statement of universal determinism is a hypothesis, that it is an empirical statement.

ROSENFELD. Laplace who is the supreme authority, I suppose, for that idea proposes it in connection with an observer; of course, his observer was an exalted being but with human qualities.

FEYERABEND. But here I would be even more radical than Laplace. I would say that the statement of determinism is a statement about the world and not about observers. Although this statement can be tested by observers – this does not mean that it asserts something about them.

ROSENFELD. The last is the only sensible way of understanding determinism but I would also contend that it was also Laplace's idea.

VIGIER. With the last statement of Professor Rosenfeld we are back to our old quarrel again: does the scientific statement have a meaning outside observers? We say Yes; he says No. Now this is a question which can only be settled by practice, it is not a question in which logic can decide. There exists the science of geology which informs us about times when no observer existed. And I think that even in the time of the big Saurians the laws of quantum mechanics did apply to molecules inside this Saurian although I cannot go and see because unfortunately they have vanished. Also the whole idea of the development of the universe on the basis of laws and of facts which we observe now, is a valid scientific edifice. This is my first point. The second point is about Laplace. Nobody wishes to return to Laplace's determinism. Bohm and I myself are very strongly against Laplace's determinism in the form which Laplace has given it. The whole point here is the question of infinity. If you find a finite ensemble of laws and an ultimate set of laws, then you are back at Laplacian determinism. But there is no such finite ensemble and closed set of laws. There is always a larger context out of which this finite ensemble of laws can be approached. This is also related to Professor Gallie's paradox – this man, this joker, is outside the context of prediction of the society of predictors.

ROSENFELD. The first point about the observers is a completely trivial point. I do not see the slightest difference between prediction about the future and retrodiction about the past: both are based on present evidence. All that we can say about those Saurians is based on what we can see of them. And all that we can say about the state of the universe when it was first being formed is also based on the known laws of physics – with some degree of idealization, of course; the reliability of those retrodictions, as well as the reliability of predictions, depending on whether the amount of idealization is excessive or not. I cannot see the least philosophical difference between descriptions of the state of the world in the secondary epoch and the description of the world as it is today. It is the same scientific procedure which is applied in both cases. But with respect to the last point about the finiteness and the infinity of levels, I would not regard this as a new point at all: it is a familiar point, and no person in the world more than Bohr would insist upon this dialectical conception of science and of our knowledge. I think I am not betraying his philosophy [184]when I describe it like that. But it is only a framework, and the question whether to fill it with subquantal monsters or with things that you know from experience is a completely different one.

BOHM. I don't quite see the equivalence between prediction into the future and retrodiction into the past. When I predict about something that will happen tomorrow then I can wait until tomorrow and see whether it happens or not. But if I retrodict that there was a monster on the earth a thousand million years ago then I have no way of going back and seeing whether it was really there. All I can do is to see that various consequences of this hypothesis continue to be verified. The second point is this: we find ourselves here in existence at a certain time and we find that we came from something that existed earlier and therefore we suppose that there was a time when there were no human beings on the earth. Now if we suppose that the only meaning of a hypothesis is what an observer might see, then we cannot understand what we mean when we say that there was a time when there was no intelligent life, after which intelligent life came into being and became capable of perceiving. This is why I want to say that we frame our hypothesis not basically on what an observer might observe, but rather, about what is in being. We then say secondarily that if a certain thing is in being, then an observer who happens to be around could observe certain consequences. If he isn't there he can't observe them, but this doesn't change the fact that they are there. Of course, we must take the whole universe in our definition of the meaning of being. We must not say that what is in being is just the universe without us. Rather, it is the universe with us in it. We are part of being and therefore when we observe something we may well change it. If this happens to a significant extent then we must remember that being includes us, our minds, our thinking, our actions and so on. Now in this sense Bohr made a very important contribution when he stressed that the observer plays both an active and a passive role – that we are both actors and observers on the stage of life. But this is not in contradiction with the notion that there is a life and there is a stage upon which we can act and observe. To say that there is no life and no stage and that there is nothing but potential or actual observation, and that this is the only thing that has meaning, I don't understand.

ROSENFELD. I protest against this complete distortion of what I have been saying.

AYER. I think that Professor Rosenfeld gave the wrong answer to Professor Vigier and that Professor Bohm gave the wrong answer to Professor Rosenfeld. Professor Rosenfeld was quite right in saying, of course, that in making conjectures about either the past or the future we do depend upon present evidence, but of course this is not at all the same thing as to say that in talking about the past or the future we are talking about present evidence. Here Professor Vigier is quite right against Professor Rosenfeld. On the other hand I don't think that the distinction which Professor Bohm made is very important. It is quite true that we shall be around to verify a few of our predictions, but we don't live very long and the great majority of the statements about the future remain unverified by us, by him or me or Professor Rosenfeld, just as do the majority of statements about the past. I suggest that if [185]Professor Rosenfeld does maintain this position he must do it in a form that would escape Professor Vigier's trouble about the Saurians. He must say that in talking about the Saurians he is talking not about anything he did observe but about something somebody might have observed, had he been there, even if he wasn't there. To make this theory work, you have got to do it in terms of the possibility of making the relevant observations and not in terms of actual observations. This, I think, might be a difficult thing to do.

ROSENFELD. I am glad that you have mentioned that – this is exactly what I meant. But I didn't mention it explicitly because I thought it was quite obvious.

FEYERABEND. I want to underline what Professor Ayer has just said. In their discussion with Professor Rosenfeld, Professor Bohm and Professor Vigier have confused a special problem of quantum mechanics with a general problem of epistemology. It is true that quantum mechanics in its present form and interpretation has some subjectivistic leanings. But in the earlier discussion about determinism Professor Rosenfeld went much further than this. He seemed to assume that the fact that we *need observers in order to test* a statement – whether of classical physics or of quantum theory – implies that the statement, every statement, *is also partly about observers*. I have tried to point out that such an inference is completely unjustified. More especially I have suggested how to formulate the problem of determinism in a way which does not make it a problem about observers although I have at the same time tried to formulate it in such a way that observers can test it and can decide whether or not it is true.

ROSENFELD. If a statement does not contain any reference to an observer – how can such a statement be tested? How can it even be understood?

GALLIE. Perhaps I can do something here. It seems to me that we want to relate determinism not to the observer but to the observable. In this sense I mean there were processes going on and there were laws before observing beings were on the earth and if there was a law before there were any sensitive beings, this law would be manifested, in Popper's language, as the propensity of certain systems to produce certain effects. But those effects we can only describe as if we could somehow, directly or indirectly, have observed them. There is another point for Dr. Feyerabend. Einstein for example propounded deterministic theories. On the other hand there are grounds, for example, von Neumann's proof, for denying that in certain fields a deterministic theory is possible. I suppose Einstein could have retorted that the difference is that in his theories the variables are well chosen, whereas in other theories they are not. Would not this rebut your disproof of determinism?

FEYERABEND. But then quantum mechanics would be false, whereas I have made the assumption that quantum mechanics is true.

VIGIER. I was very happy about what Professor Ayer said and I also agree with Professor Rosenfeld, for this is the first time I have heard Professor Rosenfeld disagree [186]with a plain, positivistic sentence. I would put the thing in a stronger form: I do not think that things which exist are things which might have been observed. This is where the split comes in a very clear form.

ROSENFELD. I don't say that. Don't continue on that line because I do not say that things only exist in so far as they could have been observed. All the statements we make about the world are necessarily descriptions of a state of affairs, of mind, of material, that an observer might perceive if he were placed in those particular circumstances.

VIGIER. Let us say then we agree that the world exists outside any observer. Did the laws of quantum mechanics apply to the world at a time when there were no observers present?

ROSENFELD. Of course.

VIGIER. O.K. If you say then, that the laws of quantum mechanics did apply at that time, then the laws of quantum mechanics are real, objective, statistical laws of nature, which have nothing to do with the observer, and are verified whether there are observers or not.

ROSENFELD. No.

VIGIER. You can't change your position and say something two minutes ago and another thing now. Let's go into the time when there were no observers at all, then you would say that the world did exist at that time in an objective way, that the laws of quantum mechanics did apply at that time in an objective, real way, and that means that the laws of quantum mechanics are real, objective, statistical laws which have nothing to do with observers or things which might have been observed, since nobody was there and couldn't possibly have been there.

BRAITHWAITE. But they could only be known by observers.

VIGIER. That is quite a different question. It is true that you get to nature through scientific practice and scientific elaboration, this is one side of the question. But what we say is that the laws that you get are real objective approximations of the real objective properties of matter. That is what is contested basically. There might have been no observers whatsoever and the laws of quantum mechanics did still work. And once this is admitted you have the right to suppose that there might be a deeper level and to admit the possibility of explaining statistical laws in terms of this level.

17

Review of John von Neumann, *Mathematical Foundations of Quantum Mechanics*, translated from the German edition by Robert T. Beyer, Princeton: Princeton University Press, 1955 (1958)

[343]Elementary quantum mechanics as presented in the early thirties was a highly successful theory, but from the point of view of rigour it suffered from various defects, mainly two: it used mathematical procedures that were not permissible, and even inconsistent; and its interpretation, viz. the Copenhagen-interpretation, although a useful guide for the theoretician, a good means of orientation for the experimentalist as well as an exciting subject for popularisation it was far from clear what exactly its presuppositions were.

The treatise at present under review, which appeared first in 1930, can be regarded as an attempt at a presentation of the theory (elementary quantum mechanics without spin) which is free from those two defects. The first defect is removed by developing an extension of the theory of the Hilbert-space and by a corresponding reformulation of the eigenvalue-problem. We shall not concern ourselves with this part of the treatise (chaps. I, II), which has found little appreciation among physicists as it "involves a technique at once too delicate and too cumbersome for the ... average physicist" (E. C. Kemble). The second defect is partly removed by an attempt to base the whole theory upon the formalism together with "a few general qualitative assumptions" (p. 295). Although this attempt has not been completely successful, it is still of importance as an attempt (*a*) to discover just what exactly is implied by quantum mechanics and what must be left to speculation and (*b*) what are the assumptions which must be added to the formalism in order to turn it into a fully-fledged physical theory.

It has sometimes been thought that those assumptions are provided by the Copenhagen-interpretation. This view is incorrect as the

Copenhagen-interpretation is neither necessary nor sufficient for connecting the formalism "with reality", that is, with the results of measurements. It is also fairly independent of the formalism, as it consists mainly in interpreting the quantum-postulate and in drawing conclusions from it which may be said to be elements of a new (and nonmechanistic) ontology. This at once raises the question to what extent this ontology is justified, not by speculation on the basis of the quantum of action, but by the theory as a whole; or, to be more precise, it raises the question which of those assertions are implied by the theory and which are based upon (ontological) speculation only.

Von Neumann's treatise attempts to provide an answer to this question. I shall use the rest of this review for a critical outline of this answer.

Interpretation. The rules that are used for connecting the formalism with "reality" are: (1) two rules (I and II on pp. 313 sq) connecting physical [344]magnitudes with operators of the formalism; and (2) the Born-interpretation (p. 210).

Rules (1) are not the only rules which are possible. Weaker rules may be used (and have in fact been used e.g., by Weyl) which do not any more enable us to prove the theorems referred to below under (*a*) and (*b*). There seem to be strong plausibility-reasons for adopting von Neumann's rules. Yet the existence of weaker alternatives shows that the so-called "Neumann-proof", that is, the attempt to prove that (A), (B), (C) below are not independent assumptions but follow from QM, is by no means as strict and straightforward as it may appear at first sight. Rules (2) are deduced from rules (1) and some assumptions about the properties of expectation-values.

Probability is introduced in two steps; firstly, in the usual way, viz. as the limit of relative frequencies within large *ensembles* (von Neumann adopts the Mises-interpretation of probability (pp. 289 sq)). This procedure, which has greatly clarified the role of probability within quantum mechanics, is suggested (*i*) by the fact that the Born-interpretation is a statistical interpretation in a straightforward and classical sense and (*ii*) by the fact that the statistical properties of large *ensembles* can be studied by experiments upon small samples, whatever happens during those experiments (p. 301). The statistical properties of the *ensembles* are then completely and unambiguously characterised by their so-called statistical operators. In the second step, which is usually overlooked by those who interpret quantum mechanics as a variety of classical statistical mechanics, it is proved (*a*) that every *ensemble* of quantum mechanical systems is either a pure *ensemble*, or a mixture of pure *ensembles* and (*b*) that the pure *ensembles* (1) do not contain *sub-ensembles* with statistical properties different from their own (they are "irreducible"), (2) are not dispersion-free. This proof (the famous *Neumann-Proof*), which, apart from a certain set of correspondence-rules (see last paragraph), also presupposes the (hidden) assumption (we may call it the *completeness-assumption*) that the statistical properties of any *ensemble* of quantum mechanical systems can be represented by a nonnegative hermitian operator (or, in other words the assumption that any *ensemble* of quantum mechanical systems is either a pure *ensemble* or a mixture of pure *ensembles*), allows for the application, *to individual systems*, of probabilities in the sense of relative frequencies; furthermore it is supposed to show that the assumption of universal determinism (even of a "hidden" determinism which underlies the apparently statistical behavior of microscopic systems) contradicts elementary quantum mechanics in its present form and (in the above) interpretation that two important elements of the Copenhagen-interpretation, viz. (A) *indeterminism* and (B) the *interpretation of Heisenberg's relations* as restrictions of the applicability of the terms involved, do in fact follow from the theory; and that it is impossible to separate the individual system from the *ensemble* to which it belongs (if this *ensemble* is a pure case); [345]this corresponds (C) to Bohr's idea of the *indivisibility of the phenomena*. But as the *completeness-assumption* is one of the hidden presuppositions of the proof all that can be said is that the Copenhagen-interpretation is consistent with QM rather than it follows that from QM interpreted in accordance with the above-mentioned rules (for this point cf. Groenewold, *Physica* 12; a similar point was made by Popper in 1935).

Here I ought to make two remarks; firstly, even if (*a*) and (*b*) were theorems of QM von Neumann's proof could not show, as has sometimes been assumed, that determinism has been eliminated once and forever. For new theories of atomic phenomena will have to be more general; they will contain the present theory as an approximation; which means that, strictly speaking, they will contradict the present theory. Hence, they need no longer allow for the derivation of von Neumann's theorem. This does not mean, however, that they will perhaps contain "hidden" variables (who would possibly defend the view that one day, perhaps, we might return to phlogiston?). Secondly, none of the attempts to refute that part of von Neumann's argument which is based upon the completeness-assumption has so far been successful. This applies also to Bohm's theory of 1951 whose "hidden" parameters are not independent of the parameters of quantum mechanics proper, but partly definable in terms of them (which is connected with the fact that statements about his "Ψ-field" and about the trajectory of a particle moving under the influence of this field, are not logically independent). Bohm's theory only refutes von Neumann's contention (326) to have shown that even the existence of *untestable* deterministic parameters contradicts QM in its present form.

Measurement. Von Neumann's theory of measurement can be regarded as an attempt at a rational explanation of the old idea of "quantum-jumps" (p. 218n). It rests upon the assumption that the state of a system may undergo two different kinds of change, viz. changes which are continuous, reversible, and in accordance with the equations of motion – those changes happen as long as the system is not observed, however strong its interaction with other systems may be; and changes which are discontinuous, irreversible, not in accordance with the equations of motion and which happen as the result of a measurement. Two arguments are presented for this assumption, an inductive argument, trying to relate the existence of the discontinuous jumps to experience (pp. 212 sqq.), and a consistency-argument which shows how they can be fitted into the theory without leading to contradictions.

The inductive argument is invalid (it is based upon an analysis of the Compton-effect which takes into consideration only those features of the phenomenon which can be understood on a classical basis). The consistency-argument is sound but it leads to a theory of measurement which is incomplete (it omits the fact that the measuring-device is [346]macroscopic and that the observer can only ascertain its classical properties) and gives rise to various paradoxes (cf. my account in Zs. *Physik*, 148, 1957): von Neumann has not succeeded in justifying the idea of quantum-jumps on the basis of the theory.

The remainder of the book deals with the theory of radiation, with thermodynamical questions (here Szilard's highly original ideas are used and extended) as well as with more specific problems of interpretation.

Trying to evaluate the book we may say that its main merit lies in the fact that (1) it gives us a "minimum presentation" of quantum mechanics, that is, a presentation which is based upon as few assumptions as possible; (2) it gives us a presentation of the theory that is formally much more satisfactory than many of the preceding expositions; (3) it tries to show how much of the Copenhagen-ontology can be justified on this basis, and thus be shown to be part of physics rather than part of philosophy; and (4) it contains valuable contributions to many special problems. In this way the book has greatly contributed to clarifying the foundations of the theory.

However it contains a theory of measurement which is incomplete and leads to paradoxical consequences. This is the more surprising as all the elements needed for completing the theory had been developed by von Neumann himself already in 1929, in his paper on the quantum-mechanical H-theorem and ergodic theorem (*Zs. Phys.*, 57), as well as in chapter V, section 4 of his book. (It was not before 1952 that these elements were used for attempting a complete theory of measurement.) Secondly the arguments which are used in order to carry out the programme referred to under (3) are by no means as straightforward as they seem to be at first sight. Reading the book one gets the impression that here *proofs* are presented to the effect that one cannot escape adopting the - Copenhagen-interpretation once one has chosen to adopt quantum mechanics. This has led to the myth that the Copenhagen-interpretation has a firm "mathematical" backing while all that could be shown was that it is compatible with QM and must therefore be defended by independent arguments (in which some of von Neumann's theorems do play an important role).[1] In this way von Neumann's book has created a good deal of confusion among physicists (cf. e.g., de Broglie's statement in the Introduction to *La Physique quantique restera-t-elle indéterministe?*) and, naturally, among philosophers.

A final remark about the formalism used by von Neumann (description of states in terms of statistical operators). This formalism has hardly ever been used in philosophical discussions of the foundations of the theory. Such discussions are either based on the formalism of wave-mechanics, or (to a lesser degree) upon the formalism of matrix mechanics. The reason for this procedure is to be found in the fact that wave-mechanics, for example, has a particularly classical appeal which makes it more plausible because [347]it suggests many seductive but often mistaken analogies. The many mistakes which philosophers have made when analysing the *content* of quantum mechanics on the basis of the formalism of wave-mechanics (as an example I may only mention Nagel's and Northrop's contention that quantum mechanics does not violate determinism as the wave-function obeys a deterministic equation) indicate clearly that a less prejudiced formulation of the theory is needed. Von Neumann's book contains such a formulation which should be known by everybody who is interested in the foundations of quantum mechanics.

---

[1] It ought to be mentioned that Bohr himself did not commit this mistake.

18

Review of Hans Reichenbach, *The Direction of Time*, Berkeley–Los Angeles: University of California Press, 1956 (1959)

[336]A purely mechanistic universe consisting of mass points which obey strictly deterministic laws presents various difficulties to the understanding, two of which are closely connected with the topic of the book under review. The first difficulty is that such a universe apparently does not provide any *physical* reason for the existence of a unidirected time. The second difficulty is that it apparently does not allow us to give an account of laws involving direction, such as the second law of thermodynamics. Both difficulties were solved, over fifty years ago, by Boltzmann's epoch-making investigations. The solution which Boltzmann proposes is very simple. According to Boltzmann the *second law* of thermodynamics is only statistically valid: given a class of closed physical systems, all of them in the same initial state of low entropy, the great majority of those systems will eventually assume a state of higher entropy, and this in spite of the fact that a *single system* behaves in a perfectly reversible way. The *direction of time* in a given part of the universe is determined by this behavior of the majority and it is therefore a property which possesses neither spatial, nor temporal universality. (It may change from one epoch to the next, and it may also change from one part of the universe to the next.) Reichenbach's essay which is mainly concerned with what we above called the first difficulty may be regarded as an exposition, enriched and enlivened by many more or less relevant asides, of this solution of Boltzmann's (which Boltzmann himself managed to condense into two pages of vol. ii of his *Gastheorie*).

Its central part, that is, the proof that time direction is a property not of a single system, but of an *ensemble* of systems with a common initial entropy, is based upon five assumptions which Reichenbach calls the *hypothesis of the branch structure* and which may be regarded as a slight clarification of Boltzmann's own presuppositions. It is interesting to note that three of these assumptions (3, 4, 5 on p. 136) have been proved to follow from classical mechanics (the proof is due to the work of Birkhoff and his school). In particular assumption 5 ("in the vast majority of branch systems the directions toward higher entropy are parallel to one another, and to that of the main system") can be shown to be a highly probable consequence of the laws of classical mechanics (it is valid for "almost all" points of the trajectory of the "universe"). This suggests that by replacing assumption 1 ("the entropy of the universe is at present low") with the assumption that the universe is dwelling in a state in which assumption 5 is strictly valid it would be possible to define a time direction for the whole of the universe. That such a procedure should be possible, at least for some stages of the development of the universe, is also supported by the fact, first made known by O. Heckmann, [337] that a classical universe is unstable and must expand, or contract, or pulsate just as its relativistic colleagues.

Whereas Boltzmann was dealing with a mechanistic universe, Reichenbach assumes "that the quantum physics of today is in need of Boltzmann's ideas just as much as the physics based upon Newton's mechanics for the very reason that this modern physics, too, did not discover irreversibility in its elementary processes" (134). This assertion touches a sore point of the interpretation of quantum theory. For few physicists would be prepared to admit that the continuous reversible changes of state implicit in Schrödinger's equation are the only elementary processes of quantum mechanics. Both Bohr and von Neumann have asserted a dualism of reversible and irreversible processes, and Landé has even gone so far as altogether to deny that Schrödinger's equation may be regarded as a process equation. On the other hand it seems that attempts, such as Ludwig's, to explain the "quantum jumps" as macroscopic phenomena which microscopically can be understood on the basis of Schrödinger's equation alone are either open to the reversibility objection, or employ a postulate which is equivalent with the assumption of irreversibility upon the macrolevel. This being the case the extension of Boltzmann's procedure to the domain of quantum mechanics seems certainly to be advisable.

The exposition of Boltzmann's theory is only part of the content of Reichenbach's essay. The remainder deals with such diverse topics as the emotive significance of time, entropy and information, quantum theory, the Feynman tracks and others. There are excellent suggestions (cf. the treatment of the Gibbs paradox on p. 63), long stretches of more or less efficient popularisation, repetitions of arguments and results which are well-known from other works of Reichenbach's. On reading the book one frequently (but not always) finds oneself confronted with the rather perplexing question as to whether that new discipline, called "philosophy of science" is really as new as it looks, or whether it is only the old enterprise of popularisation made new and fashionable by the use, in addition to the technical terms of physics, of the technical terms of the one or another "scientific" philosophy. But this question leads already far beyond the book under review, for it concerns nearly the whole of the present "philosophy of science". The fact that it does not apply to the whole of *Direction of Time* recommends this book both to scientists and to philosophers.

19

Professor Landé on the Reduction of the
Wave Packet
(1960)

[507a]In his attack on what has been called the quantum theory of meas-urement [*Am. J. Phys.*, 27, 415 (1959)], Professor Landé exhibits the ambiguities which, in more general discussions, are connected with the notion of a quantum mechanical state. He suggests to distinguish the *state* from the *expectation function* which expresses it in a given representation and which allows us to calculate the transition probabilities into other states. He therefore emphasizes, in accordance with the usual procedure, that $\langle\varphi|\alpha'\rangle$, $\langle\varphi|\alpha''\rangle$, $\langle\varphi|\alpha'''\rangle \cdots [|\alpha'\rangle, |\alpha''\rangle, |\alpha'''\rangle$, etc., the eigenstates of a complete set of commuting observables $\alpha]$, and [507b]$\langle\varphi|\beta'\rangle$, $\langle\varphi|\beta''\rangle$, $\langle\varphi|\beta'''\rangle \cdots [|\beta'\rangle, |\beta''\rangle, |\beta'''\rangle$, etc., the eigenstates of another complete set of observables with $[\alpha\beta] \neq 0]$ all represent the *same* physical state Φ. He points out, quite correctly, that this procedure is dropped as soon as the *temporal* development of a state is taken into consideration: whereas $\langle\varphi|\alpha^i\rangle$ and $\langle\varphi|\beta^k\rangle$, and even $\langle\varphi|\alpha^i\rangle$ and $\langle\varphi|\alpha^r\rangle$ are interpreted as different expectation functions belonging to one and the same state, $\Phi(t')$ and $\Phi(t'')$ $[t' \neq t'']$ are interpreted as two different states. Professor Landé regards this as an inconsistency in notation and he demands also that $\Phi(t')$ and $\Phi(t'')$ be interpreted as two different expectation functions belonging to one and the same state. If this interpretation is adopted, then Schrödinger's equation no longer plays the role of a process equation which transforms [508a]states into other states, it then rather plays the role of an equation which transforms expectation functions (or representa-tions) of a state into different expectation functions of *one and the same state*. In this interpretation states never change, unless we perform a measurement – in which case there occurs a sudden transition from a state into one of the eigenstates of the observable measured.

I would like to point out that Landé's procedure only apparently removes the temporal changes and that, strictly speaking, it is nothing but a verbal manoeuvre. By pushing the temporal changes into the representatives, this procedure makes now the dynamical variables time dependent, whereas in the usual presentation which is criticized by Landé, the variables do not change in time. However, a representation of the quantum theory in terms of stable states and moving variables is very well known: it is the Heisenberg representation. And as this representation can be shown to be equivalent to the one Landé wants to abandon, it seems that his criticism, and his suggestion completely lose their point.

20

Comments on Grünbaum's "Law and Convention in Physical Theory" (1961)

[155]There are three points in Professor Grünbaum's paper on which I wish to comment. They are: (A) his attack on Duhem's thesis that any hypothesis H can be saved in the face of adverse evidence $\sim O$ by a suitable replacement of the additional assumptions A initially needed for the derivation of O; (B) his elucidation of what is to be understood by the conventional character of space-time congruence; (C) his remarks concerning the possible conventionality of the continuum in physics.

(A) Duhem's argument results in the conclusion that, given $(HA \rightarrow O)\&\sim O$ it is always possible to find an A' such that $HA' \rightarrow \sim O$. It has been frequently assumed that Duhem has proved nothing but the possibility of introducing *ad hoc* hypotheses for the purpose of explaining away adverse evidence. Professor Grünbaum has made it very clear that this is not what Duhem intends to show. According to Professor Grünbaum's restatement of Duhem's thesis the class of the admissible A' is restricted by the following three demands:

(a) A' is not trivial (for example, it must not be the case that $A' \equiv \sim O$ or $\equiv H \rightarrow \sim O$);
(b) A' is inductively different from A;
(c) A' is such that it leaves unchanged the meaning of H.

[156]Stated in this way Duhem's thesis is very interesting, far from trivial, and definitely in need of substantiation. It is Professor Grünbaum's contention that such substantiation is still missing. More specifically, he asserts (i) that, from a purely logical point of view, the argument is a *non sequitur*; and he attempts to show (ii), that the conclusion of the argument is false as there are cases in which a modification of A is excluded as a

matter of empirical fact. It seems to me that neither (i) nor (ii) will stand up under closer examination.

To start with, it does not seem to me to be correct to assume that Duhem pretended the step from

$$(HA \rightarrow O) \& \sim O \qquad (1)$$

to

$$(\exists A')(HA' \rightarrow \sim O) \qquad (2)$$

(A' satisfying the conditions (a)–(c) above) to be justifiable on the basis of the laws of formal logic alone. I rather think that in deriving (2) from (1) Duhem would appeal to a further premise which may be stated in the following way. The set of alternative hypotheses used at a particular time by an individual scientist, or by the community of scientists, for the tentative explanation of as yet unexplained states of affairs is rich enough to provide in any case such as (1) an alternative A' satisfying (2) and (a)–(c) above. Hence, what is asserted by (2) is not, in my opinion, the *non sequitur* that as a matter of logical fact nontrivial alternatives do always exist. What is asserted by (2) is that, on reflection, the scientist will always be able to find a suitable A' because of the richness of the set of assumptions in which he operates. It is clear that this additional premise of the argument need not be true. More especially, it is to be expected that dogmatic adherence to a certain point of view to the exclusion of all alternatives will lead to a situation where alternative explanations are not forthcoming. In such a situation refutations may indeed be decisive and of the character envisaged by Professor Grünbaum in his example. But as long as different points of view are encouraged it is difficult to see how any refutation could be decisive in the sense that an alternative account which retains H, but is not *ad hoc*, could be ruled out forever. At a later point of my remarks I shall have to show that the possibility of such alternatives, and therefore the correctness of Duhem's argument is a very important presupposition of scientific progress.

To this argument Professor Grünbaum has replied, in a private discussion, that it proves nothing unless it has first been established that formula (2) above is *logically possible* for any H chosen – that is, unless it has been established that

$$(A)(H)(\exists A')\{[(HA \rightarrow O) \& \sim O] \rightarrow P(A') \& (HA' \rightarrow \sim O)\} \qquad (3)$$

[157](P(·)) being a predicate which asserts that conditions (a)–(c) are satisfied), is not logically false. That this indeed is not the case can be shown very easily: (3) is equivalent to

$$(A)(H)(\exists A')[HA \vee O \vee P(A')] \tag{4}$$

which latter formula could be logically false only if there were an A, an H, an O, and an A' such that (1) HA is inconsistent; (2) O is logically false; (3) $P(A')$ is inconsistent. As the conditions of the argument are such that (1) and (2) will never be satisfied, it follows that (3) is contingent *whatever the form of the predicate P(·)*. We conclude that Professor Grünbaum has not shown, as he claims he has, that Duhem's argument is inconclusive.

Nor has he shown, I contend, that the conclusion of the argument is false. Let us for this purpose take a look at the example chosen: what is to be tested in this case is the hypothesis H, that the geometry of the domain of investigation is Euclidean. Assume that the observational findings are such that HA (A being the assumptions that no distorting influence acts upon the measuring rods) is refuted; and that A is confirmed with the help of the criterion that within the domain in question "two solid rods of very different chemical constitution which coincide at one place ... will also coincide everywhere else". Do these observational findings exclude the possibility of a nontrivial alternative A' which would save H from refutation? I do not think they do because they cannot exclude the following A': within the domain D there exist distorting forces which act upon all chemical substances in a similar way but which make themselves noticed in a slight change induced in the transitions probabilities of atoms radiating in that region. This A' is confirmed by the evidence which also confirms A. The factual content of A' differs from the factual content of A. A' is not trivial. Furthermore, A' is possible because of the "inductive latitude afforded by the ambiguity of the experimental evidence" if we only demand that the intensity of the spectral lines be below the level of experimental accuracy reached at the time when A' is first proposed (compare this to Copernicus' reply to his opponents, that the parallax of the fixed stars is unnoticeable because they are so far away). *Yet A' enables us to retain H in the face of apparently adverse evidence.*

At this stage of the argument two remarks seem to be in order. The first remark is that A' can be chosen in such a way that it allows us save *any* geometry inside D. We only have to make the changes of intensity small enough to be undetectable even if the change in the geometrical assumptions should be drastic. The second remark concerns Professor Grünbaum's assertion that given our example he is [158]no longer able to understand what is *meant* by the assertion that all rods inside *D* experience a distortion that is independent of their chemical composition. This will, of course, be the case if the criterion of distortion which we quoted in the last paragraph is taken as an (operationalistic) *definition of the "meaning"* of the term "distortion". However, it is not at all advisable to give such a rigid account of the "meaning" of the descriptive terms of a scientific theory. Operations, like the one described above, can at best be interpreted as *test conditions* which are capable, and perhaps even in need of improvement, and which may break down under certain circumstances (the usual methods of measuring temperature will break down inside the sun, since no measuring instrument will then remain in its rigid state). Our example discusses just such circumstances, and it also explains what new tests may be available once the accuracy of the measurement of the intensity of spectral lines has been greatly increased. The charge of Professor Grünbaum that in our example either the meaning of *H* has been changed or *A* has been made meaningless is, therefore, tenable only if a narrow operationalistic theory of meaning is adopted.

If we generalize what emerges from the above discussion we arrive at the result that refutations are final only as long as ingenious and nontrivial alternative explanations of the evidence are missing. There can be no argument to the effect that the absence of such an alternative at a particular time will make the refutation final. It is very important to point out, in this connection, that the fate of many theories strongly depends upon the belief of their defenders that it will be possible, at some future time, to incorporate into them all the apparently refuting instances. Take, for example, the early belief that both the fixed stars and the planets obey the same laws of circular motion. The obvious irregularity of the motion of the planets was strong refuting evidence against this belief. Yet it was hoped that this apparently refuting evidence would somehow be explainable on the basis of the idea of circular motion, and the attempt to do so finally turned out to be successful. Similar considerations apply to Copernican astronomy (the refuting evidence was here produced by the Aristotelian theory of motion), to the kinetic theory, and the like. In all these cases the conquest of the refuting instances is carried out by vastly increasing the domain of alternative explanations which are compatible with the theory to be "saved", and then by the choice, from this domain, of a suitable *A'*. Hence, it is not much of an exaggeration if we say that the progress of science frequently depends on the belief in the validity of the conclusion of Duhem's argument.

(B) On this point Professor Grünbaum's main contention seems to be this. What is asserted by conventionality of space-time congruence [159] is the following: starting out with a definition of "congruent" that makes this word applicable either to the spatial continuum, or to the time continuum, we discover that *as a matter of empirical fact* this definition is not unique because it still allows for different metrics. According to Professor Grünübaum it is this *empirically guaranteed latitude* which is referred to by the term "conventionality" in connection with the work of Poincaré and Einstein. Professor Grünbaum emphatically denies that this latitude may be identified with the much more trivial and, indeed, universal latitude in the definition of a term which has not yet been given any meaning at all. And he points out that the existence of this kind of conventionality reveals certain properties of space and time; for example, it reveals that neither of these two entities possesses an intrinsic metric. He therefore regards as completely mistaken Eddington's contention to the effect that congruence (for line segments and time intervals) is conventional only in the trivial sense in which "the meaning of every word of our language is conventional".

What I intend to show in this part of my discussion remarks is the still unnoticed strength of Eddington's thesis. As a matter of fact, I believe that Eddington's thesis correctly mirrors the *logic* of the situation and cannot, therefore, be refuted by the *historical* remarks which Professor Grünbaum has produced against it.

In order to realize this, take the conventionality of simultaneity which Professor Grünbaum has discussed in his paper. This discussion shows that the historical process of defining the word "simultaneous" can be split into two parts. Part one is such that meaning is given to a word, "simultaneous", which previous to this action had no meaning whatever. When calling this part "purely arbitrary" and "conventional", we are indeed asserting that it is conventional in the sense in which "the meaning of every word of our language is conventional". One should also notice that in spite of its arbitrary character this first part is not without factual correlate. After all, we do not intend to introduce a term which will have no application, and we may, therefore, say that our definition reflects (our belief in) the fact that within the domain in question a relation can be defined that is symmetrical, reflexive, and transitive. Finally, it is evident that part one need not be exhausted in a single step and that it may consist in a series of stipulations, one being carried out after the other, *without thereby losing its arbitrariness and its purely linguistic character*. We now turn to part two of the process. Part two consists in the realization, on the basis of further empirical research, that there are some cases in which the arbitrary stipulations made so far do not give an unambiguous empirical result *unless some further stipulations are added*. Now the only difference between these further stipulations (which are the ones dis[160]cussed by Professor Grünbaum and which, he says, are not entirely arbitrary) and the original stipulations, which are admitted to be arbitrary, is that they have not been introduced earlier for want of some empirical information and that they must therefore (1) be consistent with stipulations already made, and (2) be such that an unambiguous answer is given to the situation in which the need for them has arisen. A little reflection will show that this difference is only an apparent one. It is apparent because the set of original stipulations will, of course, also have to be internally consistent and sufficient for the unambiguous description *of the facts then known*. And the fact that part two of the stipulational process is separated from part one by some fifty years and based upon information which was not available when part one was carried out quite obviously has nothing to do with the logic of the situation (which is the logic of an arbitrary linguistic convention). After all, the existence of a finite maximum velocity of causal chains could have been known at the very beginning, and our definition of "simultaneity" would then have consisted in part one only and would, therefore, have been admittedly verbal. Clearly, the existence of a time span between the first step and the discovery about causal chains cannot influence the logic of the situation. There is, of course, a tremendous *psychological* difference between the two parts. For when part two of the definitional procedure is carried out, the stipulations contained in part one will already have been absorbed into the behavior of those using them and will, therefore, seem to be "natural" rather than conventional. As opposed to this habitual procedure the arbitrary character of the newly introduced stipulations will be much more obvious. From a logical point of view, however, no difference can be asserted between the stipulations made in part one and those in part two. I conclude, therefore, that Eddington's account of the situation must still be regarded as correct.

(C) Continuity. The question as to whether or not continuity of space-time description is conventional may be rephrased as follows: Is it *logically possible* to conceive of a crucial experiment between continuous and discontinuous descriptions of the spatiotemporal properties of events and groups of events? If such a crucial experiment should turn out to be logically impossible, then indeed our choice of continuity rather than of discontinuity would be conventional in the sense that it could not be determined by facts, but only by considerations of simplicity and other syntactical considerations. Now the difficulty of the present situation seems to lie in the fact that discrete alternatives for the mathematics, which at the present moment is being used in physics, are missing. This, of course, does not prove anything (after all, the full theory of continuity was not developed before the second [161]half of the nineteenth century). Nor can the relative paucity of the intuitionistic approach be regarded as a serious counterargument. For what we want is not repetition, in new terms, of the *whole* of classical mathematics; what we want is a new mathematical system that may be adequate for the description of the universe which is also articulate enough for a crucial experiment to be possible. That a crucial experiment may be possible does not seem to me to be too utopian a hope. Consider, for example, a closed system containing a classical field. As is well known such a system possesses an infinite but denumerable number of degrees of freedom. If we use a discontinuous (but not dense) space-time description, then the number of events, and therefore also the number of degrees of freedom within the system, will be finite. It may well be that this difference is empirically discernible by measurements of black-body radiation of the specific heat of solids inside the cavity. In this connection there will, of course, arise the well-known difficulty that a discrete space description cannot incorporate such theorems as the Pythagorean theorem. The way out of this difficulty, however, does not seem to be impossible either. We need only assume that measurements in different directions do not commute; and then perhaps we can retain the theorem as an operator equation. This procedure is well known from quantum theory, and it has enabled physicists to reconcile the conservation laws, in individual interactions, with such phenomena as interference, penetration of potential barriers, and the like. Taking all this into account, it seems to me that it is far from obvious that we are stuck with the continuum and that we shall forever be unable to discover the *factual* reasons (and not only the reasons of convenience) behind this mode of description.

21

Comments on Hill's "Quantum Physics and Relativity Theory" (1961)

[441]In an oral prologue to his paper, Professor Hill referred to the difficulty which he as a physicist faces when attempting to discuss problems of a more general nature, or philosophical problems, as one may call them. He attributed this difficulty to the different language used by philosophers and physicists. It is extremely regrettable that an interesting attempt at a more general understanding of the present status of our physical knowledge should be handicapped by circumstances which are neither necessary nor desirable. When modern philosophers, and especially the logical empiricists, took upon themselves the task to improve the status of philosophical reasoning they took pride in the belief that in this way a closer collaboration between philosophy and the sciences would be possible. However, since those early days the situation has changed considerably. Instead of the promised collaboration we have now one more philosophical school with all the disadvantages which schools possess, such as technical terms and a jargon. The technical terms of physics as well as the formalisms of mathematics seem to be quite appropriate in this field where their sophistication is accessible to test by experience. In the field of philosophy, however, a similar degree of technicality is liable to lead to sham problems and to a situation where a jargon is used in order to solve problems created by the existence of this very jargon. There are innumerable problems which could be attacked in fruitful collaboration of philosophers and scientists. It is regrettable that the self-imposed language barrier of some philosophers should so greatly increase the difficulty of such collaboration.

I now turn to the discussion of the paper proper. Professor Hill has pointed out that mathematical ideas influence physical reasoning. This observation can be generalized. It is quite generally the case that languages influence our way of seeing problems as well as the solutions: Aristotle's physics was strongly influenced by the grammar of the [442]Greek language. However, there is a very decisive difference: whereas many traditional philosophers (some contemporary empiricists included) adopt a magical attitude toward language and assume that *what is so* (the fact that we are speaking a certain language) also *must be so* (the categories of this language have an "ontological foundation"), mathematicians are much more liable to regard a language as an artificial tool for a certain purpose which can, therefore, be modified, and exchanged, if the situation should demand it. Hence, although mathematics does influence us, it does not do so inescapably, for we can abandon it and use a different system. I am in complete agreement with Professor Hill in his belief that neither the particle concept nor the wave concept are adequate for the description of what is going on on the microlevel. In my own paper {reprinted as Chapter 6 of the present collection} I have explained how Bohr attempts to prove that our observation language will forever have to be the language of classical physics and that we shall, therefore, be forced forever to use the particle and wave terminology for the description of our experimental results. This is, of course, pure word magic. It is curious, however, to notice that many of Bohr's opponents are equally guilty of word magic, as they want to retain, not both sets of concepts, but at least one of them. It seems to me that neither procedure will do and that we shall need even a completely new set of *observational* terms in order to be able to cope with the facts. Bohm's speculations may well lead to such a new conceptual scheme.

Professor Hill's observations on the extent to which general relativity is still incompatible with Mach's principle (a fact to which Professor Grünbaum has also drawn attention) are of great importance. I find it especially fortunate that he has connected his discussion with the emphasis upon the necessity, for the purpose of prediction and explanation, of laws *and* initial conditions. For despite the fact that it has been realized – not too long ago – that explanation and prediction cannot be carried out on the basis of laws alone, the usual way of analyzing scientific theories consists in an analysis of the *dynamical laws* they contain. The properties of the space of initial conditions as well as the properties of special initial conditions used in a specific derivation are hardly ever discussed. What Professor Hill has shown is that such a discussion can lead to quite surprising results. In the case of general relativity it leads to the above observation. In the case of the quantum theory it leads to very interesting considerations concerning the universal applicability of the Born interpretation. The only thing I did not understand was Professor Hill's statement to the effect that Newton's theory is less "global" than relativity. For the class of [443]the initial conditions of Newton's mechanics does not seem to be more restricted than that of general relativity.

We now turn to the quantum theory. It is very much to be welcomed that Professor Hill views both general relativity and the quantum theory as *cosmological* theories – that is, as theories which deal with the universe as a whole and not only with a narrow part of it. Yet the empirical status of these theories seems to be very different. While general relativity is essentially supported by three facts only – and even here the support is not very strong – the quantum theory seems to be much better off experimentally. Indeed, there seems to be an innumerable array of facts which one can cite in support of the theory. I contend that this picture is somewhat misleading and that the factual support of the quantum theory may be considerably smaller than is supposed by many of its adherents. Let me start with a very trivial observation: The general theory of relativity is supported not only by the three facts mentioned but also by all the numerous facts which are in agreement with Newton's theory where this theory can be assumed to be correct. After all, one of the conditions of adequacy of the general theory of relativity is that it contain the classical theory as a limiting case. Now if I understand the situation correctly, such proof is missing in the case of the quantum theory. Neither the notorious textbook derivation for the case where $h \rightarrow 0$, nor Ehrenfest's theorem can be assumed to represent such proof. More recent investigations have taken a decisive step in the right direction without yet having solved the problem in a way that would be generally acceptable. Hence, it is by no means clear to what extent the facts of the classical domain may be regarded as evidence in favor of the quantum theory. A further consideration should be added here: An experimental fact E can be regarded as evidence for a theory T only if it is possible to express E in terms of T. In the case of the quantum theory this requires that a correlation be established between certain (observable) facts on the macrolevel and some operators of the theory. It is true that for very simple properties of very simple microscopic systems such a correlation has in fact been established. But an innumerable body of macroscopic properties is still waiting for the appropriate operator to be correlated with it. Before such correlations have been introduced the relation between the theory and "experience" is far from clear, and it is also far from clear whether the quantum theory is really as highly confirmed as is usually assumed. In view of the incompatibility between the quantum theory and the general theory of relativity which Professor Hill has shown in such a convincing way it would be of great importance to become a little clearer about this question.

22

Review of Norwood R. Hanson, *The Concept of the Positron: A Philosophical Analysis*, New York: Cambridge University Press, 1963 (1964)

[264]In Chapter IX of the present book, which is its central piece, Professor Hanson gives a detailed and most interesting account of the discovery of the positron. The philosophical intention is to show how theoretical ideas and experimental findings collaborate in a scientific discovery. The following three assertions are made: (i) "The discovery of the positive electron was a discovery of three different particles" (p. 135); (2) there existed a "profound ... resistance" against accepting a positively charged electron (p. 159); (3) this resistance was due to (a) the *conceptual* structure of contemporary "electrodynamics and elementary particle theory" (p. 159) and (b) the fact that a generation of matter out of energy seemed to be involved (p. 162).

Of these three assertions (i) is false, (3) very implausible. (2) alone has some evidence in its favor, although it seems, from the letters which Hanson himself quotes (p. 223), that the case is overstated. Still, there is enough evidence of a strong resistance against a positive electron, and against new particles in general to prompt the question: why? Hanson, dissatisfied with an explanation that is merely psychological, looks for "conceptual" reasons (p. 159) which apply not only to this or to that individual but which influence all physicists alike. He resembles Hegel in his belief that there are concepts which deeply penetrate the thinking of the physicists and which make it nearly impossible for them to consider different ideas. In the present case he suggests that "since the proton and the electron came to be thought of not only as carrying the charge, but virtually as being the charge, the very conception of a third particle beyond the proton and the electron seemed to be insupportable" (p. 159).

Neither the general belief nor the specific suggestion made by Hanson is acceptable. Let us take the latter first: an *essential* connection between charge and mass is excluded by the different transformation properties of mechanical momentum and electromagnetic momentum. On the other hand, it seems that the belief in an *accidental* correlation between charge and mass was not at all strong. Weber's ideas, which [265]Hanson mentions in his historical survey, most certainly were no longer influential. And although Rutherford had originally envisaged a neutron which consisted of a proton and an electron (1919), it was soon realized (about 1929, if not earlier) that the presence of electrons in the nucleus would lead to serious difficulties. Heisenberg therefore regarded neutrons as *primitive* particles (1932) and not as "balanced combinations of positive and negative particles" (Hanson, p. 223). Such a suggestion would in any case have been quite impossible for the neutrino which was discussed by Pauli in 1929 and publicly announced in 1931. It is true that this last hypothesis was not at all kindly received; it was regarded as exceedingly "strange" (p. 223). But my point is that the resistance was not due to a firm association, in the minds of the physicists, between mass and charge, or to their opposition to a "third particle beyond the proton and the electron" (p. 159). After all, one had been using such a "third particle" for quite some time, and without qualms – the photon. The second conceptual brake Hanson conjures up in order to explain the resistance against the positron is even more chimerical. Hanson suggests that the physical discovery of matter-energy conversion "in itself sets the positron history apart and makes it wholly different from the neutrino and neutron discoveries" (p. 162). He insinuates that the necessity to assume such a conversion was a further stumbling stone. He even says that until 1932 "it remained implicit in molar physics that ... matter could never become other than matter" and that "the discovery of the positron crushed this assumption" (p. 56). This is a surprising assertion indeed! Einstein's $E = mc^2$ had stirred the imagination for quite some time. Popular science books contained speculations on the amount of matter needed to propel a large ocean liner across the Atlantic. In the astrophysical literature of the twenties it was taken for granted that the energy output of a star was due to the conversion of matter into something "other than matter." In addition there is (in 1920) the relativistic explanation of the mass defect in composite nuclei which gave very direct evidence for the existence of mass-energy conversion. There was no "assumption" to be "crushed" by the discovery of the positron. And there was therefore no reason for resistance. Considering now the first, and more general, belief of Hanson's, it seems that the resistance against the new particle was much less well organized, much less "conceptually" conditioned, than he suggests. Indeed, that Bethe's "physicists were strictly opposed to any kind of inflation" (p. 224) seems to have been the only general attitude [266]discernible. Physicists did resist. But their reasons were vastly more complex than (3a) and (3b) above indicate, and they also varied from individual to individual. In the past this interesting and fruitful variety of physical thought was often replaced by a fictitious unity based upon "experience." Such a simplification is properly criticized by Hanson and by some contemporary historians. Unfortunately it is not realized by these historians that the unity created by their "conceptual" considerations may be equally fictitious. Assertion (i) – namely, that Dirac predicted one particle, Anderson discovered another, while Blackett identified still another – has nothing at all to do with the particular circumstances of the discovery of the positron. It is but an instance of a much more general consequence of an extreme contextualism of meaning. This general consequence implies that independent tests of a physical theory are logically impossible, and it is refuted thereby. The remaining chapters deal with Hempel's theory of explanation and with the Copenhagen interpretation of the quantum theory. The former's symmetry between explanation and prediction is attacked in various ways. There is the comment, to my opinion valid, that historically theories were often criticized as being mere predictive devices (p. 35). Hempel's theory cannot give a satisfactory account of this historical fact. Another argument against Hempel is that the quantum theory knows of events which we can explain but not predict. Thus we cannot predict where a photon will arrive in an interference experiment although we can explain why it has arrived, after it has arrived. But this assertion is due to insufficient analysis. We can explain why (and predict that) the photon will arrive at all; we can also explain why (and predict that) it will not arrive at a minimum. But we cannot explain why it arrives at a specific place on the plate – unless, indeed, we make reference to hidden variables. Hanson's account of the quantum theory has been published before, and in almost the same words. It is therefore sufficient to refer to my comments and criticisms in *Frontiers of Science and Philosophy* (Pittsburgh, 1962 {reprinted as Chapter 7 of the present collection}). I believe these criticisms refute Hanson's claim that those who try to use hidden variables ask us "to entertain what is theoretically untenable, empirically false, and logically meaningless" (p. 30; this quotation sounds almost as if it had been lifted from one of the many anti-Copernican tracts of the early seventeenth century). The future of microphysics is still a completely open matter.

23

Review of Hans Reichenbach, *Philosophic Foundations of Quantum Mechanics*, Berkeley–Los Angeles: University of California Press, 1965 (1967)

[326]Professor Reichenbach's book on the quantum theory which was first published in 1944 is now available as a paperback reprint. It (i) gives an introduction into part of the mathematics of the elementary quantum theory; (ii) discusses the familiar problems of interpretation; and (iii) suggests a new interpretation in terms of a three-valued logic. I shall comment on the last two items only.

(ii) is based on the distinction between *phenomena* ("which are determined in the same sense as the unobservable objects of classical physics") and *interphenomena* (which can be introduced "only ... within the frame of the quantum mechanical laws"). It is assumed without further argument that the inferential chains leading to the interphenomena are "of a much more [327]complicated sort than the inferential chains connecting macro-occurrences with phenomena" (21). The reason is that "in the inferences leading from macrocosmic data to phenomena we usually use the laws of classical physics" (21). This twin-assumption eliminates from the start all those interpretations which want to understand *both* phenomena and interphenomena in quantum mechanical terms (e.g., Ludwig's theory of measurement). It also shows that Reichenbach's famous "principle of anomaly" is nothing but a trivial consequence of an arbitrary definition. For according to Reichenbach the principle asserts that "the class of descriptions of interphenomena contains no normal system" (33). Now a normal system is a descriptional system for which (*a*) the laws are the same whether or not we make an observation; (*b*) the state of the object is the same whether or not we observe it (19). Also the definition of phenomena and interphenomena makes the distinction between the observed and unobserved coincide with the distinction between classical physics and the quantum theory (21). Hence, the principle of anomaly asserts only that the laws of the quantum theory are not the laws of classical physics. Some principle!

It is also pointed out that the attempt to give an "exhaustive interpretation" of the phenomena leads to difficulties, or "anomalies" as Reichenbach prefers to express himself. There are at least two non-equivalent explanations for what an exhaustive interpretation is supposed to be (cf. pp. 33 and 139) but we can safely assume that the one that is decisive for Reichenbach's argument ascribes well-defined values to all the classically definable magnitudes of an interphenomenon. A charged masspoint that has always a well-defined position and a well-defined momentum is an exhaustive interpretation of a specific interphenomenon (passage of a single electron). It is exhaustive interpretations of this kind which lead to the anomalies.

Such anomalies are: the sudden collapse of the wave accompanying a moving electron when it impinges upon a screen. The fact that particles in an interference experiment seem to "know" what is going on at distant places (par. 7). Now obviously these "anomalies" are but refuting instances of the assumption that the moving electron is a classical wave (first case) or a classical particle (second case). They show that for any theory which consists of the mathematical formalism of quantum mechanics together with some exhaustive interpretation there exist refuting instances. Reichenbach's attempt to eliminate them (which is what the whole book leads up to) is therefore, logically speaking, of the same kind as is the attempt to preserve a theory from refuting instances by a clever arrangement of ad hoc hypotheses. Let us now take a closer look at this attempt.

Reichenbach discusses four methods of removing the anomalies. *Method 1*: suggests that we should "become accustomed" to them (37), that is, it expects us not to be worried by the fact that a classical interpretation of the quantum theory turns it into a false theory. *Method 2*: advises us to use a certain interpretation only for describing those parts of the world where it works [328]and to switch over to another interpretation as soon as a difficulty arises. The only difference between this interpretation and method 1 that the former uses two or more refuted interpretations whereas the latter uses one only. *Method 3*: suggests that we stop interpreting altogether and regard the statements of the theory as cognitively meaningless instruments of prediction (40). *Method 4*, finally, suggests that we change the laws of logic so that the statements showing the inadequacy of one of the chosen interpretations "can never be asserted as true" (942). This is the method which Reichenbach himself adopts.

The objection against this method or, for that matter, against all the "methods" which Reichenbach discusses is that they remove a clash between theory and fact not by suggesting new, and more adequate scientific theories, but by an interpretation of the existing theories that removes their faults from view. Of course, it would be too much to expect from a philosopher that he "interfere with the methods of physical inquiry" (vii). But one should expect from him a little more than "conceptually clarified" (vii) recommendations for the preservation of the *status quo*. Moreover, Reichenbach contradicts the criteria of adequacy which he himself introduces. According to these criteria, every law of quantum mechanics should have either the truth value "true", or the truth value "false", but never a truth value "indeterminate" (160). Now such quantum mechanical laws or the Schrödinger equation, or the law of the conservation of energy produce, when coupled with an exhaustive interpretation, precisely the anomalies Reichenbach wants to remove. They must therefore all receive the truth value "indeterminate". (This follows from more formal considerations also). Only meta-statements, such as the principle of complementarity which no one has as yet regarded as a physical law, are exempt from this argument. Reichenbach's suggestions therefore conform neither to the quite elementary demand to take refutations seriously, nor do they conform to the criteria he has set up himself.

PART THREE

ENCYCLOPEDIA ENTRIES
(1958–1967)

24

Natural Philosophy
(1958)

[203]Natural philosophy deals with the general features of living and non-living nature, as well as with principles that can serve to explain natural phenomena. Its method is either inductive or speculative. It is inductive if it attempts to abstract these principles from the results of science and to present them in systematic connections. An inductive natural philosophy obviously assumes that a natural science already exists, and is therefore a relatively recent phenomenon. It does not always manage to avoid eclecticism, and what is sometimes referred to under this label is nothing else than a more-or-less successful popularization of scientific results. If, on other hand, general principles are introduced with the exclusive purpose of making nature understandable, completely indifferent to what other attempts exist in this direction, then we speak of *speculative* natural philosophy. A speculative natural philosophy is dogmatic if it claims to be able to prove its principles definitively and denies that a revision of its principles will ever be necessary. This brings these principles closer to statements of faith, and in fact Plato (427–347 B.C.) – who was, after Pythagoras (580–500 B.C.), the first dogmatic natural philosopher of the Western world – also wanted to defend his principles as statements of faith, namely, with the help of a sort of Inquisition (see his "laws"). A *critical natural philosophy*, [204]by contrast, acknowledges that making mistakes is human. As a consequence, it considers the most general and most enlightening principles – such as the principle of causality – as hypotheses, that is, as attempts to explain, which, just like the people who advanced them, are full of mistakes, in need of improvement, and may even be false. Critical philosophizing about nature begins with the Presocratics, and, after being nearly totally set aside by Platonism

and Aristotelianism, returns again in the late Middle Ages and in the Renaissance, and is closely connected with the development of Western science (astronomy, physics, medicine). Critical natural philosophy has never had a low opinion of *experience*. Arguments based on experience play an important role as early as the Presocratics. Everyone knows that Thales (625–545 B.C.) explained that water was the fundamental principle of the world. For him, that is, earth, fire, and air (which were viewed, together with water, as the four fundamental elements of the world) were nothing other than particular forms of water, and changes in the world were nothing other than transitions of water from one form into another. Thales doubtless reached this point of view on the basis of his experience that water can change into ice and steam (air) into water, as well as through the insight that the growth (and perhaps even the formation) of life requires moisture. But we should not forget that Thales' theory goes far beyond every possible experience, since it assumes that a unifying principle lies beneath the change we observe. Anaximander (611–545 B.C) turns against the choice of water as the fundamental principle, and he does so once again on the basis of experience: for him, fire, earth, and air play a role just as an important as water's. Thales' theory, he says, cannot justify the prominent position of water: the explanation of (observable) basic stuffs cannot itself be based on one of these stuffs. Hence, Anaximander chooses as his basic principle the "undetermined". A further example of the connection between speculation and experience in Presocratic philosophy is atomism. Atomism begins with the contradiction between the (speculative) principle of the immutability of being, which Parmenides (about 540 B.C.) forcefully advocated and the atomists did not want to surrender, and the experimental fact that things change in this world. The contradiction is overcome by multiplying beings: the atoms, and by the introduction of a new principle: empty space. What exists (that is, the atoms) [205]is still always unchangeable, but change has found its explanation. It consists in alteration of the spatial relationships between these unchangeable elements. This method of speculation tested by appealing to experience and inner unity is nothing other than the method of science. The hypotheses of science differ from the general principles of a critical natural philosophy only in their higher precision and, consequently, in their higher testability. Theoretical physics, cosmology, and theoretical biology can, therefore, correctly be called today's natural philosophy. This natural philosophy has not neutralized the older and more general principles; quite the opposite, these principles now play a decisive role as leading conceptions in the construction of scientific theories. The development of the atomic hypothesis from Gassendi to Boltzmann and Einstein is unthinkable without the simple and plausible ideas of Democritus and Lucretius, and it can also be traced back to them. The hypothesis of universal gravitation, which was first advanced by Newton in mathematical form, was, as P. Duhem has shown, prepared by some astrological ideas of sympathy among heavenly bodies. All this highlights the fundamental unity of natural science and (critical) natural philosophy, which is also thoroughly confirmed by a closer analysis of scientific methodology.

This unity is contested by Positivism, which regards as meaningful only those sentences it deems "scientific", and – in line with David Hume (1711–1776) – wishes to commit natural philosophy to the flames, as well as by a dogmatic natural philosophy, which denies science every ability to attain knowledge. Positivism wants to keep thinking out of science, dogmatic natural philosophy wants to keep experience out of natural philosophy. For the positivists, the ideal of knowledge of nature is provided by the rules adopted by Egyptian surveyors; for the dogmatic natural philosophers this ideal consists in the deeply grounded truths of Egyptian priests (who impressed Plato). Eventually, both reject the decisive achievements of early Greek philosophy, namely the rational view of nature, and, thereby, *the idea of Western science.*

As to its objects, natural philosophy deals with *space, time, matter, cosmos,* and *life.* The rest of this article presents a short description of the development of ideas about space, time, matter and [206]the world as a whole. Whenever possible, the arguments that drove this development forward will be discussed too.

Middle Ages. It has gradually become clear that the natural philosophy of the Middle Ages is not as unified as a superficial and "progressive" historiography has often depicted it. What remains unchanged is, however, that physics and the cosmology of Aristotle (384–322 B.C.), together with the astronomy of Ptolemy, were at the center of natural philosophical discussions. Let us give a closer look at this world picture, which Dante described so wonderfully. The world is finite (its diameter is about the same as the diameter of the orbit of Venus in modern astronomy). That means, by the way, that *space* is finite, too. Beyond the sphere of the fixed stars, which limits the world, is "neither space nor place nor body". Space cannot therefore be the abstract, flat, Euclidean space, as this is necessarily infinite. Space, for Aristotle, is the real variety of places and borders of bodies. There is no empty space. He tries to show that by appealing to various arguments. According to one of these arguments, later found in Descartes, void and nothing are one and the same thing. But nothing cannot exist, and not even God can make it exist. Therefore, void cannot exist, and not even God can make it exist. (This argument had been condemned by the Archbishop of Paris in 1277.) The Aristotelian theory of motion appeals to a more physical argument. According to this doctrine, finite power without resistance leads to infinite speed. Therefore, if void exists, there must also be infinite speed, which is considered absurd. The world is not homogenous. In the vicinity of the Earth, which rests at the center, there are the four elements: earth, water, air, and fire (they contain, so to say, the essence of what we normally call earth, water, air, and fire). Each of these elements strives for its "natural place". The natural place for earth is the center; for water, the next spherical layer; for air and fire, still higher layers. This hierarchy is not only a physical order, but also a value order. Earth is common stuff, fire is far nobler. Were the elements in the order of their natural hierarchy, that is, if the area of the globe, the "sublunary" area, were already in its natural order, then nothing could change. But the elements are mixed; each piece of matter – this piece of paper here, for example – contains earth (namely, the [207]ashes that remain when it is burned) as well as water (burning produces moistness), fire, and air. The changes in the sublunary region are a consequence of this lack of order and of the striving of the elements to take their natural order. In the area of heaven, the final order already exists. (This order was interpreted in a social sense: the firmament is an exemplary, total order, in which each element has and retains its assigned place. This is the reason why totalitarian social philosophers, such as Plato, had such a great interest in the proof that the orbits of the planets, which prima facie appeared to behave as ill-manneredly as undesirable "vagabonds" on the Earth, are ruled by strict laws.) Here no change is possible, at least not through radial movement, away from the Earth or toward it. Circular movement is the essential movement of heavenly bodies and of the crystalline spheres to which they are attached. (There have been various theories about these crystalline spheres. They were sometimes conceived as coarsely material, and the planets as opaque, as reflecting knots in them. But we also find a more "mathematical" conception of the spheres, which sometimes is quite similar to the original conception of Eratosthenes.) On the basis of Aristotelian physics, in which not only accelerated motion, but constant speed and generally every kind of change has to be explained, a mover must be introduced for the sudden changes of the heavenly bodies, just as for the movement of a thrown stone or a pushed object. In the last two cases, the air traveling along is taken to be the mover – an assumption that would be criticized in the 14th century by appealing to simple facts of experience. The sudden changes of heavenly bodies are explained by appealing to a spiritual mover (there is in no other way since air belongs to sublunary areas only): superlunary spirits shove the crystalline spheres and interfere with the sublunary region (which favors astrology and provides it with a theoretical basis); spiritual dispositions and urges are introduced to explain the movement of earthly bodies. This doctrine of ghosts is the second reason for the attacks on Aristotelian physics that were advanced by various sides in the middle of the 14th century. Nicole Oresme remarked that God could indeed have wound up the world like a clock and then, without the help of spirits, left it to itself. Such a point of view naturally required a new theory of motion. [208]This theory about movement was gradually constructed by Buridan, Oresme, Albert of Saxony, by the philosophers of the Merton College in Oxford, and is the direct predecessor of Galilean dynamics. It attributes an impetus to a pushed body that remains, or slowly declines, when the moving body meets with no resistance. This impetus is compared with the continued glowing of a piece of iron after removal from a coal fire. But the theory is still incomplete, and allows for a "cooling down", that is, a gradual disappearance of the impressed force. (Even Newton must still appeal to God to keep the universe in motion!) The cosmos had to be turned inside out before further development became possible. This took place with the introduction of the Copernican hypothesis. For a proper assessment of the accomplishments of Nicholas Copernicus (1473–1543), we must not lose sight of the reasons that motivated him, which are all of a philosophical nature, for example, his preference for circular motion as the most noble movement, his insight that the Ptolemaic system did not take this idea seriously – many of the circles introduced by Ptolemy have only an imaginary center – and his thoroughly Neo-Platonic worship of the Sun, which possesses absolute power, resting "on a royal throne [to] govern the family of planets revolving around it".[1] Copernicus never racked his brain with observations: he simply took over the facts transmitted by Ptolemy's followers, and he did so with a naïvety that appeared thoroughly unbelievable by his more empirically minded followers, such as Tycho Brahe (1546–1601). Furthermore, the Copernican cosmos cannot be reconciled with Aristotelian

---

[1] Copernicus, *De revolutionibus orbium coelestium*, Nuremberg: Johann Petreius, 1543, fol. 9*v*, 9–10; English translation by Edward Rosen, *On the Revolutions*, Baltomore: The Johns Hopkins University Press, 1992, p. 22.

physics. (In order to make this irreconcilability plain to see, one should simply try to reverse it, and construct Newton's theory of gravitation in a Ptolemaic universe!) But then, how can we explain changes on the Earth itself? How is it possible that the globe of the Earth moves circularly, when all movements in the sublunary region are radially directed, either toward or away from its center? It is not true that the immobility of the Earth in Ptolemaic astronomy was nothing other than a prejudice, or an imitation of the Bible; it was a necessary consequence of Aristotelian physics! And the arguments against the motion of the Earth that Ptolemy offers are all physical arguments, namely: first, the only motion possible for Earth, as a standard sublunary body, is downward. The speed of this motion is proportional to the weight of the falling body. [209]As a consequence, "due to the colossal overweight of its size", the globe of the Earth would "run ahead of all bodies as it fell in the deep, and the detached living things and heavy bodies would be swinging in the air while it would end by breaking through the firmament with enormous speed. But already the mere thought of such possibilities appears incredibly ridiculous".[2] Second, the assumption of a movement of the globe of the Earth toward somewhere "under" is in any case absurd, "because in the universe there is neither 'above' nor 'below' with respect to the Earth".[3] Third, the rotation of the Earth cannot be reconciled with the principles of physics: a stone let loose falls perpendicularly to the Earth, but if the Earth rotates, it must lag behind its starting point, and do that with high speed, so that "all bodies ... would apparently have to travel in one course toward the west, that is, toward the place the Earth has left behind".[4] In this way the construction of the Aristotelian cosmos was also explained physically, and all appearances were organized in a harmonious system. This harmony, this inner connection, is destroyed by the Copernican system. "[The world is] all in pieces, all coherence gone, / All just supply, and all relation", wrote John Donne sometime later.[5] Understandably, many astronomers saw the Copernican hypothesis as a "calculation trick" belonging to higher astronomy, which simplifies the order of phenomena but has nothing to do with reality. Contemporary

---

[2] Ptolemy, *Syntaxis Mathematica*, I, 7; in *Claudii Ptolemaei Opera quae Extant Omnia*, edited by Johann L. Heiberg, vol. I, Leipzig: Teubner, 1898, pp. 23.20–24.4; *Ptolemy's Almagest*, translated and annotated by Gerald J. Toomer, Princeton: Princeton University Press, 1998, p. 44.

[3] *Ibid.*, pp. 22.22–23.1; English translation, p. 44.

[4] *Ibid.*, pp. 25.13–14; English translation, p. 45.

[5] John Donne, *An Anatomy of the World*, London: Samuel Macham, 1611, 213–214.

textbooks in astronomy mention the Copernican teaching in their forewords, they call it a difficult mathematical hypothesis that would confound beginners and, therefore, must be dealt with in a later volume (which then normally never appeared). (The same thing happened to the theory of relativity, about 350 years later.) Anyway Copernicus himself is completely satisfied with his turning of the Aristotelian–Ptolemaic system inside out. He retains a number of its fundamental tenets: the fixed stars, and thereby the finiteness of the cosmos (although not of space), as well as its centrally symmetrical structure. The actual spatial and material *infinity of the world* as well as the homogeneity of physical space was first announced (not without predecessors!) by Giordano Bruno (1548–1600), a thought that was first taken up in Newton's time. Then Kepler (on the basis of a homogeneity postulate, which plays a key role, but with false empirical data, in the cosmology of the 20th century) argued against an actual infinite world; Galileo left the question undecided; Descartes [210]ascribed only to God actual (or, as he called it, "positive") infinity, and deemed the world borderless (that is, without end to human understanding), or undetermined. The actual infinity of space later returns in physics through the detour of Neo-Platonic theology and the spiritual theory of Henry More (1614–1687). For More, space (and not matter, as for the Cartesian Spinoza) is an attribute of God, and is therefore actually infinite. Newton expressed himself very cautiously on this issue. But we can say that his conception of space stands in close relationship to More's.

The theory of motion necessary to round off the Copernican cosmos is nearly simultaneously offered by Galileo and Descartes. It is amazing how little Galileo uses new facts in his arguments for the Copernican system. The experience on which he bases his arguments is, in fact, the very same everyday experience Aristotelians had interpreted. But, he orders this experience in a completely new way, and so makes it understandable. It is a great mistake to believe that the "scientific revolution" of the Renaissance consisted in the replacement of speculation by the description of facts. "It was supremely difficult", wrote Herbert Butterfield in his extraordinary study *The Origins of Modern Science*, "to escape from the Aristotelian doctrine by merely observing things more closely".[6] And also the hectic collecting of facts really does not begin until later, with Francis Bacon (1561–1626), whose writings become the Bible of the Royal Society. What happened in Galileo's time is the development of a

---

[6] Herbert Butterfield, *The Origins of Modern Science: 1300–1800*, London: Bell, 1949; new edition, 1957: p. 4.

new standpoint for viewing and for better understanding the same old facts. The discovery of new facts, such as the new star of 1572, whose appearance so dramatically showed the changeability of the superlunary sphere; the moons of Jupiter, which constitute a small model of a Copernican system; and the Sun spots merely had the function of strengthening this new standpoint. Alchemy, in which, at the time, facts were collected with great industry, has scarcely any progress to record.

The law of inertia is nowhere to be found purely formulated by Galileo (1564–1642). Galileo still had a predilection for circular motions – a thrown stone describes, according to him, a circular orbit. He did not accept Kepler's ellipse theory. As a consequence, the world picture he advocated in his *Dialogues* does not have the unity of the Aristotelian picture. The nature of space and forces is unclear. The relationship between matter and space is touched only fleetingly. The man we must consider [211]as the creator of the new natural philosophy is Descartes (1596–1650). This natural philosophy once and for all banned spirits from the world. The essence of matter consists in extension, the essence of spiritual principles, such as the soul, in self-consciousness. As a consequence, the presence of spirits and demons in the created world is logically impossible. The contribution of this theory to the disappearance of witch-hunting, which in the 15th and 16th centuries reached an unbelievable scale in Catholic as well as in Protestant areas, is not to be underestimated. Also, according to Descartes God did not act in the world; he is nowhere in it – *nullibi*. Hence, the behavior of the world is now to be explained by itself. Descartes reached this result with his law of inertia, according to which the sum of all motions in the world remains constant; *his vortex hypothesis*, according to which all central motion, such as the motion of the Earth around the Sun, is to be explained through the rotation of world matter around the center (the planets are just like dust particles swept around the Sun by the rotation of world matter; they are surrounded by smaller vortexes, which bear their moons; each fixed star has its vortex), as well as through a detailed atomic hypothesis. Fontenelle popularized this world picture in his dialogues between a beautiful lady who is thirsty for knowledge and her teacher, but otherwise unsuccessful lover. In the 17th century it was the common knowledge of educated people on the European continent.

Not so in England. Henry More turned sharply against Descartes' identification of space with matter. His arguments, which are partly physical in nature, are perfectly clear as to the main point. But his aim was above all that of incorporating penetrable, unbreakable, yet extended beings – that is, spirits – into the world. His description of such entities perfectly fits with light (here is a continuation of the astrology of light and the metaphysics of light of the Middle Ages!) as well as with the modern concept of field. More's space, free from matter, is an infinite, homogeneous attribute of God. The space used by Isaac Newton (1643–1727) has very similar properties. It is infinite, homogeneous, and substantial; it contains matter (without collapsing with it), which is made of small, hard elements: the atoms. Newton offered strong arguments against the fullness of the Cartesian world: the ease with which comets shoot in all directions through the solar system, without concerning themselves in the least with Cartesian vortexes (nearly 100 years earlier, Tycho Brahe offered comets against the existence [212]of crystalline spheres!), as well as the impossibility of explaining the laws of planetary motion, Kepler's laws, hydrodynamically. Instead of the vortexes, Newton appealed to gravitation. But he hesitated to take gravitational forces as new principles: indeed, he viewed the existence of physical principles with the features of gravitation as really absurd. He suggests – and Bentley has made full use of these suggestions – that gravitation must necessarily be caused by a spiritual being. Matter in the form of impenetrable atoms, forces most likely caused by spiritual beings, both embedded in empty, but real, absolute space and in empty, but real, absolute time – this is Newton's universe. "A Frenchman", writes Voltaire in his *Letters Concerning the English Nation* (1733), "who arrives in London, will find Philosophy, like every thing else, very much chang'd there. He had left the World a *plenum*, and he now finds it a *vacuum*".[7] (It is worth noticing that it was just these "letters" of Voltaire which introduced Newton's conception of the world, together with Locke's philosophy, into the continent.)

At the beginning of the 18th century this conception of the world is attacked from two sides: by Leibniz, who partially carries further the Cartesian arguments, and partially develops his own theological, metaphysical, and natural philosophical ideas; and by Berkeley, a subtle predecessor of modern positivism. The motive of both thinkers is the fear that Newton's natural philosophy threatens religion and will necessarily lead to materialism. Apart from this motive, the approach of these two philosophers is very different. George Berkeley's (1685–1753) arguments have essentially an epistemological nature. He does not try to show, that is, that Newton's principles of absolute space, absolute time, and gravity

---

[7] François-Marie Arouet (Voltaire), *Letters Concerning the English Nation*, London: C. Davis and A. Lyon, 1733: Letter XIV, p. 109.

do not exist; he goes much further and questions whether the very words employed to express such principles make any sense. For him, the only meaningful words are those that denote *sensations*. Neither gravitation nor matter refers directly to sensations. Insofar as these words are meaningful at all, they are shortened references that order a definite class of sensations. Ontologically put, this means that only sensations exist. Gravitation, matter, atoms do not exist; but the words "gravitation" and "matter" help to order representations of sensations economically. This method of "semantic analysis" would eventually dissolve natural [213]philosophy in the 19th century. Berkeley's theory of space is of great interest, quite independently from his positivist standpoint (with which it is otherwise connected). His short essay titled *De Motu*, in which this theory about space is developed, can count as a masterpiece of analysis: if the purpose of science is to coordinate sense perceptions, then it can only make use of spatial relationships, as far as these are nothing but relationships among perceptible bodies.

For upwards, downwards, leftwards and rightwards, all locations and regions are grounded on a relation, and necessarily describe and assume a body which is different from movement. So, for example, if we assume that only a ball exists, all other bodies having been destroyed, it would be impossible to perceive any motion of it.... That is, since motion, by its very nature, is relative, it is impossible to perceive it before other bodies, related to it, are given.[8]

The relativity of space and time is emphasized even more radically by Gottfried Wilhelm Leibniz (1646–1716). "Without matter there is no space" – that is, space and also time are nothing other than "related structures of existing things": "Once we have decided about these things and their relations, there is no longer any choice about time and place, which indeed in themselves contain nothing real, nothing that differentiates them, nothing that can be differentiated in any way".[9] (Descartes had said the same, in a similar way: the interregnum between two worlds, one of which is destroyed, and the other created "sometime later", is timeless.) The arguments Leibniz offered in support of his position are different from Berkeley's: namely, if there existed something such as an

---

[8] George Berkeley, *De Motu: Sive, de Motus Principio & Natura, et de Causa Communicationis Motuum*, London: Jacob Tonson, 1721, nos. 58–59.

[9] "Mr. Leibnitz's Fifth Paper. Being An Answer to Dr. Clarke's Fourth Reply", in Samuel Clarke, *A Collection of Papers, Which Passed Between the Late Learned Mr. Leibnitz, and Dr. Clarke, in the Years 1715 and 1716. Relating to the Principles of Natural Philosophy and Religion*, London: James Knapton, 1717, p. 197, no. 47 and p. 219, no. 57.

empty, absolute, homogeneous space, then there would be no sufficient reason for the hypothesis that God created the first piece of matter here and not there. The principle of sufficient reason was for Leibniz the foundation stone of philosophy. (According to this principle, every judgment must have a reason from which it can be correctly inferred.) Prima facie, Leibniz's and Berkeley's arguments are most plausible. However, we cannot overlook the fact that neither of them satisfactorily explained how the force of inertia – Newton's most important argument for the existence of absolute space – can be explained on the basis of the relations among material elements. Such an explanation was first delivered by Albert Einstein (1879–1954) – but not by disposing of, but rather by modifying Newton's idea of abso[214]lute space, different from matter (but now no longer unaffected by it). The discussion of the concept of absolute space, in which Leibniz is, in fact, unwillingly involved, is only part of his criticism of Newton's conception of the world. Leibniz criticizes Newton's need for God's intervention in order to keep his cosmos in motion. (A physical argument offered by Newton's followers in support of the assumption that the universe would stop if a supramundane force did not intervene steadily, or at some given intervals, is the disappearance of impetus in a nonelastic push.) He demands, quite in the sense of Descartes, that the world be fully explained in accordance with its own laws. Leibniz, or still better Descartes, but not Newton, is also the predecessor of Laplace's world machine. For Newton, space and time are attributes of God. And the famous "*Scholium Generale*" he added to the second edition of the *Principia*, in which he discussed in detail his metaphysical ideas, contains at the end the following argument:

A few things could now be added concerning a certain very subtle spirit pervading gross bodies and lying hidden in them; by its force and actions, the particles of bodies attract one another at very small distances and cohere when they become contiguous; and electrical [i.e., electrified] bodies act at greater distances, repelling as well as attracting neighboring corpuscles; and light is emitted, reflected, refracted, inflected, and heats bodies; and all sensation is excited, and the limbs of animals move at command of the will, namely, by the vibration of this spirit. . ... But these things cannot be explained in a few words.[10]

Newton's God works effectively in this world, as did the biblical God in the first six days of creation, and this work is required for the

[10] Isaac Newton, *The Principia: Mathematical Principles of Natural Philosophy*, translated by I. Bernard Cohen and Anne Whitman, assisted by Julia Budenz, Berkeley–Los Angeles–London: University of California Press, 1999: pp. 943–944.

maintenance of the cosmos. The God of Leibniz and Descartes is "the Biblical God on the Sabbath Day, the God who has finished his work and who finds it good" (A. Koyré).[11] In the 18th century the conviction developed that natural philosophy and physics did not require the hypothesis of God's intervention (Laplace's answer to Napoleon, "I do not need these hypotheses", has exactly this sense). This leads to the conception of the world offered by mechanical philosophy.

Mechanical Philosophy. According to this world picture, which quickly becomes generally dominant, gravitation and all other forces are accepted as physical principles. Atoms and forces in empty space – that is the catchphrase of mechanical natural philosophy of the 18th and 19th centuries, the third great dogmatic natural philosophy after Aristotle and Descartes. [215]In the 19th century mechanical philosophy is the creed not only of physicists (with the exception of the genius Faraday and his followers), but also, and above all, of biologists. The grand development of all sciences in this century has provided mechanical philosophy with a host of arguments. But it also prepared the way for its collapse. Present-day natural philosophy is the result of this process of collapse, which is very complicated and by no means over; furthermore, various concerns mingle with it, above all those doctrines about spirits for which Cartesianism and then mechanical philosophy made life so difficult. In what follows, we will try to describe as well as possible some ideas that play a role in this process of collapse.

The undertaking immediately meets with the difficulty that, in the 19th century, epistemological arguments start to play an important role in natural philosophy. (This epistemological criticism of natural philosophical positions and the following partial collapse of natural philosophy is quite comparable to the Sophists' criticism of Presocratics' natural philosophy, and the partial collapse of this trend that followed.) The process begins with Immanuel *Kant* (1729–1804). Kant's main philosophical question is not "How is the world made?", but "What can we know about the world?" He is deeply convinced of the correctness of the Newtonian world picture (which he decisively extended with his cosmological hypothesis), and he wants to justify it. His justification has an epistemological nature, not a natural philosophical one: the principles belonging to general natural science, such as the principle of causality and the conservation laws (and later almost the whole Newtonian theory) are absolutely true, because our intellect is prepared to think Newtonian.

[11] Alexandre Koyré, *From the Closed World to the Infinite Universe*, Baltimore: The Johns Hopkins University Press, 1957, p. 240.

First of all, there are therefore rules of reasoning, and nature is made of (chaotic) sensations ordered by these rules of reasoning. That explains why we are capable of understanding nature: its general traits are indeed our own work. According to Ernst Mach (1838–1916), who proceeds more radically, the concept of nature has no longer any basis in experience, and must be removed. For science (besides which no independent natural philosophy is allowed) can and should do nothing else but systematically order experience; a concept, or a theory, is allowed as long as it contributes to this ordering. Such an epistemological principle must now dispose not only of the general concept of nature, but also of the special concepts of substance, absolute space, absolute [216]time, and atoms; but this means that it must also rid itself of the very heart of the mechanical concept of the world, that is, mechanics. Here Berkeley led the way; he was the first to condemn Newton's natural philosophy from a positivistic standpoint, and, as we saw, also delivered a decisive criticism of absolute space, absolute time, gravitation, and matter. Mach's criticism, as also Kant's philosophy did before him, initiates the epistemological dissolution of mechanical philosophy.

In the 19th century, mechanical philosophy confronts physical obstacles too. Maxwell's electrodynamics, whose correctness is secured by Hertz's researches, works with the concept of field, which allowed no reasonable and clear mechanical interpretation. The releases of heat show directionality, which is foreign to mechanics (temperature differences equalize each other, but do not arise without outside intervention), and which, as can be shown, virtually contradicts it. This contradiction was resolved by Boltzmann and Einstein – but only at the price of introducing new, nonmechanical principles, such as objective probabilities. Mechanical philosophy, however, is crucially attacked by the special theory of relativity (1905), according to which all properties that mechanics attributed to matter, space, or time (inertia, extension in space, and time) are conceived of as relationships between courses of events and systems of coordinates, which therefore change as soon as the state of motion of the system of coordinates changes; and according to which every form of energy, and therefore also the energy of the gravitational field, possesses inertia, and therefore a material character. But relativistic space-time (which no longer breaks cleanly into space and time) is still always objective, and the same goes for the relations among events in it. In addition, it possesses a well-defined structure: events that can be connected through light signals are simultaneous; events in the lower half of the light cone belong to the past; events in its upper half belong to the future. (In Newtonian space-time, the light cone collapses in the doubly occupied level of space. The structure of Newtonian space-time consists, then, in a vertical fabric and a horizontal stratification.) All in all, the special theory of relativity knows three objective entities: space-time, matter, and the (electromagnetic and gravitational) field. The transition to the general theory of relativity (1913) consists, roughly, in the blending together of space-time and the gravitational field. In order to produce the blending, the general [217]theory of relativity appeals to curved Riemannian space-time rather than the flat space-time of the special theory. Everyone knows that curved space-time or quickly accelerated motion brings the inertial properties of matter to light: forces appear (centrifugal force, Coriolis force). In this way, indeed, Newton had tried to distinguish between relative and absolute motion. Therefore, a curvature of space-time, which cannot be eliminated by a transformation of coordinates, produces "genuine" forces (in contrast to "apparent" forces, such as the centrifugal force), namely gravitation, which thus is shown to be a manifestation of inertia. And conversely, the curvature depends also on the distribution of matter. It is important to highlight that the curved, nonhomogeneous, and anisotropic space-time of the theory of general relativity plays a thoroughly independent or, if one so wishes, "absolute" role: Newton's absolute space and absolute time are by no means eliminated by the theory of relativity; they are only blended and remodelled in Riemannian space-time, which now takes over the role of the old "ether". "To deny the ether", says Einstein, "is ultimately to assume that empty space has no physical qualities whatever. The fundamental facts of mechanics do not harmonize with this view".[12] Modern positivists have often described the elimination of absolute space and absolute time as one of the great achievements of their philosophy. What has just been said shows that this claim is not only historically incorrect (Einstein never dealt with the meaning of physical claims, only with their correctness), but also physically false; just as before, physics needs space and time as independent (albeit influenced by matter) entities! Attempts by Weyl and Kaluza to incorporate the electromagnetic field have not had much success. A second attempt, however, is more important: it has been taken up most recently, in a different connection, by de Broglie and his school. It is the attempt, by Einstein and Grommer, to interpret particles of matter as individual entities, that is, as a sort of knot in

---

[12] "Ether and the Theory of Relativity", in Albert Einstein, *Sidelights of Relativity*, translated by George B. Jeffrey and Wilfrid Perrett, London: Methuen & Co., 1922, pp. 3–24: p. 16.

curved space-time. This attempt, which has progressed to the deduction of the equations of motion from the field equations (Einstein, Infeld, and Hoffmann), leaves only two entities as ultimate principles, space-time and the electromagnetic field. This grand simplification can count as the high point of classical physics – the physics, that is, which sees the world as an entity independent from the subject, and which seeks to find the ultimate principles [218]of this world. According to Newton, these principles are those of (divine) space, time, (atomic) matter, and gravitation. Later came the field, in two forms: deterministic (Faraday), to which various thinkers (Thomson, Lorentz) sought to reduce matter; and probabilistic (statistical mechanics). Einstein unifies space and time to space-time and reduces gravitation and perhaps even the electrical field and matter to it. Now, the curvature of space-time allows tackling the cosmological problem in a new way, too. The infinite world of Newton is unstable (it seems it was O. Heckmann, who first proved it); but in Newton's space a finite amount of matter must disperse (Einstein's evaporation objection). Curved space-time is likewise unstable, but allows for meaningful dynamics, it can pulsate (that is, it can return to prior states), and it is also possible to form a picture of a world without boundaries, but nevertheless finite, and which contains only a finite amount of matter. (A good comparison, in two dimensions, is the surface of the Earth: it is without boundaries – a rambler wandering on a smooth Earth would stop nowhere – but finite.) These possibilities have invited speculation and led to models of the world, all of which account more or less for experience, but which – all this notwithstanding – present only a limited similarity. The idea of a constant creation of new matter is the most recent, and daring, hypothesis in this field. But all these attempts, and the proud edifice of Einstein's theory on which they rest, are seriously threatened by new knowledge about matter, which goes under the name of quantum theory, and by a philosophical movement supported by its creators, which requires no less than the renunciation of its founding principles, namely, the principles of objectivity and determinism. We turn now to a description of this movement.

It soon turned out that the elements of matter are not as simple as they were thought to be. Atoms and their core have a structure. On the one hand, this increases the number of elementary particles from which matter and fields can be made (today we know about a dozen of them). But it was, above all, the knowledge of the nature of these elementary particles that demanded a radical change of the classical standpoint. Since around the mid-19th century, it seemed to have been established that light has a wave-like behavior, and spreads out in concentric spheres from the light source. Experiments conducted [219]in the first third of the 20th century led to the opinion that the energy transported in the field of light has to be concentrated in packets, and that the interaction between light and matter is accompanied by a directed momentum exchange (Compton effect). This and other experiences can only be explained if one attributes an internal structure to light, and interprets it as something like a swarm of particles. But the latter assumption does not give us an understanding of the phenomena related to interference, which obey the laws of wave optics exactly. Hence, we face the fact that all experience of light (and, as it turns out, of elementary particles in general) falls into two groups. The experience of one group can be exactly explained by appealing to wave theory; but they contradict our picture of particles, which assumes that they can be spatially localized. The experience of the second group can be exactly explained by appealing to our picture of particles, but contradict wave theory. We call these two facts the duality of light and matter. What is crucial is that the two sides of radiation and matter correspond to the two sides, or better, the two halves, of space-time. From the dualism of the particle picture and the wave picture follows a dualism of configuration and motion, in which the configuration is given by position in space-time, motion by the current direction of the path. Furthermore, the two cannot exist together: space-time falls into a directional space and in a configuration space, and the concept of path becomes unusable. But this entails, as Poincaré noticed very early, and as Heisenberg later set out in detail, the dissolution of both space-time and the concept of motion of classical physics.

Now, it might be supposed that this dissolution is to be followed by a synthesis based on a new and more comprehensive principle. This synthesis would conceive the character of particles and the character of waves as two sides of a more abstract object, and lead to the conjecture that this dualism forces on us a further, and indeed objective, generalization of space, time, matter, and motion. Such an idea is rejected by the majority of contemporary physicists. They maintain that dualism is a basic fact, for which no further explanation can in principle be given. "It would be a misconception to believe", writes Niels Bohr, "that the difficulties of the atomic theory may be evaded by eventually replacing the concepts of classical physics by new conceptual forms".[13] That means that we have

---

[13] Niels Bohr, *Atomic Theory and the Description of Nature*, Cambridge: Cambridge University Press, 1934: p. 16.

to content ourselves with the debris of the classical world picture, since [220]a deeper unity does not exist; and that we can do only one thing, that is, to employ these fragments usefully in the prediction of events. Natural knowledge in the sense of knowledge from universally valid principles is no longer possible.

From this follows, first, that determinism can no longer be allowed (determinism is linked to the classical concept of path, and this is abandoned without being replaced), and second, that things are no longer independent of the observer (in fact, the employment of the description in terms of waves or particles depends on which experiments are conducted, and since there is no deeper reality underlying either description, it is no longer possible – just as with the theory of relativity – to objectify the relations involved in that dependence).

We must be aware of one thing: this picture of nature is the outcome of the epistemological turn that began with Kant, and is the result of a direct application of Mach's philosophy (which is also acknowledged). Why is it maintained that the dualism is a fundamental fact, which cannot be further explained? The reason is not a natural philosophical insight into the nonexistence of deeper principles, but the epistemological (today one would say "semantic") demand not to allow any explanation with the help of such principles, since it would be "metaphysical". Therefore, not only has the classical picture of nature been abandoned, but also the much more general precept (which the development of physics from Galileo to Einstein uses as a foundation), not to stop at phenomena, but to understand them by way of deeper reasons: the revolution of modern physics, so loudly proclaimed by the advocates of quantum theory, consists, in fact, in the transition to a positivist epistemology and in a corresponding change of the principles of physical explanation. And for the first time in the history of physics, it was possible to advance a theory that, as it seems, does not allow for any realistic interpretation.

It is completely obvious that such a change allows principles that Cartesianism and mechanical philosophy wanted to see removed to give access to the world, namely, spiritual movers, fine spirits, "free will" (not in Kant's sense!): the more physics distances itself from natural philosophy, the more it limits itself to "pure facts", the greater the freedom it allows obscurantists – who, indeed, are always looking for an opportunity to introduce their principles into the world. That is, apart from its [221]methodological impossibility, the mistake of Positivism: it does not overcome metaphysics. It only weakens good and critical philosophizing, and invites unbridled speculation, which is no longer subject to the slightest limitations in its field. And the literature shows how it benefits from the retreat of physics from natural philosophy.

For some time Einstein and Schrödinger were the only physicists who stood against this new conception of the world (which, by the way, has not developed any satisfactory theory of space and time). Over the last decade or so, however, a reaction against it has appeared that aims to return to the clear and objective ideas of classical physics, and rejects the solution of difficulties in physics by way of a "recourse to epistemology" (Schrödinger).[14] In this reaction we count, on the one hand, ideas developed in quantum electrodynamics without polemical intentions, and according to which we have to represent the world as filled with a fine ether (not the relativistic one!) capable of various excited states. These excited states are identified with elementary particles. Also, an initial discrete theory of space and time has been developed. Generally speaking, however, quantum electrodynamics progresses in a positivist fashion, too, as an instrument for prediction, whose internal coherence is of minor importance. Here begins the "counterrevolution", associated with the names of Bohm and Vigier, to which de Broglie also adhered, and which had a long preparation even in Russia. These physicists have deliberately put the idea of an objective description of nature at the top of their researches. And they have set themselves the task to develop microphysics from the beginning in a relativistic way, and to construct a synthesis of quantum theory and relativity theory, long overdue. According to Bohm, all the laws we discover in a given field are merely approximations, on whose limited validity accurate investigations must shed light. The fluctuations discovered by such an investigation, and which the aforementioned laws render "uncertain" to some extent, can be explained if one reaches a deeper layer of entities (as in the theory of Brownian movement). The world is split up in endlessly many layers of that kind, which have relative autonomy and which, however, also influence and disturb one another. As a consequence, every uncertainty, every indeterminism can be explained by reverting to one or more of the deeper layers, in which, however, a qualitative new uncertainty [222]comes in. Certain regularities hold for all layers, and they allow Bohm to infer the quantum conditions (which have universal validity, too) by way of simple reflections on time lapses in the elements of a certain layer (each element has an internal time – a pulse rate, so to speak – which is the result of an overlapping of

[14] Erwin Schrödinger, "Die gegenwärtige Situation in der Quantenmechanik", *Die Naturwissenschaften*, 23, 1935, pp. 807–812, 823–828, 844–849: p. 823.

the pulse rates of endlessly many elements of deeper layers). The construction of layers itself should admit of explanation through set-theoretical considerations, which can also serve, among other things, to indicate the number of degrees of freedom, that is, the "space" of a system, from its motion. But this means that the concept of motion (which is no longer continuous, but still the ergodic theorem suffices) is derived fundamentally from the concept of extension. Bohm explains that this world picture is a development of ideas we basically find in Hegel. Positive results, which go beyond the success of the orthodox theory, are not yet at hand (although the theory is so developed that decisive experiments should be possible very soon), and many physicists view these "speculations" with scepticism, if they do not reject them altogether. But they still have the merit that they make us philosophize about nature once again, and once again we defend a view according to which physics is pursued not only in order to make atomic weapons, but above all to know.

Theory. We do not really need to emphasize that this description of the development of ideas about inanimate nature is very sketchy, and that important developments had to be overlooked. The question now arises whether a classification of natural philosophy should not be considered along with its history, although the latter always depends on accidental elements, such as the death of an influential thinker, or temporary philosophical and academic power relations. This question makes sense in the context of dogmatic natural philosophy, which believes in the existence of eternal and demonstrable principles; or in a philosophy of history, which holds fast to the principle that the path of history leads completely and automatically in the direction of the growth of rationality. A critical natural philosopher knows, however, only two things: first, the history of natural philosophy – and this is for him the history of the development of natural philosophical ideas, of the arguments that have driven this development forward, of the measures (adopted, say, by a powerful organisation, such as Plato's inquisition or that of the Middle Ages), which tried to steer it in a certain direction with force or astuteness – and, second, the currently accepted theories together with the [223]reasons that led to their adoption and to the rejection of alternative theories.

The philosophy of life. Theories about matter and the cosmos have long freed themselves both from the influence of Aristotelian habits of thought and problems and from Aristotelian terminology. (We explicitly say "theories", because in philosophy the situation is less hopeful). Since the high point of Aristotelian philosophy, there have been at least two further comprehensive natural philosophies in the Western World,

namely, Descartes' natural philosophy and mechanical natural philosophy. Both removed teleology, once and for all, from the realm of inanimate nature. Furthermore, Descartes and the later mechanical
philosophers tried also to understand the behavior of living organisms
and even of the living human body on the basis of purely mechanical
principles. "It is an error to believe", writes Descartes in his psychology,

that the soul is the principle of the movement . . . of the body.. . . So as to avoid this
error, we must note that death never occurs because of the soul, but only because
one of the principal parts of the body decays. And we can say that the difference
between a living body and a dead body is just like the difference between a
watch, . . . when it is wound up . . . and, on the other hand, a broken watch or a
watch whose (mechanical) principle of movement ceases to be active.[15]

The fundamental principle of Descartes' biology is, then, that the causes
and principles of the behavior of living organisms are none other than the
principles that generally control the behavior of matter. In spite of the
great progress to which this principle has led, above all in the 19th
century, and in spite of the fact that the later mechanical philosophy
was extended to psychology, and not without success, there still always
were, and are, philosophers and biologists who believe that we will never
be able to explain important properties of living organisms in this way.
They are convinced that the forms of organisms, the fact of their purposeful behavior as well as the fact of their adaptation to the environment,
require an explanation on the basis of nonmechanical principles. In the
attempt to find such principles, thinkers repeatedly fell back on ideas that
were originally developed by Plato and Aristotle. According to Plato,
every earthly thing possesses an (indestructible) form, or idea, which the
thing copies with greater or lesser success. At times, Plato thinks of the
relation between the idea and the thing in a very abstract way, at times
he describes it once again in terms of [224]the relation between the father
of a family and the children, who see in him a model and try to imitate
him. The discrepancy between the thing and the idea is traced back to the
inertia of matter, from which the thing is made, as well as to certain
contrary tendencies, which are inherent to matter. If contrary tendencies
take over, the thing distances itself from the idea, dissolves, and disinte-
grates. This is how Plato, in the *Republic*, explained the development of
government forms from original patriarchy, to oligarchy, democracy,

[15] René Descartes, *Les Passions de l'Ame*, Paris: Henry Le Gras, 1649: nos. V–VI, pp. 6–8;
reprinted in *Œuvres de Descartes*, edited by Charles Adam and Paul Tannery, vol. 11,
Paris: Vrin, 1947: pp. 330–331.

and, finally, to tyranny: the first stage is so close to the idea of state that it can hardly change, and yet it carries in itself the seed of decay. The mixing of classes leads to a contamination of the gold of the leaders' nature, the noble and wise guides of the state, by the common metal that characterizes the members of other classes. This mixing distances the original state from the idea and causes its dissolution and decline. Analogously, Plato explains the origin of species (in *Timaeus*). The first individuals are strong and wise men, who are still very close to the idea of man. But here degeneration sets in too: from men, who are inclined to talkativeness and are unable to restrain this tendency, come women, then birds, fish, and finally beasts, "which lack any wisdom".[16] In both cases paradise is in the past, whereas the present is a degenerate and worthless condition. This metaphysics of unchangeable forms and ideas and changeable things, which more or less successfully imitate the forms in earthly stuff, links Plato with a theory of science according to which only the unchangeable can be an object of science, whereas the changeable eludes the reach of possible knowledge. Only forms, therefore, are accessible to scientific examination. So much for Plato's theory. It is taken over almost unchanged in idealistic morphology and in its two great predecessors, Goethe and Cuvier. The idealistic morphology sets itself the task of finding, behind the variety of concrete organisms (or fossils), the forms they embody to a greater or lesser degree. According to idealistic morphology, an individual is relevant only to the extent to which it corresponds to its form. Each deviation from the form counts as a deformity completely devoid of scientific interest. That leads to a kind of mathematical theory of forms, in which what really matters is not the coexistence relations that actually exist, but the necessary matching of ideal elements, and [225]which in the end totally abandons experience or rather uses experience only to establish how incompletely the forms are realized. Of course, a causal explanation of concrete forms can never be achieved in this way. Today Platonism no longer plays any special role in biology: it has changed its operational field and now operates in certain trends in Gestalt psychology, and above all in depth psychology (archetypes, symbols) and the philosophy of culture.

Biologists and philosophers, who were less interested in the forms of organisms than in the causes of their behavior, found no great help in Platonism. They have for the most part taken over the Aristotelian theory

[16] Plato, *Timaeus*, 92 A 7-B 2.

and have adapted it to their goals. Aristotle's theory postulates a direct influence of the forms on concrete individuals. In Plato's theory, things orient themselves toward their forms and try to imitate them. In Aristotle's theory, everything is a compound consisting of stuff in which no formative tendencies are inherent, together with a form, which shapes this stuff and hence creates a concrete and especially shaped individual. The forms of Aristotle play the role of active principles, which steer the behavior of complex things. All kinds of varieties of vitalism take this theory as their basis. Vitalism aims at explaining the behavior of organisms by assuming that not only physical principles but also immaterial "entelechies" direct their motion and growth. To account for such an assumption, sometimes philosophical and sometimes empirical reasons are offered. If we consider the empirical reasons that vitalism accepts for the existence of its vital forces, we soon notice that they rest either on a dissatisfying analysis or on an overly narrow, "mechanical" conception of physical systems. How is it possible, the perplexed vitalist asks himself, that in organisms all chemical–physical processes work so perfectly together, that organisms remain alive, even in case of serious troubles such as wounds, shock, hunger, drought, and so forth? He believes the answer can only be provided by appeal to an entelechy that ensures that the physical processes proceed in an orderly way. In so doing, it is assumed implicitly (and completely in the sense of Aristotle!) that matter alone has no inherent ordering principles, so that matter without entelechy can be nothing but chaos. This standpoint overlooks the simple fact that every stone contains a host of atoms, which retain all of their order when the stone is thrown, [226]heated, or otherwise "disturbed". Laws such as Pauli's exclusion principle coordinate events that occur in widely separated places of one and the same physical system, and thereby give the system the "character of an integrated whole". Physics does not only know the laws that govern the course of individual processes; it also knows structural laws that guarantee that numerous processes work together, thereby guaranteeing the stability of macroscopic structures. The simplest laws of this kind are the laws of probability: 6,000 throws of a normal die give 1 a thousand times, 2 a thousand times, 3 a thousand times (only approximately, of course), and so forth – not because an entelechy lying beyond space and time steers the die at the 5,001st throw, so that it fulfils its probability target – each individual throw is influenced only by those causes that directly act on it: the shaking method employed, for example, or the movement of the players' hands, air currents and so forth – but because the *initial conditions* of each individual throw stand in specific relations to one another. So far, no reason has been given why the character of "integral unity" of living beings should not be treated in an analogous way, that is, without control by "life police"! Quite apart from the fact that vitalism makes by far too poor a picture of physical laws, however, it also contradicts them. For modern vitalism does not limit itself to postulating entelechies, but further identifies the inert matter of Aristotelian theory with the matter of physics. As a consequence, it would be required to prove that entelechies steer the behavior of organisms without affecting any known physical laws. Hans Driesch (1867–1941), who developed vitalism in this very specific form conjectured (as Descartes did, too, and, in a more undetermined form, Epicurus before him), that the entelechy works orthogonally to the direction of the movements of the organism, and therefore without any effort. In so reasoning, however, one neglects the fact that the law of energy is not the only law that matter must obey; the law of impetus in its vectorial form (which Descartes did not yet know) must also be introduced – and this prohibits any change in the direction of a particle's motion without material influence. Neither Driesch nor his followers was able to show how this difficulty could be overcome until quantum theory apparently offered a way out: [227]if the motion of a material system is not strongly determined, then there is some leeway, a gap in physical regularity, and in this gap the entelechy can work. However, this attempt to save vitalism quantum mechanically also must fail. For quantum theory does not say that we cannot offer any detailed physical cause for the behavior of a well-determined system, but that no system is ever in a completely determined condition. What is lacking, then, is not a cause, but a system on which this cause could work. The conclusion of our considerations is, therefore, that vitalism has, on the one hand, a poor idea of matter, and that, on the other, its modern version, in which Aristotelian matter is identified with the (no longer formless or lawless) stuff of physics, cannot be made compatible with physical laws.

25

Philosophical Problems of Quantum Theory
(1964)

[1]1. INTRODUCTION

Philosophical ideas are connected with the more recent development of microphysics in three different ways. First there are those fairly concrete ideas which guided research between 1900 and 1927 and which contributed to the invention of the earlier forms of the quantum theory. Secondly there are the ideas which were used in order to understand the finished products of the preceding period. Here the principle of complementarity plays a most important role. The discussion of this principle has led to a detailed examination of theories, of the scientific method, and to much technical argument. The particular interpretation of microphysics which it entails is called the Copenhagen interpretation. The discussion of complementarity has also led – and this introduces the third set of ideas – to considerations of a much more general kind which were no longer guided by the requirements of physics but by the standards of the one or the other school of philosophy. It was hoped that complementarity might lead to new solutions of old philosophical [2]problems and that it might also give an improved account of the *a priori* elements in our knowledge. Returning from such ventures into pure thought, some thinkers, physicists as well as philosophers, even assumed that now *a priori* demonstrations could be given of the very same physical phenomena which had led them to their speculations in the first place. This starts already very early, at any rate before the invention of wave mechanics. The literature is vast and rather amusing. It will be dealt with only cursorily.

The threefold division is arbitrary only in part. As we proceed in time the ideas tend to become more general, less elastic, less concrete, less useful for the *advance* of physics, but very useful for the defence of a certain interpretation of its *results*. This is especially true of Bohr's philosophy, which in one way or another forms the background of almost all "orthodox" thinking on the matter. This philosophy is not easy to trace. It emerges only after a detailed analysis of the published material (including, of course, Bohr's paper on physics) *as well as* of the much less accessible unprinted sources of influence. Such analysis suggests that Bohr did not initially recognize the customary sharp distinction between physics and philosophy, fact and speculation. Research in physics was for him a subtle and delicate interaction of general principles [3]and technical procedures, the principles shaping the procedures and being in turn modified, and made more concrete, by the results obtained with the help of these procedures. It is only later, after 1927 (or 1953) that the interaction terminates and that the philosophical element petrifies, first into a doctrine and then, with the younger generation of physicists, into a "presupposition" whose existence remains increasingly unnoticed. This process was accelerated by the emergence of "vulgar" versions which, while paying lip service to Bohr's ideas, either simplified them beyond recognition, or else identified them with the tenets of some fashionable and more easily comprehensible philosophy (positivism, Kantianism, dialectical materialism, neo-thomism). Today – 1964 – the old worries have largely ceased to exercise the minds of the physicists. The S-matrix theory, Heisenberg's unified field theory, the symmetry considerations which dominate contemporary particle physics seem to be self-sufficient and not in need of an interpretation in terms of complementarity. The continued discussion of this notion is therefore regarded by many physicists as an unnecessary projection of antediluvian (1927!) troubles into a time which has outgrown them. It is in this spirit that D. D. Iwanenko (1956) invites us "at once to commence with the philosophical discussion of the most recent problems" and [4]not to waste time with "tackling opponents moving far behind the front lines of research". Such an invitation overlooks that scientific progress and corresponding solution of topical problems has frequently proceeded, and may again proceed, from changes in the deepest hinterland. Nor is it true, as P. A. M. Dirac has insinuated (1963), that the further development of technical physics will automatically take care of the more general or "philosophical" difficulties (the problem of measurement is mentioned by Dirac as an example) which are therefore not in need of separate treatment. The reason is that technical developments obtain coherence, simplicity, and purpose only by being connected with a comprehensive point of view, and such a point of view often emerges from the discussion of philosophical problems. It follows that a critical examination of the Copenhagen interpretation is now as topical as it was thirty years ago.

2. THE QUANTUM OF ACTION

A physicist invents theories, develops them so that they give a detailed account of the relevant phenomena, and tries to replace them by better theories once they are refuted. The discovery of the quantum of action showed that a revision was needed of the theories, and perhaps even of the philosophical background of the "classical" physics of Newton and Maxwell. The following example illustrates the problem.

[5]Classical physics and common sense tell us that a physical system such as a pendulum swinging with the energy E between the extreme points A and B will adopt any intermediate position, will swing for any length of the supporting thread, and will swing for any smaller amount of energy. It is also taken for granted that the pendulum remains an objective thing whether or not it is being observed. Disturbances it might experience in the course of a crude observation, such as touching, may change its objective structure; they may break it, but they will not turn it into a dream.

On the other hand we have the assumption, first introduced by Max Planck (1900) that the frequency v of a vibrating system that interacts with a radiation field is connected with its energy E by the equation

$$E = hv \tag{1}$$

where $h = 6.6 \times 10^{-27}$ CGS, the so-called quantum of action. It was soon realized (by Einstein, 1905) that this equation is not restricted to the specific situation (blackbody radiation) for which Planck devised it, but is valid generally. Applying it to the pendulum in a "naive" fashion, without any further analysis, the reader will note (i) that a "kick" of energy $<hv$, where $1/v$ is the period of the pendulum, will not affect it at all; (ii) that the attempt to slow down the motion of a pendulum swinging with hv will either be unsuccessful, [6]or will lead to an abrupt stop and release of the whole "quant" hv; (iii) that a pendulum which is receiving energy will swing for a discrete manifold of lengths only; (iv) that a pendulum, while receiving energy from a source (and we assume, in accordance with experience, that such a process takes a finite amount of time) cannot be separated from it, not even in principle (if the transfer takes Δt, then even a "conceptual" separation after $\Delta t/2$ will leave the

pendulum with only part of hv absorbed, which contradicts equation (1); (v) since (iv) applies to every interaction, it applies to every observation: observation unites observer and observed object in a manner that resists further analysis; (vi) assuming that a given interaction has more than one possible result, it follows that the emergence of a particular such result is not in reality connected with the situation before the interaction started: the idea of *determinism* cannot be universally valid; (vii) considering that every part of the universe interacts with every other part we see, finally, that the universe is an indivisible block and that the observer cannot be separated from it: the idea of *objectivity* cannot be universally valid. It also seems that the very possibility of a physical science is threatened.

The development of the quantum theory from 1900 to about 1927 is characterized by various attempts to "evade" (Bohr's term) such consequences. Almost always the [7]"naive" application of Planck's fundamental formula is replaced by something more complex. Thus it may be assumed that (iv), (v), (vi), and (vii) are correct only as long as certain complicated and not easily accessible "hidden" processes are neglected and that consideration of these processes will at once restore objectivity, determinism, and perhaps all the laws of classical physics. Or it is assumed that while some *particular* classical theories may have broken down, the more *general* ideas of determinism and objectivity stand still unassailed. Physicists subscribing to this latter belief tried to find new models for elementary particles and their interactions. Louis de Broglie for some time (before 1926) considered extended wavelike entities containing a punctiform core (theory of the pilot wave, theory of double solution). He was anticipated by Einstein who had suggested (1909) that Maxwell's equations might possess pointlike, "singular" solutions in addition to waves, and who later on applied the same hypothesis to the equations of the general theory of relativity. In this connection Einstein and his collaborators obtained the encouraging result (1927) that *if* the equations of general relativity produced singularities, then these singularities would move along geodesic lines and therefore behave exactly like material objects. These and many other ideas were discussed in the belief that some classical theories had been refuted, that refuted theories must be eliminated, [8]that it was necessary to look for new theories to fill the vacuum, and that these new theories, while giving a correct account of experimental results, would not remove (determinism and) objectivity.

Bohr's procedure differs in an interesting manner from all these attempts. He too hoped, at least initially, that some day a new and more satisfactory theory would be found; but he also believed that the older views already contained a distorted picture of such a theory. He therefore insisted that they be used as a guide for research and this not only occasionally, as was Einstein's habit, but in a systematic fashion. Rather than discard them completely, he tried to isolate and to arrange in an orderly manner their still valuable remnants. The discovery of a mathematical formalism which unites these remnants, exploits their predictive power to the utmost, incorporates such relations as Planck's fundamental equation seemed to him to be a necessary, or at least a very useful first step to a new and unified picture of atomic processes.

It is clear that a formalism of this kind, or a "*rational generalization* of the classical mode of description*", as Bohr used to call it, is not yet a theory in the full sense of the word. Theories, such as the kinetic theory of matter, or Newton's theory of gravitation, or the general theory of relativity, combine two [9]functions: they establish connections between observational results, and they also give an account of these connections in terms of certain entities, of their properties and their interactions. They are predictive device and ontological system in one. Thus the general theory of relativity not only gives excellent predictions of the behavior of planets, satellites, double stars, photons, clocks, but it also indicates the mechanism (bodies, or singularities moving along geodesics in a four-dimensional Riemannian continuum whose regional properties depend on the region as well as on the global distribution of matter) which brings about such observable features. Not unlike a rational generalization, the general theory of relativity *corrects* the earlier notions and demands that they applied under certain conditions only. (e.g., it relates our spatio-temporal assertions to coordinate systems and thereby restricts their classical usage.) But it also suggests *new descriptive concepts* (the four dimensional interval as an example) which admit of universal application and which can therefore embody a new view of reality. These new concepts can be used to *explain* the restrictions by pointing out that what was thought to be reality (absolute distances, absolute time intervals) is just an appearance, and by specifying the conditions (velocity relative to the frame of reference chosen) under which a particular appearance appears. A rational generalization [10]carries out the first, the corrective function; it does not carry out the second, the synthetic function. The classical concepts it contains are restricted and are incapable of universal application. The nonclassical concepts have the express purpose to exploit the predicative power of what is left. They do not describe anything. Seen from the point of view of past history of science, a rational generalization is only a first step toward a new theory of microcosm. It is not yet such a new theory.

We may now characterize the point of view of Bohr and of his collaborators, of the so-called Copenhagen school of quantum mechanics as saying that in microphysics we must rest content with rational generalizations, or to speak more dramatically, it is contented that the concept of objective reality breaks down in the domain of elementary particles. Why is this view held and what are the arguments in favor of it?

3. THE COPENHAGEN INTERPRETATION

For a positivist, it is a direct consequence of his principle that physics deals with predictions and that the idea of objective reality is unscientific. Now this principle was actually used by some members of the Copenhagen School, for example Heisenberg, Jordan, and Pauli, who also allowed it to enter their more technical arguments (especially in connection with the uncertainty-relations). It appears, [11]then, that the strange consequences which have been drawn from the quantum theory are not forced upon us by indisputable experimental results, but are put over on us via a physics that has been infected by positivism in the first place. Such an impression receives support from the fact that idealistic tendencies in physics have increased since the end of last century, that many of the younger physicists were strongly influenced by these tendencies, and that contemporary physics is almost completely submerged in them. Reacting to such clues, some philosophers thought that what they regarded as the ailings of quantum theory were capable of simple treatment, and they hoped to restore reason (*their* reason) by a resolute crusade for realism. They overlooked that while the quantum theory and its history may be wrapped up in positivistic trappings there still exist features which are independent of these trappings and which cannot be easily reconciled with a realistic point of view.

To see this, consider again item (iv)–(vi), Section 2: two systems, A and B interact by exchanging a single quant. The assumption that during the interaction each system retains a well-defined, though possibly very complex and therefore unknown *and perhaps even unknowable*, structure leads to an inconsistency with Planck's fundamental formula as interpreted by Einstein (the presence [12]of the italicized clause shows that we are now dealing with a problem not accessible to a positivist). The difficulty exists even for the new models which were devised by Einstein and de Broglie: these models are still definite structures. A definite structure, however, means a definite intermediate energy during interaction, and this is not allowed. It seems, then, that there are circumstances (interaction, for example)when we must (1) separate the energy of a system from its remaining features (assuming these remaining features are still well defined); and (2) relax its precision. Now it was soon found (refutation of the theory of Bohr, Kramers, and Slater) that (2) cannot be satisfied by merely changing from determinism to statistics. Measured energy always has a well-defined value and this value satisfies the observation laws. The relaxation of precision must, therefore, be restricted to intervals *between* measurements and, as a measurement can be derived out of an arbitrary time, chosen by the observer, must be made dependent on the measurement. It must also be more than a relaxation of predictability that leaves untouched the state of the system. To satisfy these demands Bohr framed the following two hypotheses: (i) hypothesis of the *relational character of state descriptions*: state descriptions are relative to experimental conditions; they change as soon as these [13]conditions change. (The dependence is noncausal and may be compared to the sudden change of the relation "*a* is longer than *b*" when *b* expands.) (ii) *Indefiniteness hypothesis*: there are situations when ascription of a definite energy is not only *false*, but *meaningless*.

The indefiniteness hypothesis can at once be applied to other magnitudes. For example, the assumption that a particle has always a well-defined position is not easily reconciled with the laws of interference which it also obeys (the fact that elementary particles have position *and* obey laws of interference is known as *duality*). These laws assert (a) that the pattern created by the joint action of various coherent channels is different from the algebraic sum of the patterns created by each single channel; (b) that interference patterns do not change with intensity. (a) entails dependence between events in different channels, (b) entails that such dependence exists *even for a single particle*. This is the phenomenon which De Broglie's theory of the pilot wave intended to explain. The difficulty is that on measurement the particle is always found to possess the whole energy and the whole momentum so that nothing is left to the field. The alternative suggestion (discussed, a. o., by Landé and Popper) to regard the interference pattern as a condition-dependent stochastic process is refuted by the very same evidence [14]that refuted the original theory of Bohr, Kramers, and Slater. The hypothesis that there are situations (interference) when it is not only false but meaningless to ascribe a definite position to the particle is again seen to be our best bet. In the present case we can even obtain a rough estimate of the limitations of the position-momentum ontology.

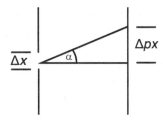

A slit with width Δx restricts the position of a particle passing it to the interval Δx. The idea of a trajectory must be given up to the extent to which interference occurs that is at least up to the first minimum. The first minimum is characterized by the angle α where Δx·sinα = λ (λ the wave length of the monochromatic process correlated with the particle). Using the equation

$$\lambda = h/p \tag{2}$$

(which does for momentum what equation (1) does for the energy) as well as sinα = $\Delta p_x/p_x$ we obtain

$$\Delta x \Delta p_x > h \tag{3}$$

as an estimate of the domain in which definite assertions concerning position and momentum are not allowed. Considering that the laws of interference are valid for all particles and in all circumstances, we must conclude that (3) is generally valid also. This is a form of the famous *uncertainty relations*.

[15]The uncertainty relations have played a great, but on the whole unfortunate, role in recent philosophical discussions. Let us, therefore, make it quite clear what is entailed by the above estimate and especially by the indefiniteness hypothesis. The simple and ingenious physical hypothesis has so often been misinterpreted that a few words of explanation seem to be needed. First it must be pointed out that its reference to meaning has nothing whatever to do with the now quite customary attitude of preferring semantic analysis to an investigation of physical conditions. There are well known classical examples of terms which are meaningfully applicable only if certain physical conditions are first satisfied and which become inapplicable and therefore meaningless as soon as these conditions cease to hold. A good example is the term "scratchability" (Mohs scale). It is applicable to rigid bodies only and loses its significance as soon as the bodies start melting. Secondly, it should be

noted that the hypothesis does not refer to *knowledge*, or *observability*. It is not asserted, for example, that two interacting systems, A and B, may be in some state, unknown to us, or that interfering particles have some unknown position. For equation (1), the laws of interference and the conservation laws do not merely exclude knowledge, or observability of well-defined (classical) states; they exclude these [16]states themselves. Nor must the argument be read as asserting, as is implied in many presentations by physicists and especially in Heisenberg's treatment of the matter that the states in question (trajectories; sharp energy) do not exist *because* they cannot be observed. For it refers to laws which deal with existence, and not merely with observability. The third remark concerns a suggestion for getting around the kinematics of ill defined states which has often been made in connection with wave mechanics. According to this suggestion, the difficulties which arise when we try to explain cases like those discussed above are due to the fact that the classical point mechanics is not the correct theory of dealing with atomic systems, and the state descriptions of classical point mechanics not the adequate means for describing states of systems upon the atomic level. The advice is not to retain the classical notions, such as position and momentum, and make them less specific, but rather to introduce completely new notions which are such that when they are used states and motions will again become well defined. Now, if any such new system is to be adequate for the description of quantum phenomena, then it must contain means for expressing the quantum postulate and it must therefore also contain adequate means for expressing the concept of energy. However, once this [17]concept has been introduced, in the very same moment all our above considerations apply again with full force: while being part of A+B, neither A nor B can be said to possess a well-defined energy from whence it follows that also the new ingenious set of concepts will not lead to unambiguous kinematics. The same is true of other variables. The uncertainty relations therefore teach us that for all we know *indefiniteness is a universal and inherent property of matter*. This is the *first argument* in favor of the assertion at the end of the last section. It is seen that the argument goes beyond the positivist's insistence to admit only what is observable. It is an attempt to make sense of a long series of refutations, all of them involving definite hypotheses concerning the structure of microentities.

These refutations convinced Bohr that what was at stake was not just a particular view about microprocesses but the much more general idea of objective spatio-temporal existence. At the same time it was clear that a wholesale elimination of the "classical mode of description" would not do.

Some classical laws (the conservation laws, the laws of interference) were still valid and classical concepts were needed to express them: Bohr's main objection against Einstein's photon theory of 1905 was that it made nonsense of its own basic equation, viz. equation (1). In this equation v refers to a frequency. Frequencies are determined by [18]interference experiments. The content of v therefore depends on phenomena which (see the above argument) cannot be reconciled with a universal application of the photon idea. Considerations of this kind, but especially the breakdown of the theory of Bohr, Kramers, and Slater, seemed to indicate that while the classical concepts have limits we must yet retain them if we do not want to eliminate what is known to be true. This result was accepted by Bohr the more readily as it was in agreement with philosophical ideas which seem to have guided him from the very beginning: for one thing, Bohr believed in the existence of a priori concepts. Like Kant before him, but on different grounds, he held that our observation statements, and even the most ordinary descriptions are, and must be formulated with the help of theoretical terms. Examples are our spatio-temporal notions and the idea of causality. These concepts which have been refined beyond the possibility of further improvement by classical physics (relativity included) are the basis of objective description and perception of reality *and* they *are a priori*. But Bohr also believed (perhaps under the influence of W. James) that all our concepts, including the a priori ones, are *limited* and that we discover their limitations in the course of physical research. He was never tired of demonstrating to his pupils the way in which even the most ordinary [19]notions might break down and of preparing them in this manner for experimental revision. These two elements – the existence of a priori concepts which make objective description possible and the existence of limitations for these concepts form the philosophical background of the *idea of complementarity*. Physics adds to them the particular way in which the quantum of action restricts the classical concepts. The consequences are as follows.

(i) If classical concepts are not universally applicable to microprocesses, if they work only under certain conditions; if on the other hand they are the only concepts for objective description; then the idea of objective existence must be given up on the microlevel and must be replaced by the idea of a condition-dependent set of appearances. (This corresponds to (vii), Section 2, and to the hypothesis of the relational character of quantum mechanical states.) The idea of determinism cannot be retained either, *not* because there are not sufficient antecedent causes, but because there exists no unambiguous description of the processes that might be influenced by them. (This corresponds to (vi), Section 2. It also eliminates the quantum theoretical "solution" of the problem of free will, the quantum mechanical defence of vitalism, and similar jokes.) One should note the "transcendental" character of the derivation. (ii) The classical concepts are valid on [20]the macrolevel. We are therefore able to give an intersubjective account of our experimental findings. There are no other descriptive statements. This solves the problem posed immediately after (vii), Section 2. (iii) We may now project particular experimental results into the microdomain, reading them as features of entities there if we only realize that such ascription has no validity apart from the situation in which it arises. Thus we may say that an electron was *found* in position P, but we must not imply that an electron is therefore a particle in the sense that it *always has* some position. We may even retain the concept of a microscopic *object* as long as we regard its dynamical properties (position, momentum, but *not* the charge) as classical *appearances* which come forth in particular circumstances and interpret the object itself as a bundle of appearances, each appearance relative to certain conditions, without substratum. The principle of complementarity then states how the various "sides" of the bundles are related to each other and under what conditions they appear. The "bundle theory" is of course nothing but the idea of a rational generalization presented in the material mode of speech. This is a *second argument* for the assertion at the end of Section 2. (iv) A bundle theory of the kind just described need not be restricted to the domain of physics. Thus the properties of a live organism, its material aspects, [21]and those gross phenomena which we commonly connect with life may be regarded as various "sides" which appear under different experimental conditions but have no validity apart from these conditions. The same may be said about the relation of consciousness to its material concomitants. It is clear that a point of view like this will regard the physiologists' attempt to give a uniform, "materialistic" account of life, or of consciousness, as based upon a naive conceptual error (something very similar to a category mistake, as Professor Ryle and his school would express it). But it is also clear that the definite reasons which led to the bundle theory in physics are entirely absent here and that the analogies that may remain are altogether too vague to justify such a radical restriction of research.

4. WAVE MECHANICS

The ideas we have been discussing so far were developed in connection with the older quantum theory, that strangely successful collection of classical and nonclassical features which centered around the atomic model of Rutherford and Bohr. Despite its power of uniting many otherwise disconnected results, the older theory was regarded as unsatisfactory by many physicists. Its main fault was seen to lie in the fact that it combined disparate assumptions in an *ad hoc* fashion, with a view to prediction, but without a coherent theoretical background. It was therefore [22]frequently believed to be but a stepping stone on the way to a point of view that would truly deserve the name of a theory. Of course, the members of the Copenhagen school saw in those very same features a promise of things to come. But for the majority of physicists such confidence was an unjustified generalization from an accidental intermediate stage of scientific knowledge. This attitude seemed to receive support from the new wave mechanics of de Broglie and Schrödinger. When this theory was finished it was hailed by many as the long expected coherent account of the microcosm. The indefiniteness hypothesis and the hypothesis of the relational character of quantum mechanical states, so it was thought, had only reflected the indefiniteness and incompleteness of the early theory, and was no longer needed. More especially, it was assumed either that the states were now new, but well-defined entities (the Ψ-waves); or it was assumed that whatever incompleteness occurred was due to the statistical character of the theory (these are the two outstanding nonorthodox interpretations of the theory). The *third* and perhaps most powerful *argument* in favor of the sentence at the end of Section 2 and of the Copenhagen interpretation is provided by the refutation of these assumptions and by the discovery that the wave mechanics which was produced by a realistic philosophy, untarnished by the "Copenhagen spirit", [23]which gave an as yet unpreceded account of the microlevel; which explained what was postulated in the old theory and whose experimental success was quite astounding, was just another "rational generalization of the classical mode of description". This discovery, and its dramatic demonstration at the Fifth Solvay Congress in Brussels, October 1927, convinced most physicists and turned the Copenhagen interpretation from a possible hypothesis into the foundation of a new orthodoxy. The result could, of course, have been foreseen: Bohr's argument presupposes equations (1) and (2), duality, the conservation laws, and it therefore applies to any theory containing these laws. Still, the detailed examination has considerable interest of its own.

Let us first analyze the suggestion that the quantum theory provides a new type of complete and well-defined states, viz. the Ψ-waves which are to replace the descriptions, in terms of trajectories, of the classical point mechanics. This assumption has been made by Planck. It is also similar to Schrödinger's original interpretation of his theory. It derives support from the fact that the Ψ-function obeys a superposition principle for ordinary, spatio-temporal waves. According to Planck "the material waves constitute the primary elements of the new world picture. . . . In general the laws of the matter [24]waves are basically different from those of the classical mechanics of material points. However the point of central importance is that the function characterizing matter waves, that is, the wave function . . . is fully determined for all places and times by initial and boundary conditions" and is also fully objective. In order to see the inadequacy of this interpretation, consider the case of two interacting systems, A and B (cf. (iv)–(vii), Section 2, as well as the review of this example in Section 3). Assume that the state of the combined systems, $A + B$, is represented by a Ψ-wave Ψ_{AB} which is not a product $\Psi_A \cdot \Psi_B$. It can be shown that in this case the assumption that both A and B are in a definite state, that is, are characterized by a definite wave function, leads to a contradiction. A and B cannot be said to be in any state whatsoever. Considering that according to the present interpretation all properties are supposed to be contained in the wave function, neither A nor B can be said to have definite properties. Even a statistical interpretation which views the Ψs as referring to classical collectives is in difficulty here. For there is nothing in the idea of a collective of interacting pairs that would prevent the elements of the pairs from being members of some other collectives. The Copenhagen interpretation which asserts that dynamical properties are possessed by a system not absolutely, but only in [25]certain experimental conditions, is in agreement with the case: the conditions Ψ_{AB} are incompatible with the conditions which would allow A and B to have definite states. We may, of course, observe A and obtain a state $\Phi(A)$ for it. Such an observation introduces new conditions. For the reasons just given, the state obtained by it cannot have been present while Ψ_{AB} was present.

The last statement describes in a nutshell the process of *measurement* in the quantum theory. The peculiarity of this process lies in the fact that the transition from the initial state Φ of a system S to be measured to the final state Φ', the so-called *reduction of the wave packet*, can be explained neither on the basis of the equation of motion (applied to either S itself, or to the combined system $S+A$, where A is the measuring apparatus), nor as one of the various possible compositions of collectives we know from probability theory. For example, it would be incorrect to assume that the transition from Φ to Φ' is a transition, due to selection, from a less definite description to a more definite one (as is the transition, in classical statistics, from "the probability that the outcome is α is $1/2$" to "the outcome is actually α"). For consider a measurement whose possible outcomes are represented by the states Φ' and Φ'' and which occurs when the system is in state $\Phi = \Phi' + \Phi''$. Here Φ cannot be interpreted as asserting [26]that one of two mutually exclusive alternatives, Φ', or Φ'' occurs, for Φ leads to objective physical processes which do not occur, neither when Φ' is realized, nor when Φ'' is realized (cf. Section 3, the laws of interference). The transition, on measurement, from Φ to, say, Φ'', is therefore accompanied by a physical change; it cannot be merely the result of a selection. Even worse: what the Schrödinger equation yields when applied to the system Φ + measuring apparatus is, strictly speaking, another pure state $\Psi = A\Phi' + B\Phi''$ so that now a *mere look* at the apparatus (after which we assert that we have found Φ'') seems to precipitate a physical change, namely the destruction of interference between $A\Phi'$ and $B\Phi''$ (Schrödinger's paradox). All these difficulties disappear when we carefully restrict our assertions about states to the relevant conditions and regard a measurement as a transition to other conditions which cannot be further analyzed.

We are now ready to deal with the second suggestion according to which "the description of the quantum theory" must be regarded as "an incomplete and indirect description of reality", such that the Ψ-function "does not in any way describe a state which would be that of a single system; it rather relates to many systems, to an 'ensemble of systems' in the sense of statistical mechanics" (Einstein). According to this interpretation the single [27]system is always in a well-defined (classical) state, and any indeterminacy that is introduced is due to our lack of knowledge of this state. Quite obviously this assumption is inconsistent with indefiniteness hypothesis and with the considerations leading up to it. However let us not forget that for anybody unfamiliar with the qualitative arguments of the last section there are strong reasons to assume that it is correct. These reasons consist above all in the fact that within wave mechanics the connection between the theory and "reality" is established by *statistical* rules (Born's rules): "It has not till now been taken sufficiently into account" writes Popper "that to the mathematical derivation of the [uncertainty relations] there must correspond, precisely, a derivation of the *interpretation* of these formulae". And he points out that, given Born's rules and *nothing else*, we must interpret Δx and Δp_x in formula (3) as the standard deviations, within large ensembles, of quantities *which are otherwise well defined*, not as "statements imposing limitations upon the attainable precision of measurement", and certainly not as statements dissolving the clear outlines of the classical universe. This is completely correct reasoning *provided* we already know what are the elements of the ensemble to which reference is made. [28]*If* these elements are classical objects with well-defined, classical properties, then the argument is valid, but unnecessarily laborious. If we do not yet know what to regard as an element, then the mere fact that we are dealing with a statistical theory will not help us. We may still consider at least two alternatives: (1) the elements possess their values before the measurement and retain them throughout the measurement; (2) the elements do not necessarily possess their values before measurement, at least not in the general case, but obtain them by the measurement itself. To decide between these alternatives we need additional considerations – and this brings us back again to the informal arguments outlined in Section 3 and to the Copenhagen interpretation (which favors 2 over 1).

(Incidentally, it is not only impossible to use the statistical character of wave mechanics as an argument in favor of 1; it is also impossible to use it as an argument in favor of 2. The usual interpretation of von Neumann's "proof" is therefore incorrect.)

Popper's argument achieved an apparent success by silently assuming the very point at issue, viz. the principle P that physical systems have objective properties which may be changed by active physical interference, but are independent of all other processes. A much more startling argument of exactly the same kind was given, in 1935, by Einstein, Podolsky, and Rosen. [29]These authors utilized certain correlations inherent in the formalism of wave mechanics to calculate the state of a system S via a measurement which did not involve any physical interaction with S. Using, in addition, the above principle P (what is discovered by such a measurement cannot have been created by it: alternative 2 above goes into 1 for measurement without interaction) they could infer that every physical object must always possess all the classical properties it is capable of. In this case the proper interpretation of equation (3) would, of course, be the one suggested by Popper. While this argument refutes some primitive interpretations of (3) which ascribe the uncertainty to a disturbance caused by a physical interaction (Heisenberg), it does not touch Bohr's point of view which denies P. Its result may also involve a violation of some conservation principles.

Reviewing these and other, more technical arguments one realizes that wave mechanics is a theory of rather unusual kind. In the past the issue between realism and instrumentalism was discussed by considering the content of a theory, and not by its truth value. The instrumentalists pointed out that theoretical terms, interpreted as descriptions of actual situations in the universe, would transcend experience and violate empiricism. The realists, on the other hand, insisted that such transcending was essential to knowledge and that a mere [30]summary of observational results would not do. Both parties assumed that the *truth value* of the theory was not affected by the change of interpretation. Wave mechanics seems to be the first theory which excludes a realistic interpretation not only by a philosophical maneuver, but by the very physical assumptions it contains (duality; the conservation laws, equations (1) and (2); the laws of interference). It is, of course, possible to introduce variables which restore a semblance of realism. But as long as we retain the remainder of the theory these variables will be as superfluous as the mechanical models which were at some time used to give substance to the laws of the electromagnetic field. Does this show that a return to realism and to definite models is impossible?

5. THE FUTURE OF MICROPHYSICS

According to the majority of contemporary microphysicists it does. Nobody denies, of course, that the collection of techniques and bits of theory which constitute the quantum theory of today may be in need of revision. Such revision is going on all the time. Occasionally the strife between alternative points of view is as fierce, as was the much older controversy between Copenhagen and Princeton. But despite all their differences the opposing parties seem to share the belief that (3) defines a restriction which will never again be crossed by physics. [31]However large future changes, they will always be untouched the indefiniteness hypothesis, the relational character of states and those parts of the formalism (Hilbert space or suitable extension of it; commutation relations between operators defined by this space) which most directly express them. How is this attitude supported?

Bohr's original arguments are hardly used any longer except in simplified positivistic versions (which seem to have originated with Heisenberg). What is used is (a) the empirical success of the quantum theory; (b) the fact that alternative interpretations would destroy the symmetries which form such a decisive part of the theory; (c) much less importantly, von Neumann's proof. As regards (c) we have already indicated that it involves an illegitimate transition from properties of collectives to property of individuals. Besides it uses the completeness of the quantum theory (the Ψ-function contains all that can be consistently said about a system) as an *explicit* premise: it is circular. Finally, it presupposes the correctness of a certain very specific theory (the elementary quantum theory as formulated by von Neumann) which is now recognized as being of approximate validity only. (b) is often used against the attempt to approach the problems of microphysics from the point of view of general relativity, or of classical [32]point mechanics. The symmetry properties of general relativity are different from those of the orthodox quantum theory and it is evident that, initially at least, a combination of the two would have rather awkward formal properties. This, by the way, was the reason why the connection with general relativity was severed in the first place, and why the latter theory has receded into the background. However, it is clear that even the most beautiful and the most symmetrical theory cannot survive argument showing its inadequacy. But – and we move to (a) – no such argument has been found yet. Quite the contrary, so the defender of the orthodox point of view will continue, the history of the quantum theory is an uninterrupted series of successes showing again and again the correctness of the fundamental scheme. But it is of course also a series of failures, each of them necessitating modification, adjustment, improvement, and this process has by no means come to an end. Now if the successes are supposed to support the orthodox view, must we not say that the failures undermine it and suggest that the search for a new theory is indicated? Moreover, it is somewhat naïve to expect that a theory which has so many possibilities to adapt itself to difficulties and to evade refutation will ever admit defeat *unless* it is confronted with a radically different theory which succeeds where the orthodox have succeeded; [33]explains in a straightforward manner what the orthodox can explain only by strenuous adaptation; makes definite predictions beyond the reach of the orthodox point of view (within formula (3), for example), and thereby offers a tangible alternative choice with different advantages. Such a theory cannot be excluded in advance, not even by Bohr's own reasoning. In order to see this, let us repeat the three arguments in favor of complementarity which have been developed in the text above. It is important to emphasize that while these arguments *refute* certain ideas about elementary particles, they do not thereby *prove* the Copenhagen interpretation. They show perhaps that this interpretation is *superior* to the rest; they do not show that it is *correct*. This was the position we have described so far. But this position is still much too strong. Consider the *first argument*. It assumes the universal validity of the conservation laws, of the laws of interference, of the quantum of action; and it rejects positions such as the idea of the pilot wave, or the statistical interpretation of Einstein and Popper as not being in agreement with these laws. This certainly means going too far. All the laws mentioned have been tested in restricted domains only. We have not the slightest guarantee that they will continue to be valid in other domains. Moreover our remarks a few lines [34]above indicate that a test of the validity of the laws on which the first arguments rests will need theories which, while being in agreement with the available (finite) evidence *contradict* the laws where they have not yet been tested and thereby suggest new and decisive experiments. Nobody can say in advance how the suggested new experiments will turn out. The *second argument* rests essentially on the difficulty to transcend the spatio-temporal framework of classical physics. However, instead of seeing this difficulty for what it really is, viz. the rather stable result of an adaptation that has gone on for a long time, it turns it into an insuperable limit for the human mind. This means excusing lack of imagination by the postulation of transcendental limits for the human mind – a procedure which some Aristotelians and some Kantians have practiced *ad nauseam*. "Scientific philosophy" has long ago looked through these manoeuvres and has heaped scorn and contempt upon those who defended the. It is time that the more "modern" apriorism be also recognised for what it is. Finally, the *third argument* is based upon the elementary quantum theory and certain features of the field theories and it is valid only (a) as long as the Copenhagen interpretation is retained – but this interpretation is based upon the first argument and loses its uniqueness with the removal of this argument; (b) as long as the corresponding theoretical assumptions can be taken for [35]granted – which we want to find out by the very same move the third argument forbids us to use. Result: the future of physics is still a completely open matter. Orthodox opinions to the contrary [notwith-standing,] there is not a single *argument* that excludes for example the future usefulness of determinism and objectivity. Which of course should not prevent anyone from having *faith* in his favorite interpretation and from developing it further. For it is only through the comparison of results achieved by a somewhat unreasonable faith that our knowledge grows and advances.

LITERATURE

The reader is advised not to start with Bohr's philosophical writings as they appear in *Atomic Physics and the Description of Nature*, Cambridge 1932, or in *Atomic Physics and Human Knowledge*, New York 1958. As an introduction, they hardly make sense. He is invited first to familiarize himself with the many suggestions, the difficulties, the tentative hypotheses which characterize the period from about 1900 to 1927. This will give a proper impression of the richness of the material utilized by the Copenhagen Interpretation. For details and literature cf. an early edition (for example the third edition) of Vol. 1 of Sommerfeld's *Atombau und Spekrallinien*, or Geiger-Scheel, *Handbuch der Physik*, first edition, Vols. 4 (1929), 20 (1928), 21 (1929), [36]23 (1926) as well as Wien-Harms, *Handbuch der Physik*, Vols. 21 (1927), 22 (1929). A historically not quite trustworthy survey which is however interesting in itself is P. Jordan, *Anschauliche Quantentheorie*, Leipzig 1936. The so far best historical presentation which does not only register "facts" but also deals with the interplay of ideas is K. Meyer-Abich's study, *Korrespondenz, Individualität, und Komplementarität. Eine Studie Zur Geistesgeschichte der Quantentheorie in den Beiträgen Niels Bohrs*, Dissertation, Hamburg 1964. The state of the theory in 1927, when it faced its most important test as well as the discussions leading to its almost general acceptance can be found in *Rapports et discussions du 5ᵉ Conseil*, Institut International de Physique Solvay, Paris 1928. For an evaluation of this conference cf. de Broglie, *La Méchanique quantique, restera-t-elle indeterministe?* Paris 1953, Introduction. After the conference Einstein was almost the only physicist to continue criticizing the Copenhagen Interpretation. For his view cf. "Physics and Reality", first published in 1935, reprinted in *Ideas and Opinions*, London 1954. The long drawn out debate between Bohr and Einstein is beautifully reported in Bohr's contribution to the Einstein volume of the *Library of Living Philosophers*, Illinois 1949, and commented upon by Einstein in the same volume. The argument of Einstein, Podolsky, and Rosen [37]appears in *Phys. Rev.*, Vol. 47 (1935). Bohr's reply which according to Einstein (Einstein volume, p. 681) comes "nearest to doing justice to the problem", in *Phys. Rev.*, Vol. 48. The, by now, quite voluminous discussion of Einstein's example is reviewed in my essay "Problems of Microphysics", *Frontiers of Science and Philosophy*, ed. Colodny, Pittsburgh 1962 {reprinted as Chapter 7 of the present collection}. For a clear and quite original presentation of Bohr's philosophy and an application to concrete examples and to the formalism of wave mechanics cf. D. Bohm, *Quantum Theory*, Princeton 1952. For an excellent and brief discussion of the formalism of elementary quantum mechanics cf. G. Temple, *Quantum Mechanics*, London 1952.

Among alternative versions of the Copenhagen interpretation, Heisenberg's positivistic version has achieved greatest notoriety. It is found in practically all textbooks. The origin is *The Physical Principles of the Quantum Theory*, Chicago 1930. For the best criticism cf. Ch. IX of K. R. Popper's *Logic of Scientific Discovery*, New York 1959 (first published 1935).

The present situation of fundamental discussion is well reflected in the symposium *Observation and Interpretation*, ed Körner, London 1957 {see Feyerabend's own paper, in *PP1*, ch. 13, as well as the excerpts of the discussions, reprinted as Chapter 16 of the present collection}. Here Bohm, a pupil of Einstein, and Vigier, a pupil of de Broglie, develop and defend their causal model of microphysics. One should also consult *Philosophische Probleme der Modernen Naturwissenschaft. Materialen der Allunionskonferenz zu* [38]*philosophischen Fragen der Naturwissenschaft*, Moskau 1958, Akademie Verlag Berlin, 1962; "The Foundations of Quantum Mechanics", report on a conference at Xavier University, October 1962, *Physics Today*, Vol. 17/1 (1964), pp. 53ff; and Dirac's lecture at this conference (to which reference was made in Section 1) which appeared in *Scientific American*, May 1963. The reasons why alternative points of view should be developed even at a time when orthodoxy seems to be without blemish are developed in Popper's "Aim of Science", *Ratio*, Vol. 1 (1957) as well as in my paper, "How to be a Good Empiricist", Vol. II of the Publications of the *Delaware Seminar*, New York 1963 {reprinted in *PP3*, ch. 3}.

26

Ludwig Boltzmann, 1844–1906
(1967)

[334a]Austrian physicist and philosopher of science, was born in Vienna and received his Ph.D. in physics from the University of Vienna in 1866. He held chairs of mathematics, mathematical physics, and experimental physics at Graz, Munich, and Vienna. In addition, he taught courses on the methodology and general theory of science.

Boltzmann combined a strong and unerring philosophical instinct with a sharp intellect, a great sense of humor, a somewhat violent temperament, and an exceptional mastery of presentation. His lectures in theoretical physics were attended by many nonphysicists, who could understand the problems which Boltzmann took care to state independently of the mathematical arguments. "The true theoretician", he wrote, "makes only sparing use of formulae. It is in the hooks of the allegedly practical thinkers that one finds formulae only too frequently, and used for mere adornment". His lectures on experimental physics were exquisite performances which he prepared with care and presented with flamboyance. Boltzmann often invited his academic opponents, such as Friedrich Jodl, to his philosophical lectures and debated with them in front of the students. Music was a special interest: he had studied with the composer Anton Bruckner and played the pianoforte; and he periodically arranged evenings of chamber music and more lighthearted parties at his home.

The effect of his teaching upon the younger generation of natural scientists can hardly be exaggerated. "All of us younger mathematicians were then on Boltzmann's side", Arnold Sommerfeld wrote about the Lübeck discussions on energetics in 1895, where the "bull" Boltzmann, supported by the mathematician Felix Klein, defeated the "torero Ostwald despite the latter's expert fencing". Svante Arrhenius and Walther Nernst studied with Boltzmann in Graz, Paul Ehrenfest attended his Vienna lectures, and Einstein – not inclined to listen to lectures – read his published work and was strongly influenced both by the physics and by the philosophy it contained. Wilhelm Ostwald called Boltzmann "a man [334b]who was superior to us all in intelligence, and in the clarity of his science". On the occasion of Boltzmann's sixtieth birthday, thinkers from many different countries – among them O. Chwolson, Pierre Duhem, Gottlob Frege, Max Planck, and Ernst Mach – contributed to an impressive *Festschrift* (Leipzig, 1904). Two years later Boltzmann, who was subject to severe depression, committed suicide.

...

PHILOSOPHY OF SCIENCE

Physics amid philosophy are inseparably connected in Boltzmann's work. He was one of those rare thinkers who are content neither with general ideas nor with simple collections of facts, but who try to combine the general and the particular in a single coherent point of view. He felt scorn and even hatred for the "philosophers of the schools" (he mentions Berkeley's "crazy theories", Kant, Schopenhauer, Hegel, and Herbart), who, according to him, offer but a few "vague and absurd ideas". He admitted that they had "eliminated theories of a still more primitive nature", but he criticized them for not going further and for believing that the final truth had already been obtained (see *Populäre Schriften*, chs. 18 and 22). The same criticism was applied to those contemporary physicists – and especially to their philosophical leaders, Ernst Mach and Wilhelm Ostwald, who thought that they had achieved a purely phenomenological physics which did not transcend physical experience and which could therefore be retained in all future developments.

Hypothetical Character of Knowledge

The reasons Boltzmann gave for his dissatisfaction with dogmatic systems of thought were partly biological, partly logical. Boltzmann's biological argument took the following lines: Our ideas are the result of a process of adaptation by trial and error (Boltzmann was an enthusiastic follower of Darwin, whose theory he extended both to the origin of life – thus anticipating Aleksander Oparin – and to the development of thought). Some of these ideas, such as that of the Euclidean character of space, may be "inborn" in the sense that the individual is endowed with them through the development of the species. Such an origin explains their force, the impression of incorrigibility, but "it would be a fallacy to assume, as did Kant, that they are therefore absolutely correct" (ibid., ch. 17). The future development of the species, aided by scientific research, may head to further modifications and to further improvement.

The logical argument (ibid., chs. 10, 14, 19) was as follows: It is to be admitted that one can try to adhere as closely as possible to experience and that one might in this way obtain a physics that is free from hypotheses. But, first of all, such a physics has not yet been achieved – "the phenomenological physics only apparently contains less arbitrary assumptions. Both [that is, atomism and the phenomenological theory] proceed ... from experimental laws which are valid for macroscopic objects. Both derive from them the laws which are supposed to be valid on the microlevel. Pending further examination, the latter laws are therefore equally hypothetical". Second, it is very doubtful whether a physics without hypotheses would be desirable – "the bolder one is in transcending experience, the greater the chance to make really surprising [335a]discoveries ... the phenomenological account of physics therefore really has no reason to be so proud for sticking so closely to the facts!" Boltzmann's conclusion was that "the edifice of our theories does not consist of ... irrefutable truths.... It consists of largely arbitrary elements ... so-called hypotheses" (ibid., ch. 8).

In his realization of the hypothetical character of all our knowledge, Boltzmann was far ahead of his time and perhaps even ahead of our own time. Apriorists and empiricists both claimed to have shown the existence of incorrigible principles, the former supporting their claims by reference to "laws of thought", the latter by reference to "experience" and "induction"; and the majority of thinkers supported either one group or the other. Therefore, the particular consequences that Boltzmann drew from his views were recognized only slowly and rather reluctantly. These consequences are the following.

Scientific Theories

Theories transcend experience in two ways. They express more than is contained in our experimental results, and they represent even the latter in an idealized manner. They are therefore only partly determined by the facts. The rest, which is usually almost the whole theory, must be regarded as "an *arbitrary* invention of the human mind" and can he judged only by its simplicity and by its future success. This being the case, it is "quite conceivable that there are two theories, both equally simple, and equally in agreement with experiment, although they are both completely different, and both correct. The assertion that one of them gives a true account can in this case be only the expression of our subjective conviction" (see ibid., chs. 1, 5, 9, 10, 14, 16, 19). Einstein wrote in *On the Method of Theoretical Physics* (Oxford, 1933), concerning the idea that theories are "free inventions of the human mind", that a clear recognition of the correctness of this notion really only came with the general theory of relativity which showed that we can point to two essentially different principles, both of them corresponding to experience to a large extent.

Deductive and Inductive Methods

The method of presentation which is best suited to such a situation is the deductive method. Boltzmann discussed both the deductive method and the inductive method in the introductory chapter of his treatise on mechanics (1897).

The deductive method begins with a formulation of basic ideas and establishes contact with experience only later, by specific deductions. It emphasizes the arbitrariness of the basic ideas, clarifies their content, and exhibits the inner consistency of the theory. Its disadvantage is that it gives no account of the arguments that led to the theory in the first place.

The inductive method only apparently removes this disadvantage. It tries to demonstrate how theoretical ideas are obtained from experience, but its proofs cannot stand up under closer examination. Despite considerable effort the gaps in the derivation have not been closed. The inductive method's peculiar way of mixing theoretical ideas and experimental notions leads to a loss in clarity and makes it impossible to judge the consistency of the edifice. "The lack of clarity in the principles of mechanics seems to be connected with the fact that one did not at once start with hypothetical pictures, framed by our minds, but tried [335b]to start from experience. The transition to hypotheses was then more or less covered up and it was even attempted artificially to construct some kind of proof to the effect that the whole edifice was free from hypotheses. This is one of the main reasons for the lack of clarity".

The historical development of a physics consisting entirely of hypotheses would be "by leaps and bounds", by cataclysmic changes which would wipe out almost all that is believed at a certain time without regard for even the most fundamental principles; it would not be a gradual process of growth which adds to, but never takes away from, an already existing body of facts and theories.

Criticism

Finally, the hypothetical character of our knowledge makes criticism the most important method of investigation. Boltzmann developed important elements of the point of view which is today connected with the name of K. R. Popper. In his obituary for Josef Loschmidt, Boltzmann suggested, half jokingly, half seriously, a journal dealing only with experiments that had failed (note that the Michelson-Morley experimente, e.g., was long regarded as a dismal failure). Criticism can be by comparison with facts, but it can also be by comparison with other theories. "To attack problems from different sides furthers science and [I welcome] each original, enthusiastic piece of scientific research". Boltzmann even encouraged the defenders of the inductive mode of presentation to attempt to reveal the errors in his description so that both sides might profit from the controversy.

PHYSICS

Boltzmann maintained the critical attitude even in actual research. Almost all of his better-known investigations were based upon classical mechanics and on the idea of localizable mass points in Euclidean space. Yet he was fully aware that these notions were not final and tried to formulate them as clearly as possible in order to make future criticism easier. He speculated on the possibility of an electromagnetic explanation of mass, force, and inertia; he doubted the continuity of time, even the notion of continuity itself, and he suggested that the laws of mechanics and perhaps all laws of nature "might be approximate expressions for average values, and not differentiable in the strict sense"; he pointed out that mass points need not be individually distinguishable; he regarded the concept of Euclidean space as a matter in need of further examination. These were striking anticipations of later developments. But Boltzmann did not elaborate them further. He was convinced that the resources of the existent mechanics and of the existent atomic theory were riot yet exhausted, and his work in physics was guided by this conviction. He was largely responsible for the central position which the atomic theory and statistical considerations have assumed in contemporary thought.

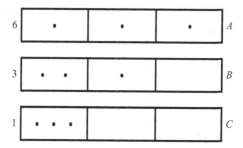

Toward the end of the last century, the atomic theory was all but abandoned. Boltzmann was regarded as a brilliant defender of a lost cause. The reasons given by most French and German scientists for opposing atomic theory were in part methodological. The idea of small and invisible particles seemed metaphysical in comparison with the theories of thermodynamics that postulated measurable [336a]quantities only. In part these reasons were based on real physical difficulties: the Second Law of Thermodynamics asserts the existence of irreversible processes which apparently cannot be explained by the atomic theory. However, for any motion that is in accordance with the laws of mechanics, the reverse motion is equally in accordance with the same laws. Now Boltzmann pointed out that while this argument correctly describes the properties of the laws of mechanics, it has not considered the initial conditions or "microstates" the distribution of which will certainly play an important role in systems containing a great number of constituents. It may well be that initial conditions leading to a reversal in motion are almost never realized in nature. In order to explain this, consider, first, a container with three molecules (1, 2, and 3) and assume that our macroscopic measurements (on which the phenomenological account is based) allow us only to determine density differences in volumes of the indicated size (see diagram). Situation A will then be described as a gas having equal density α all over the container. Situation B is a gas with density 2α on the left, α in the middle, o on the right. Situation C has density 3α on the left and o everywhere else. Now A may be realized in six different ways, by the following six different arrangements of the individual molecules: 123, 132, 213, 231, 312, 321. Situation B can be realized in three ways only: [12]3, [32]1, [13]2. And there is only one way to realize C. Hence, if all arrangements are equally likely, it is more likely to find equal density than unequal density. For three molecules the probabilities are 6:3:1 for A, B, and C. For N molecules, the relation between A and B will be about 2^N, a tremendous number for real gases. Therefore, while the Second Law of

372 Encyclopedia Entries (1958–1967)

Thermodynamics correctly represents the statistically most probable distribution, the possibility of minor fluctuations in the microstates is still open.

We must realize, however, that so far we have allowed the molecules to move freely and to take their positions according to the laws of chance. In actual fact, the molecules obey the laws of mechanics, which means that there exist relations of dependence between their positions at different times and – via collisions – between the positions of different molecules. The assertion that all arrangements are equally likely and that A is therefore vastly more probable than B is no longer guaranteed. Boltzmann's so-called H-theorem (which he had to reformulate various times as the result of criticism) solves this problem to some extent by showing that the laws of mechanics, combined with certain probabilistic assumptions about collisions, would lead to the very same result we have just obtained: what[336b]ever the initial conditions, there is an overwhelming probability that the magnitude H will decrease until the case of maximum probability according to the above reckoning is realized.

Entropy

In connection with the above investigations, Boltzmann also established a relation between the number of arrangements W leading to a certain macroscopic situation and the entropy S of that situation:

$$S = k \cdot \log W$$

This equation provides a definition of entropy that is more general than the thermodynamic definition (which is applicable only to systems in a state of equilibrium). It is also seen that the newly defined entropy does not satisfy the strict Second Law but admits fluctuations between the situation of maximum probability and other, less probable but by no means impossible, situations of a lower entropy. "As soon as one considers bodies which are so small that they contain only a few molecules the [second] law must cease to be valid". (This was written in 1878, in the *Vorlesungen über Gastheorie*, and was severely criticized. The existence of the predicted fluctuations of 1905 had been confirmed by Jean Perrin and Theodor Svedberg.) The problem of irreversibility is not yet completely solved by this account, however, for the probability that a state of low entropy is followed by a state of higher entropy is identical with the probability that a state of high entropy is followed by a state of lower entropy. In order to solve this difficulty, Boltzmann, in *Vorlesungen über*

Gastheorie, compared local developments to the overall development of large regions of the universe. The large-scale development provides a reference point for small-scale fluctuations and in this way determines a local direction of time in a "timeless" universe.

Considering the universe as a whole both directions of time are still indistinguishable just as space knows no above and below. Yet at a specific point of the surface of the earth we call the direction towards the center the downward direction; in the same way a living being dwelling in a certain time-phase of such a local world will distinguish the direction towards the more improbable state from the opposite direction, calling the former the past, and the latter the future. This seems to be the only method which will allow us to conceive the second law, the unidirected development of each partial world without postulating an unidirected development of the whole universe.

All the above ideas have led to further developments. The statistical physics of today, quantum statistics included, is inconceivable without the work of Boltzmann, who provided a framework that could be retained despite the need to change some more specific assumptions. The same is true of the highly technical developments in connection with the *H*-theorem and the ergodic hypothesis. It is unfortunate that Boltzmann's general philosophy, which is intimately connected with his physics, is practically unknown, for his ideas are still relevant to contemporary discussions and present a promising field for further study.

[337a]Bibliography

Boltzmann's research papers were collected in three volumes by F. Hasenöhrl and published as *Wissenschaftliche Abhandlungen* (Leipzig, 1909). His larger works are *Vorlesungen über die Prinzipien der Mechanik*, 2 vols. (Leipzig, 1897–1904); *Vorlesungen über Maxwells Theorie der Elektricität und des Lichtes*, 2 vols. (Leipzig, 1891–1893); and *Vorlesungen über Gastheorie*, 2 vols. (Leipzig, 1896–1898). Boltzmann's more general writings have been collected in *Populäre Schriften* (Leipzig, 1905).

The biography by E. Broda, *Ludwig Boltzmann* (Vienna, 1955), contains an excellent account of Boltzmann's physics and his philosophy. For a more technical discussion of the *H*-theorem and of related questions, see Paul and Tatiana Ehrenfest, *The Conceptual Foundations of the Statistical Approach in Mechanics* (Ithaca, N.Y., 1959); Hans Reichenbach, *The Direction of Time* (Berkeley, 1956); and Dirk ter Haar, "Foundations of Statistical Mechanics", in *Reviews of Modern Physics* (July 1955).

27

Werner Heisenberg
(1967)

[466a]German physicist, was born in Würzburg in 1901. Heisenberg studied physics in Munich under Arnold Sommerfeld and received his doctorate from Munich in 1923. He became a lecturer and assistant to Max Born at Göttingen in 1924. He continued his studies at the University of Copenhagen, where he collaborated with H. A. Kramers. He succeeded Kramers in 1926 as lecturer in physics there. Heisenberg was professor of physics at Leipzig from 1927 to 1941 and professor at Berlin and director of the Kaiser Wilhelm Institute for Physics from l941 to 1945. He was named honorary professor and director of the Max Planck Institute for Physics at Göttingen in 1946 and has been honorary professor and administrative director of the Max Planck Institute for Physics and Astrophysics in Munich since 1958. He was awarded the Nobel Prize for Physics in 1932.

Heisenberg's contributions to physics are contained in over 120 papers covering a great variety of topics. We shall here deal with two topics only, with the invention of matrix mechanics and with Heisenberg's more recent theory of elementary particles.

MATRIX MECHANICS

The *older quantum theory* of Bohr and Sommerfeld had tried to combine classical physics with the new quantum laws and to utilize the predictive power of both. The resulting theory was a mixture of classical notions – some useful, others apparently redundant – of new ideas and of *ad hoc* adaptations. Thus, for example, transition probabilities and selection rules were calculated, or guessed at, by examining the Fourier coefficients of the motions

$$\Phi_i(t) = \sum_{n=-\infty}^{+\infty} X_i(n, \omega_i)\exp(in\omega_i t)$$

of the independently vibrating parts of the atom, while the motion Φ_i itself had to be denied any physical significance. In addition, the theory had failed in important respects. It clearly was but an intermediate step on the way to a satisfactory mechanics of the atom. The final theory is essentially due to the efforts and the very different philosophies of two men, Heisenberg and Schrödinger.

According to Heisenberg we must abandon all attempts to give a detailed description of the unobservable internal motions of the atom. Such motions are but the result of the continued use of classical ideas in a domain that is inaccessible to direct experimental examination. Considering that these ideas may be in need of revision it would seem [466b]to be wise to construct a theory that is expressed solely in terms of such "outer" magnitudes as frequencies and intensities of spectral lines. Speaking formally this means that we want to predict by using the X_i *directly* and without appeal to the Φ_i. Now Bohr's investigations had already gone a long way toward determining the required properties of the X. His idea of a rational generalization corresponds exactly to what Heisenberg had in mind. Heisenberg himself provided additional rules of calculation which were sufficient for solving some simple problems, such as the problem of the harmonic oscillator. It was not known to him at the time that the rules were those of an algebra of noncommuting matrices; this was soon recognized by Born, who, together with Pascual Jordan and Heisenberg, completed the formalism a few months after Heisenberg's first paper had appeared. A new atomic mechanics was at last in sight. Its meaning, however, was far from clear. Macroscopic objects whose positions and momenta could be ascertained with a higher degree of precision were represented by infinite arrays of complex numbers, none of them corresponding in a simple way to visible properties. "Can you imagine", objected H. A. Lorentz at this stage, "me to be nothing but a matrix?" It was again Heisenberg who, after the theory had been completed in a somewhat unexpected fashion by Schrödinger, made an essential contribution here by showing, in his *uncertainty relations*, to what extent classical notions could still be used in the interpretation of microphysical theories.

Heisenberg was to use the principle to rebuild a theory by working "from the outside in" once more in 1943, in order to eliminate certain difficulties in the quantum theory of fields. Believing these difficulties to be due to the disappearance of the ordinary space-time relations below 10^{-13} cm, he tried to replace field theory by a formalism which for any interaction transforms asymptotic anterior states into asymptotic posterior states without dealing with the details of the interaction. This so-called S-*matrix theory* was taken up by Geoffrey Chew and others for the calculation of the properties of strongly interacting particles. This has led to what some physicists regard as the beginning of a "third revolution" of twentieth-century physics, to the idea that particles are composites and that the properties of all of then can be obtained in a step by step procedure, starting with the interaction of any small subset ("bootstrap hypothesis"). Spatiotemporal relations are alien to this scheme, which therefore cannot develop a theory of measurement. Nor does there seem to be any possibility of extending it to other types of interaction.

ELEMENTARY PARTICLE THEORY

Heisenberg, who had been the first to stress the nonexistence of a criterion for distinguishing "elementary" particles from composites, has in the meantime developed a different theory in which elementary particles are stationary states of a single physical system, "matter". The field operators refer no longer to particles lilt to this basic matter (which Heisenberg sometimes compares to Anaximander's *apeiron*). The masses of the particles arise wholly from the interactions due to the nonlinearity of the basic field equation. There are no "bare particles". Other properties are supposed to follow from the symmetries of the field equation. Strange [467a]particles of spin 0 and 1/2 have been dealt with, to a certain extent, on the basis of approximation methods (this refers to 1962). There are only programs, no exact predictions, or weak interactions.

Heisenberg's philosophical speculations have always been intimately connected with his physics. They have been original and exciting. The same cannot be said about his more general observations on philosophical matters. However, he should not be blamed for this disparity, as it is at any rate only in close connection with reality that philosophy can be both interesting and fruitful. (For a discussion of Heisenberg's contributions to quantum theory, see QUANTUM MECHANICS, PHILOSOPHICAL IMPLICATIONS OF.)

Bibliography

Heisenberg's early work and its relation to wave mechanics and to experiment is described in *Die physikalischen Prinzipien der Quantentheorie* (Leipzig, 1930), translated by Carl Eckart and Frank C. Hoyt as *The Physical Principles of the Quantum Theory* (Chicago, 1930). The theory of the S-matrix is explained in "Die ,beobachtbaren Größen' in die Theorie der Elementarteilchen", in *Zeitschrift für Physik*, Vol. 120 (1943), 513–538 and 673–702. For a survey of Heisenberg's new field theory, see "Quantum Theory of Fields and Elementary Particles", in *Reviews of Modern Physics*, Vol. 29 (1957), 269–278, and "Die Entwicklung der einheitlichen Feldtheorie der Elementarteilchen", in *Naturwissenschaften*, Vol. 50 (1963), 3–7. *Die Physik der Atomkerne* (Braunschweig, 1943) explains nuclear physics and Heisenberg's contributions to it. *Wandlungen in den Grundlagen der Naturwissenschaften* (Leipzig, 1935) is a survey of advances and discoveries from 1900 to 1930. *Physics and Philosophy* (New York, 1959), which contains the Gifford lectures given in 1955/1956, deals with the same subject on a broader historical and philosophical basis and also discusses current objections to the Copenhagen interpretation of quantum theory.

For the atmosphere at Göttingen in the "golden twenties", see F. Hund, "Göttingen, Kopenhagen, Leipzig im Rückblick", in Fritz Bopp, ed., *Werner Heisenberg und die Physik unserer Zeit* (Braunschweig, 1961), pp. 1–7, and Robert Jungk, *Brighter Than a Thousand Suns* (New York, 1958).

28

Max Planck, 1858–1947
(1967)

[312b]German physicist, discoverer of the quantum of action, also called Planck's constant. Born in Kiel, Planck studied physics and mathematics at the University of Munich under Philipp von Jolly and at the University of Berlin under Hermann von Helmholtz and Gustav Kirchhoff. After receiving his Ph.D. at Munich (1879), he taught theoretical physics, first in Kiel, then (starting in 1889) in Berlin, as Kirchhoff's successor. "In those days", he wrote later, "I was the only theoretician, a physicist *sui generis*, as it were, and this circumstance did not make my *début* so easy". At this time Planck made important, and indeed quite fundamental, contributions to the understanding of the phenomena of heat, but he received hardly any attention from the scientific community: "Helmholtz probably did not read my paper at all. Kirchhoff expressly disapproved of its contents". The spotlight was then on the controversy between Ludwig Boltzmann and the Wilhelm Ostwald–Georg Helm–Ernst Mach camp, which supported a purely phenomenological theory of heat. It was via this controversy, and not because of the force of his arguments, that Planck's ideas were finally accepted. "This experience", he wrote, "gave me an opportunity to learn a remarkable fact: a new scientific truth does not triumph by convincing its opponents and making them see the light, but rather because its opponents eventually die". Nevertheless, the discovery of the quantum of action in 1900 (see QUANTUM MECHANICS, PHILOSOPHICAL IMPLICATIONS OF), for which Planck received the Nobel prize in physics (1918), was a direct result of these earlier studies. In 1912 Planck became permanent secretary of the (then) Prussian Academy of Sciences, a post which he retained with only minor interruptions for the rest of his life. He used this position with excellent judgment for

furthering the international collaboration of all scientists. From 1930 to 1935 he was president of the Kaiser-Wilhelm Institut, which later became the Max-Planck-Institut.

Politically Planck was conservative, loyal to the Prussian ideas of the state and of honor, and loyal to Wilhelm II. During World War I he more than once expressed his devotion to the cause of the German people united in battle, and he received the order of "pour le mérite", one of the highest orders of Wilhelm's Germany. However, he opposed the Nazi regime. He defended Einstein, first against his scientific opponents, then against his political enemies. Despite severe criticism by Johannes Stark, Phillip Lenard, and Ernst Müller, he continued to defend Einstein and other Jewish scientists (such as Walther Nernst) even after 1933. He later personally demanded of Hitler that those scientists who had been imprisoned be freed; as a consequence he was removed as president of the Physical Society, was refused the Goethe Prize of the city of Frankfurt (he was awarded it after the war, in 1946), and finally was forced to witness the execution of his only son, who had been connected with the German resistance. Antiquated as some of his political ideas may have been, he [313a]nevertheless put individual justice above all and defended it even at the risk of his own life. At the end of the war he was rescued by the Allied Forces. He spent the last years of his life in Göttingen.

APPROACH TO SCIENCE

Planck's research was guided by his belief "of the existence in nature of something real, and independent of human measurement". He considered "the search for the absolute" to be the highest goal of science. "Our everyday starting point", he explained, "must necessarily be something relative. The material that goes into our instruments varies according to our geographical source; their construction depends on the skill of the designers and toolmakers; their manipulation is contingent on the special purposes pursued by the experimenter. Our task is to find in all these factors and data, the absolute, the universally valid, the invariant that is hidden in them".

This point of view was not allowed to remain a philosophical luxury, without influence upon the procedures of physics. One of the main objections which Planck raised against the positivistic creed was its sterility in the promotion of theory. "Positivism lacks the driving force for serving as a leader on the road of research. True, it is able to eliminate obstacles, but it cannot turn them into productive factors. For ... its glance is directed backwards. But progress, advancement requires new associations of ideas and new queries, not based on the results of measurement alone".

SCIENTIFIC DISCOVERIES

Of new ideas Planck himself produced essentially two. He recognized and clearly formulated those properties of heat which separate it from purely mechanical processes, and he introduced and applied to concrete problems the idea of an atomistic structure not only of matter but of radiation also. In his doctoral dissertation he had already separated thermodynamic irreversibility from mechanical processes and had interpreted Rudolf Clausius' entropy as its measure. Later he showed (independently of Willard Gibbs) that "all the laws of physical and chemical equilibrium follow from a knowledge of entropy". His conviction that the principle of the increase of entropy was a genuine and independent physical law and his belief in the universal (or, to use his term, "absolute") validity of all physical laws led him to apply thermodynamic reasoning in domains which until then had been regarded as inaccessible to it. For example, he determined that the lowering of the freezing point of dilute solutions could be explained only by a dissociation of the substances dissolved, thus extending the science of thermodynamics to electrically charged particles. This tendency to strain laws to the limit rather than to restrict them to the domain of their strongest evidence caused a temporary clash with Boltzmann, who was quite unperturbed by the fact that in his approach the entropy of a system could both increase and decrease. But it also led to Planck's greatest triumph – his discovery of the quantum of action. He was the only one to correlate the relevant features of radiation with the entropy, rather than the temperature, of the radiant body. "While a host of outstanding physicists worked on the problem of spectral energy distribution, both from the experimental and theoretical aspect, every one of them directed his efforts solely towards exhibiting the dependence of the intensity of radiation on the temperature. On [313b]the other hand I suspected that the fundamental connection lies in the dependence of entropy upon energy. As the significance of the concept of entropy had not yet come to be fully appreciated, nobody paid attention to the method adopted by me, and I could work out my calculations completely at my leisure". These calculations furnished a formula which agreed with experiment and contained the existing theoretical results (Wien's formula and the Rayleigh-Jeans law) as limiting cases. In the attempt to find a rationale for this result, Planck utilized Boltzmann's statistical interpretation of entropy and was thus led to the discover of the "atomic", or discontinuous, structure of action (energy).

REALISM, DETERMINISM, AND RELIGION

The discovery of the quantum of action was brought about not only by the specific physical arguments used but also by the philosophical belief in the existence of a real world behaving in accordance with immutable laws. The intellectual climate of the late nineteenth century was opposed to such a belief (Boltzmann was almost the only other figure to uphold it). This climate not only found expression in the philosophical superstructure but influenced physical practice itself. Laws were regarded as summaries of experimental results and were applied only where such results were available. However, it was the "metaphysics" of Planck, Boltzmann, and, later on, Einstein (whom Planck interpreted as a realist from the very beginning) which made possible many of the theories that are now frequently used to attack realism and other "metaphysical" principles.

Planck never accepted the positivistic interpretation of the quantum theory. He distinguished between what he called the "world picture" of physics and the "sensory world", identifying the former with the formalism of the Ψ waves, the latter with experimental results. The fact that the Ψ-function obeys the Schrödinger equation enabled him to say that while the sensory world might show indeterministic features, the world picture, even of the new physics, did not. His belief in the existence of objective laws also provided him with an important steppingstone to religious belief. Planck argued that the laws of nature are riot invented in the minds of men; on the contrary, external factors force us to recognize them. Some of these laws, such as the principle of least action, "exhibit a rational world order" and thereby reveal "an omnipotent reason which rules over nature". He concluded that there is no contradiction between religion and natural science; rather, they supplement and condition each other.

Works by Planck

Theory of Heat Radiation, translated by Morton Masius, Philadelphia, 1914; 2d ed., New York, 1959.
Eight Lectures on Theoretical Physics, translated by A. P. Wills, New York, 1915. Lectures given at Columbia University in 1909.
The Origin and Development of the Quantum Theory, translated by H. T. Clarke and L. Silberstein. Oxford, 1922. Nobel Prize address.

A Survey of Physics; A Collection of Lectures and Essays, translated by R. Jones and D. H. Williams, London, 1925. Reissued as *A Survey of Physical Theory*, New York, 1960.

Treatise on Thermodynamics, translated by Alexander Ogg, London, 1927; 3rd rev. ed., New York, 1945.

Introduction to Theoretical Physics, translated by Henry L. Brose, 5 vols., London 1932–1933; New York, 1949. Includes *Gen[314a]eral Mechanics, The Mechanics of Deformable Bodies, Theory of Electricity and Magnetism, Theory of Light*, and *Theory of Heat*.

Scientific Autobiography and Other Papers, translated by Frank Gaynor, New York, 1949.

The New Science, translated by James Murphy and W. H. Johnson, New York, 1959. Includes *Where is Science Going?* (a defense of determinism with a preface by Albert Einstein), *The Universe in the Light of Modern Physics*, and *The Philosophy of Physics*.

Works on Planck

Schlick, Moritz, "Positivism and Realism", in A. J. Ayer, ed., *Logical Positivism*, Glencoe, Ill., 1959. This essay was a direct reply to the criticism of positivism that Planck expressed in *Positivismus und reale Außenwelt*, Leipzig, 1931.

Vogel, H., *Zum philosophischen Wirken Max Plancks*, Berlin, 1961. Excellent biography with detailed bibliography.

29

Erwin Schrödinger, 1887–1961
(1967)

[332b]Austrian physicist, was born in Vienna and studied physics and mathematics with Franz Exner, Rudolf Hasenöhrl, and Wilhelm Wirtinger. After brief appointments at Jena, Stuttgart, and Breslau, he became (1922) professor of mathematical physics at Zurich. It is here that he developed his wave mechanics in the fall of 1925. Schrödinger succeeded Max Planck at Berlin in 1927, only to leave in 1933, shortly after Hitler's rise to power. After a few years in Oxford and Graz (which he had to leave in 1938, when Austria was made part of the Third Reich) he accepted an invitation by Eamon de Valera to Dublin's newly founded Institute for Advanced Studies. Schrödinger received the Nobel prize in physics (jointly with Paul Dirac) in 1933. He returned to Austria in 1956.

THOUGHT

Though the ideas of Werner Heisenberg and Schrödinger are now usually regarded as complementary aspects of one and the same point of view, they arose from different motives and were originally incompatible. Schrödinger was strongly influenced both by Boltzmann's physics and by his philosophy. "His line of thought", he said in his address to the Prussian Academy in 1929, "may be called my first love in science. No other has ever thus enraptured me and will ever do so again". In 1922, five years before Heisenberg's uncertainty relations saw the light, Schrödinger criticized "the *custom*, inherited through thousands of years, of thinking *causally*", and he defended Boltzmann's conjecture (which had been elaborated by Exner and Hasenöhrl) that the observed macroscopic regularities might be the result of an interplay of inherently indeterministic microprocesses. (This, he reports later, "met with considerable shaking of heads".) Yet "the inherent contradictions of atomic theory" (which played such a central role in Bohr's work and which were to some extent regarded by him as something positive) sounded harsh and crude to Schrödinger "when compared with the pure and inexorably clear development of Boltzmann's reasoning". He even "fled from it for a while" into the field of color theory. It was mainly the work of Louis de Broglie which encouraged Schrödinger to return to the theory of the atom and which finally led him to develop his wave mechanics.

There can hardly be a greater disparity than that between the procedures of Heisenberg and Schrödinger. Heisenberg had participated in the various attempts to adapt the classical theories to the new experimental situation and to arrive at what Bohr called a "rational generalization of the classical mode of description". In the course of these endeavors he concentrated on observable magnitudes and had explicitly refrained from giving an account of the internal processes of the atom. Schrödinger, on the [333a]other hand, proceeded to do just this. His reasoning is realistic throughout. Briefly, the argument is as follows: Classical geometrical optics can be summed up in Fermat's principle, according to which a beam of isolated light rays transverses an optical system in such a manner that each single ray takes the shortest possible time to arrive at its destination. This formal principle can easily be explained by reference to the way in which wave fronts are accelerated or retarded when passing media of different refractive index and turn in consequence. The classical mechanics of mass points can be formulated in a manner that makes it formally identical with Fermat's principle: The path of a system of mass points is such that a certain integral, depending on this path, assumes an extreme value. "It seems" (so Schrödinger describes his train of thought in his Nobel address) "as if Nature had effected exactly the same thing twice, but in two very different ways – once, in the case of light, through a fairly transparent wave mechanism; and on the other occasion, in the case of mass points, by methods which were utterly mysterious, unless one vas prepared to believe in some underlying undulatory character in the second case also". Now, if this analogy is correct, then we have to expect diffraction phenomena in regions comparable to the wave length of the hypothetical matter waves. The breakdown of the ordinary Hamiltonian formalism inside the atom indicates that such regions may be of the size of the atom, so that the atom itself becomes "really nothing more than the diffraction phenomena arising from an electron wave that has been intercepted by the nucleus of the atom". Calculations carried out on this basis led to all the features known from experiment and later on produced a coherent theory that could in principle be applied to any possible situation. In this theory the collection of particle coordinates forming the initial condition of classical point mechanics is replaced by a "wave function" whose development in time is given, in a perfectly causal fashion, by the so-called time dependent *Schrödinger equation*. The similarity to classical mechanics where we start out with an objective state of affairs which then develops causally in time cannot be overlooked. However, this appearance of objectivity and causality soon turned out to be deceptive. For example, the theory cannot give a satisfactory account of such phenomena as the tracks of particles in a Wilson chamber. There are other difficulties, too, which after considerable discussion led to a new interpretation of the wave function (Max Born's interpretation) that closed the gap that apparently existed between matrix mechanics and wave mechanics (both theories were also shown to be mathematically equivalent). In this new interpretation the waves are not real and objective processes but only indicate the probabilities of the outcome of certain experiments. The general transformation theory, then, altogether robbed the wave Function of its realistic connotation and turned it into a purely formal instrument of prediction.

Schrödinger, though aware of the difficulties of his original ideas, never acquiesced in the interpretation transferred upon it by the Copenhagen school. He attacked this interpretation both on physical and on philosophical grounds. On the physical side he doubted the consistency of the various approximation methods which were used to [333b]establish a connection between theory and fact. Philosophically, he objected to the tendency "to forgo connecting the description [of what is observable] with a definite hypothesis about the real structure of the universe". He also pointed out that strictly speaking the connection between theory and experiment had been established only at very few points and that "the tremendous amount of empirical confirmation" could be accepted only at the expense of consistent procedure and mathematical rigor. It is unfortunate that his criticism had only very little influence and that the clarity, simplicity, and consistency of a Boltzmann, which it intended to achieve, has now largely become a thing of the past.

Bibliography

All the writings of Schrödinger which are of interest for a wide audience are contained in *Science, Theory, and Man* (New York, 1957). For a more recent criticism of orthodox quantum theory, see "Are There Quantum Jumps?", in *The British Journal for the Philosophy of Science*, Vol. 3 (1952), 109–123 and 233–242. See also *"Die gegenwärtige Lage in der Quantentheorie"*, in *Naturwissenschaften* (1935). Schrödinger's views on various philosophical and religious topics not directly related to physics are presented in *My View of the World* (Cambridge, 1964).

Name Index

Adam, Charles, 342
Agassi, Joseph, xxix, 47, 105, 126, 159
Aharonov, Yakir, xxiii
Albert of Saxony, 327
Alder, Berni J., 263
Alexander, Peter, 101
Alfraganus, 256
Anaximander, 211, 324, 376
Anaximenes, 211
Archimedes, 243
Aristarchus, 212
Aristotle, 243–4, 257, 312, 325, 334, 342, 344
Arrhenius, Svante, 366

Bacon, Francis, 198, 280, 329
Balmer, Johann Jakob, xiii, 55, 119, 165
Bartley, W. W. III, x, xxvi
Baumrin, Bernard, xxix
Bentley, Richard, 331
Berkeley, George, 11–12, 34, 53, 74, 116,
     170, 191, 204, 331–3, 335, 367, 373
Besso, Michele, 246
Bethe, Hans, 317
Beyer, Robert T., xix, xxxiv, 23, 294
Birkhoff, George, David, 232, 251, 255,
     264–5, 300
Blackett, Patrick, 317
Blokhintsev, Dmitry, 86, 123, 134, 139, 223
Blumenberg, Hans, 244
Bohm, David, xx, xxii–xxiv, xxvi–xxvii,
     xxxi, xxxiv, 226, 257, 275–7, 279,
     281, 283, 285–93, 296–7, 312, 340–1,
     364–5

Bohr, Niels, xiii–xv, xvii–xviii, xxi,
     xxiii–xxiv, xxviii, xxxii–xxxiii, 192–3,
     204–6, 221–2, 226–7, 229, 241,
     246–7, 273, 276, 279, 281, 290–1,
     296, 298, 300, 312, 338, 347,
     349–52, 354–5, 357, 360–4, 374–5,
     384
Boltzmann, Ludwig, xxxv, 252, 254–5, 262,
     280, 299–300, 325, 335, 366–73, 378,
     380–1, 383–5
Bondi, Herman, 263–4
Boole, George, 232
Bopp, Fritz, 279, 377
Born, Max, xiv, xvi–xx, 216, 225, 259,
     271–3, 275, 277, 295, 313, 359,
     366–7, 374–5, 378, 383, 385
Borrini-Feyerabend, Grazia, ix
Brahe, Tycho, 327, 331
Braithwaite, Richard, 288, 293
Brillouin, Léon Nicolas, 235, 238
Broda, Engelbert, 373
Broglie, Louis de, xiv–xv, xxii–xxiii, xxvi,
     xxxi, 298, 336, 340, 349, 351–2, 357,
     364–5, 384
Brose, Henry L., 382
Brown, Robert, 340
Bruckner, Anton, 366
Bruno, Giordano, 243–4, 329
Brush, Stephen, 252, 255, 262
Budenz, Julia, 333
Buridan, Jean, 327
Butterfield, Herbert, 329
Butts, Robert, 248

Subject Index

Printed in the United States
By Bookmasters